Carbon Nanomaterials for Biological and Medical Applications

Carbon Nanomaterials for Biological and Medical Applications

Sekhar Chandra Ray
Department of Physics, University of South Africa
1 Preller St, Pretoria, 0002
South Africa

Nikhil Ranjan Jana
Centre for Advanced Materials
Indian Association for the Cultivation of Science
2A & 2B Raja S C Mullick Road
Kolkata 700032, India

ELSEVIER

elsevier.com

Elsevier
Radarweg 29, PO Box 211, 1000 AE Amsterdam, Netherlands
The Boulevard, Langford Lane, Kidlington, Oxford OX5 1GB, United Kingdom
50 Hampshire Street, 5th Floor, Cambridge, MA 02139, United States

Notices

Library of Congress Cataloging-in-Publication Data
A catalog record for this book is available from the Library of Congress

British Library Cataloguing-in-Publication Data
A catalogue record for this book is available from the British Library

ISBN: 978-0-323-47906-6

For information on all Elsevier publications visit our website at
https://www.elsevier.com/books-and-journals

Publisher: Matthew Deans
Acquisition Editor: Simon Holt
Editorial Project Manager: Sabrina Webber
Production Project Manager: Susan Li
Designer: Greg Harris

Dedication

Dedicated to all the people for the betterment of their life.

Contents

About the Authors

SEKHAR CHANDRA RAY (PhD, University of the North Bengal, India) is a National Research Foundation (NRF) B-rated researcher in physics. He is currently a professor of physics at the University of South Africa (UNISA) in experimental condensed matter physics. Prof. Ray hails from India where he completed his doctoral studies which focused mainly on photovoltaic solar grade materials. He has worked as a research fellow and visiting scientist in Italy (INFM Fellow), Taiwan (NSC Fellow), Spain (ICMM, CSIC, Foreign Researcher Fellow, Ministry of Science and Technology, Spain), South Korea (Brain Pool Research Fellow, Government of South Korea), and India (IACS, visiting scientist). His research group focuses on carbon nanostructure material in electronic structure/magnetic properties for the possible fabrication of spintronic devices application and bioimaging process. At present Prof. Ray is working on different 2D structure materials such as *graphene*, MoS_2, *stanene*, *silicene*, and *germanene*. During his 20-year research career, he has published 90 peer-reviewed research articles, including four in *Nature Publishing Group (NPG)* Journal with more than 2500 citations in internationally recognized journals. Prof. Ray also acts as an editorial board member of Scientific Reports—NPG journal.

NIKHIL RANJAN JANA (PhD, Indian Institute of Technology, Kharagpur, India) is currently an associate professor at the Center for Advanced Materials, Indian Association for the Cultivation of Science, India. Prof. Jana worked as a postdoctoral research fellow and research scientist at University of South Carolina, Columbia (USA), University of Arkansas (USA), and Institute of Bioengineering and Nanotechnology (Singapore). His research group focuses on biomedical application of functional nanomaterials. He has 25 years of research career with 125 peer-reviewed research articles in internationally recognized journals, which have 15,000 citations.

Preface

This book has been written to meet the basic requirement of researchers getting training in carbon nanomaterials for biological and medical applications.

The contents of the book are from different articles published in different journals from different research groups worldwide. We, the book authors, are very much thankful to those publishers and authors from whose publications we have collected all useful information that are presented in this book. This book consists of six chapters, *viz.*, synthesis process of carbon nanomaterials, removal of biologically toxic materials, biosensors, bioimaging probe, drug and gene delivery carrier, and toxicology/biosafety.

We hope that this book will be useful for all researchers who are working in this research field.

Any suggestions toward its further improvement will be thankfully acknowledged and incorporated in the next editions.

Professor Sekhar Chandra Ray
University of South Africa
Johannesburg, South Africa

Professor Nikhil Ranjan Jana
Indian Association for the Cultivation of Science
Kolkata 700032, India

Acknowledgments

Professor Sekhar Chandra Ray and Professor Nikhil Ranjan Jana wish to extend their thanks to their friends and relatives for their support toward writing this book.

Introduction

Rapid development in the field of nanomedicine is bringing novel opportunities for improved removal of biologically toxic materials, biosensors, bioimaging probe, drug and gene delivery carrier, and toxicology/biosafety. Among various nanomaterials involved in biological and medical applications, carbon nanomaterials possess a unique 0D, 1D, 2D structure with interesting intrinsic mechanical, physical, and chemical properties that have been extensively explored for a wide range of applications in biology and medicine. This book provides an overview of how carbon nanomaterials are used in different aspects of biomedicine, including drug and gene delivery and cancer treatment, bioimaging as well as biosensing. The recent developments, future perspectives, and major challenges in these fields are discussed.

CHAPTER 1

Different Synthesis Process of Carbon Nanomaterials for Biological Applications

CONTENTS

1.1 INTRODUCTION

Carbon nanomaterials (CNMs) have a unique place in nanoscience owing to their exceptional electrical, optical, thermal, chemical, and mechanical properties and have found application in diverse areas such as composite materials, energy storage and conversion, sensors, drug delivery, medical, field emission devices, and nanoscale electronic components. Conjugated CNMs cover the areas of carbon nanoparticles (CNPs), carbon nanotubes (CNTs), fullerenes, graphene, graphene oxide (GO), and reduced graphene oxide (r-GO). CNPs, CNTs, graphene, and GO/r-GO continue to gain attention and have impacted many fields, and the number of potential applications continues to grow. The chemistry of CNPs, CNTs, graphene, and GO/r-GO to control over electronic properties and the assembly of graphene devices are particularly active areas in this field of research work. Research work in different CNMs has reviewed vigor with significant advances in the field of bioapplications and supramolecular assembly over the 20 years. Graphene/GO is perhaps the newest of the CNMs and promises to be a very active and challenging research area. Already since its "isolation" in 2004, it has grabbed the attention of the chemistry, materials, and physics communities. It promises to rival CNTs in terms of properties and potential applications with the number of publications rising day by day. CNMs cross many disciplines and therefore make an ideal subject for review and discussion on various biological and medical applications. We have deliberately identified three key areas of CNMs: (1) CNPs, (2) CNTs, and (3) graphene and GO/r-GO, which although look very similar, often realized on common ground. Much of the work on CNTs and CNPs has original research and now graphene and GO/r-GO are building on CNM work. The aims are to gather knowledge for researchers from different backgrounds and disciplines, such as biochemists, biomaterial scientists, and biophysicists, and to interact with each other disciplines.

The emergence of CNPs shows high potential in biological labeling, bioimaging, and other different optoelectronic/electronic device applications (Gruber et al., 1997; Neugart et al., 2007; Batalov et al., 2009; Glinka et al.,

Carbon Nanomaterials for Biological and Medical Applications. http://dx.doi.org/10.1016/B978-0-323-47906-6.00001-1

1999; Zyubin et al., 2009; Yu et al., 2005; Sun et al., 2006; Zhou et al., 2007; Fu et al., 2007; Wee, 2007; Lim et al., 2009; Gao et al., 2007; Liu et al., 2007a,b, 2009a,b; Zhao et al., 2008; Selvi et al., 2008; Bourlinos et al., 2008a,b; Mochalin et al., 2009; Ushizawa et al., 2002; Cahalan et al., 2002; Huang et al., 2004; Kong et al., 2005a,b). The CNPs are also being explored widely for use in various biological applications such as drug delivery and cancer treatment. Studies reveal that cancer treatment using radio waves can heat and destroy a tumor, lymphoma, or metastasized cancer. These particles can be used in humans. These CNPs are biocompatible and chemically inert and can be surface functionalized with organic molecules, or polymers can be chemically bound to the particle surface. Nanotechnology has enabled the use of these engineered CNPs with diameters of <100 nm in industrial applications, cancer treatment, medical imaging, disease diagnoses, gene therapy, drug delivery, and many other areas.

CNT is another key area of CNMs. About 30 years ago with the discovery of the football-shaped CNPs, called fullerenes (Kroto et al., 1985), a new development in carbon research began. It will lead to lasting effects on many areas in technology and everyday life. Once their principle of design had been uncovered, more variations of the original molecules were developed. As a result of these developments the "elongated" form of fullerenes, called CNTs, were discovered by the Japanese scientist Iijima in 1991. These nanomaterials are allotropes of carbon, made of graphite, and have been constructed in cylindrical tubes with nanometers scale in diameter and several millimeters in length (Hirlekar et al., 2009; Singh et al., 2012). CNTs are categorized into two structural forms, including single-walled carbon nanotubes (SWNTs) and multiwalled nanotubes (MWNTs). Motivated by the unique one-dimensional (1-D) structure with impressive structural, mechanical, and electronic properties (due to their small size and mass); incredible mechanical strength; and high electrical and thermal conductivity (Usui et al., 2012; Zhang et al., 2010a,b,c), they have been introduced in different applications. CNTs have been extensively explored in the field of biology and medicine. CNTs have been first used as additives to various structural materials for electronics, optics, plastics, and other materials of nanotechnology fields. Since the beginning of the 21st century, they have been introduced in pharmacy and medicine for drug delivery system in therapeutics. Thanks to their high surface area, excellent chemical stability, and rich electronic polyaromatic structure, CNTs are able to adsorb or conjugate with a wide variety of therapeutic molecules (drugs, proteins, antibodies, DNA, enzymes, etc.). They have been proven to be an excellent vehicle for drug delivery by penetrating into the cells directly and keeping the drug intact without metabolism during transport in the body (Hirlekar et al., 2009; Singh et al., 2012; Usui et al., 2012; Zhang et al., 2010a,b,c). Many studies have demonstrated that when bonded to CNTs, these molecules are delivered more effectively and safely into cells than by traditional methods (Singh et al., 2012; Usui et al., 2012; Zhang et al., 2010a,b,c). This fantastic discovery has opened a new way for drug preparations that is completely different from traditional techniques used in pharmaceutical industry and radically changed anterior concepts of pharmacology (Singh et al., 2012; Usui et al., 2012). It has been first applied to bind antineoplastic and antibiotic drugs to CNTs for cancer and infection treatments, respectively. Then, other linkages of biomolecules (genes, proteins, DNA, antibodies, vaccines, biosensors, cells, etc.) to CNTs have been also assayed for gene therapy, immunotherapy, tissue regeneration, and diagnosis of different ailments (Kateb et al., 2010; Liu et al., 2007a,b; Zhang et al., 2011; Rosenand and Elman, 2009; Bekyarova et al., 2005). Therefore, CNTs have become the focus of attention by scientists in a wide variety of disciplines. They may be promising antioxidants for health protective effect and ailment prevention in the future (Galano, 2010). Besides these applications of CNTs, they have been shown as a powerful tool for enantiomer separation of chiral drugs and chemicals in pharmaceutical industry as well as in laboratory and for extraction of drugs and pollutants in different media by solid phase extraction before analysis (Yu, 2011; El-Sheikh and Sweileh, 2011). Different research groups have already contributed to the development of different novel-functionalized CNT techniques for drug delivery as well as for drug analysis and also to the study on the interactions between CNTs and albumin or drugs (Li, 2008; Li et al., 2010a,b,c,d, 2013; Chen et al., 2011a,b; Zha et al., 2011; Jiang et al., 2012a,b; Xiao et al., 2012, 2013). Moreover, CNTs with excellent mechanical properties have also found applications as potential tissue engineering scaffold materials (Tran et al., 2009; Veetil et al., 2009).

The other key areas of CNMs are graphene and GO/r-GO. Graphene has sp^2-hybridized planar honeycomb lattice structure with zero-energy band gap and remarkable properties such as high conductivity, large surface area, and high mechanical and thermal stability (Geim et al., 2007; Stankovich et al., 2006; Zhi et al., 2008; Geim, 2009; Debgupta et al., 2014; Rao et al., 2009). These properties offer graphene valuable materials in various applied research such as in fuel cell catalysis, supercapacitor, photocatalysis, heterogeneous catalysis, water purification, drug delivery, and biosensing (Huang et al., 2012; Hou et al., 2011; Dua et al., 2010; Seger et al., 2009; Xiaoqiang et al., 2011; Upadhyay et al., 2014; Stoller et al., 2008). In particular, high charge carrier mobility (up to 10^5 $cm^2\ V^{-1}\ s^{-1}$), high intrinsic strength (130 GPa), high thermal conductivity at room temperature (10^3 $Wm^{-1}\ K^{-1}$), and large surface area (>2000 $m^2\ g^{-1}$) compared to graphite (10 $m^2\ g^{-1}$) or CNT (1300 $m^2\ g^{-1}$) have been used to make graphene-based nanocomposites for optical and electronic applications (Hou et al., 2011; Lee et al., 2008; Balandin et al., 2011; Novoselov et al., 2004; Chen et al., 2010). In addition, it is reported that the efficiency of graphene-based materials is generally higher as compared to other forms of carbon such as CNT and fullerene (Machadoab et al., 2012). Graphene is highly hydrophobic in nature, does not dissolve in hydrophilic solvent, and forms large aggregates via $\pi-\pi$ interaction/van der Waals interaction. To overcome this issue, GO is commonly used as an alternative. The chemical exfoliation of graphite powder by strong oxidizing reagent produces highly colloidal stable exfoliated sheets of hydrophilic GO. The GO contains carbons with sp^2 (C=C, C=O, etc.) and sp^3 (C–C, C–OH, C–O–C, etc.) hybridization and with oxygen functionality such as hydroxyl, epoxy, carbonyl, and carboxyl groups (Dreyer, 2010). However, unique properties of graphene are compromised in GO, which limits its use particularly, in electronic- or energy-based applications. In contrast, colloidal GO offers opportunities for functionalization with molecules/polymers/metal nanoparticles and extends the application potential (Chen et al., 2012; Jung et al., 2014; Xu et al., 2013; Yangab et al., 2013; Huang et al., 2011). Again, GO can be transformed into r-GO by providing thermal/chemical/photochemical/electrochemical reducing atmosphere (Loryuenyong et al., 2013; Mikhailov, 2011; Some et al., 2013; Pei et al., 2011; Zhang et al., 2013). These processes increase the number of sp^2-hybridized carbon atoms and greatly recovered the lost properties of graphene. The r-GO has enhanced self-aggregation processes and lowered the solubility but enhanced the conductivity, light absorption property, and mechanical and thermal stability (Kumar et al., 2014; Park et al., 2009; Becerril et al., 2008). Quality of r-GO depends on the processes and nature of reducing agents (Dreyer et al., 2010; Gao et al., 2010a,b; Guo et al., 2009; Dreyer et al., 2011; Nethravathi et al., 2008; Fan et al., 2011; Dey et al., 2012; Thomas et al., 2013). Similar to GO, the r-GO is also widely used in preparation and processing of graphene-based composites for different applications.

Moreover, other nanoparticles including metal/metal oxide nanoparticles (e.g., Pt, Pd, Co_3O_4, Au, Ag), semiconductor nanoparticles (e.g., TiO_2, CdSe, ZnS, ZnO), and magnetic nanoparticles (e.g., Fe_2O_3, Fe_3O_4) are also widely used in optical, biomedical, and electronic applications (Saha et al., 2009; Kleijn et al., 2014). Metal/metal-oxide/hydroxide nanoparticles act as active center for chemical/electrochemical reactions (e.g., O_2 reduction, biofuel oxidation, organometallic reaction) (Qiao et al., 2011; Sarina et al., 2013; Guo et al., 2013). The semiconductor nanoparticles have definite energy band gap suitable for absorbing/emitting light in the UV–visible range and act as a photocatalyst (Mills et al., 1993). The plasmonic nanoparticles absorb/scatter light in UV–visible region and are used in surface-enhanced Raman spectroscopy, electrochemical, and other detection applications (Wang et al., 2012; Jana et al., 2008; Manimaran et al., 2007; Mohammed et al., 2014). The magnetic nanoparticles have been used in magnetic separation and magnetic resonance imaging applications (Shena et al., 2009). In particular, Ag-, Au-, and Pt-based nanoparticles are widely used in biosensing applications; Pt, Pd, Si, SnO_2, MnO_2, and $Ni(OH)_2$ nanoparticles are widely used in energy-based applications; TiO_2 and ZnO are widely used in photocatalytic applications; and Fe_3O_4- and γ-Fe_2O_3-based nanoparticles are mainly used in magnetic separation applications. But carbon-based nanomaterials and CNP composites offer the property of both components and remove some of the limitations of individual components and in some cases enhance the performance of individual components (Muszynski et al., 2008; Yogeswaran et al., 2007; Yu et al., 2011; Li et al., 2011a,b; Fei et al., 2014; Kamat, 2010) compared to other nanoparticles. In continuation on focusing of nanoparticle-based composites with other carbon-based materials such as CNT and fullerene, the graphene-based similar composites as discussed earlier also offer better performances in most of the applications (Guo et al., 2012; Li et al., 2010a,b,c,d; Yang et al., 2013). The nanoparticles are linked with graphene through hydrophobic, electrostatic interaction or covalent bonds, and the synergistic effect prevents the $\pi-\pi$ stacking between graphitic sheets as well as aggregation between nanoparticles (Huang et al., 2012; Dreyer, 2010; Sarma et al., 2011). Recently, several reviews appear on graphene-based composites with nanoparticle and polymer, focusing mostly on synthesis and property, and limited attention is paid on application potential (Xiaoqiang et al., 2011; Upadhyay et al., 2014; Chen et al., 2012; Fei et al., 2014). In this chapter, we will focus mostly on the synthesis and fundamental basic properties of CNMs (CNPs, CNTs, graphene, and GO) and their composites toward the biological/biomedical and environment-based applications.

1.2 PREPARATION/SYNTHESIS PROCESS OF DIFFERENT CARBON NANOMATERIALS

A common route in making CNP includes high-energy ion beam radiation–based creation of point defect in diamond particle followed by annealing (Gruber et al., 1997; Neugart et al., 2007; Batalov et al., 2009; Yu et al., 2005), laser ablation of graphite followed by oxidation and functionalization (Sun et al., 2006; Gao et al., 2007), thermal decomposition of organic compound (Selvi et al., 2008; Bourlinos et al., 2008a,b), electrooxidation of graphite (Zhao et al., 2008), and oxidation of candle soot with nitric acid (Liu et al., 2007a,b). A wide range of fluorescent carbon particles (CPs) of different colors can be prepared by these approaches: octadecylamine-functionalized diamond nanoparticle showed blue fluorescence (Mochalin et al., 2009), nitrogen-doped diamond showed red fluorescence (Gruber et al., 1997), and candle soot–derived particle (Fu et al., 2007) or thermal decomposition method (Bourlinos et al., 2008a,b; Liu et al., 2009a,b) or laser ablation method (Sun et al., 2006) produced particles with multiple colors. However, quantum yield of most of these particles are too low (<1%) (Liu et al., 2007a,b; Zhao et al., 2008), but have few exceptions (Sun et al., 2006; Liu et al., 2009a,b). In addition, the synthetic methods are cumbersome and inefficient. For example, in the high-energy ion beam radiation–based method, it is difficult to introduce a large number of point defects into ultrafine nano-CPs (<10 nm) for bright luminescence (Gruber et al., 1997; Neugart et al., 2007; Batalov et al., 2009; Yu et al., 2005). Thermal decomposition–based methods produce a low yield of soluble and fluorescent particles with a significant fraction of insoluble product (Selvi et al., 2008; Bourlinos et al., 2008a,b). Soot-based synthesis produces a particle mixture of different colors, and isolation of different colored particles by gel electrophoresis is a difficult task (Liu et al., 2007a,b). A report showed that surface passivation can lead to a significant increase in fluorescence quantum yield (4–15%); however, the exact mechanism is not yet clear (Sun et al., 2006; Bourlinos et al., 2008a,b). Thus, simple, efficient, and large-scale synthesis of fluorescent CNPs and their isolation, purification, and functionalization and hence uses in biological and biomedical applications are very challenging.

In case of CNTs, the growth control of CNTs is mostly done by chemical vapor deposition (CVD) approaches using hydrocarbon sources (CO, methane, ethylene, acetylene), while the arc-discharge method (using arc-vaporization of two carbon rods) and laser ablation methods (using graphite) produce only tangled nanotubes mixed randomly with by-products. After preparation, CNTs are submitted to purification by acid refluxing, surfactant-aided sonication, or air oxidation procedure to eliminate impurities such as amorphous carbon, fullerenes, and transition metals introduced as catalysts during the synthesis (Singh et al., 2012; Usui et al., 2012; Digge et al., 2012). The chemistry involved in the growth and controlling the catalytic nanoparticles could eventually allow for the control of the diameter and chirality of nanotubes. However, we will discuss the most useful synthesis process/techniques for the CNTs. Implications of nanotube functionalization, molecular sensors, and interfacings with biological molecules are also discussed.

The high-quality graphene is generally synthesized by micromechanical exfoliation on silicon substrate or CVD on transition metal surfaces, but these methods have low production efficiency. In addition, insufficient functional groups generated on graphene surfaces limit the preparation of graphene-based nanocomposites (Novoselov et al., 2004; Sarma et al., 2011; Compton et al., 2010; Mkhoyan et al., 2009). Thus, colloidal GO, prepared by Hummer's method or modified Hummer's method, is commonly used for large-scale synthesis of graphene-based composites (Hummers et al., 1958; Marcano et al., 2010).

1.2.1 Synthesis of Carbon Nanoparticles and/or Fluorescence Carbon Nanoparticles

CNPs and/or fluorescence carbon nanoparticles (FCNs) have been synthesized through physical methods, such as high-energy radiation-based creation of point defects in diamond (Gruber et al., 1997; Yu et al., 2005) and laser ablation of graphite (Sun et al., 2006), or through chemical methods, such as oxidation of candle soot (Liu et al., 2007a,b; Ray et al., 2009), carbonization of carbohydrates (Selvi et al., 2008; Peng et al., 2009; Zhang et al., 2010a,b,c; Yang et al., 2011, 2012), thermal decomposition of small molecules (Bourlinos et al., 2008a,b; Wang et al., 2010a,b; Pan et al., 2010; Zhang et al., 2010a,b,c), pyrolysis of polymers (Liu et al., 2009a,b), microwave-based pyrolysis (Liu et al., 2009a,b; Li et al., 2011a,b; Wang et al., 2011; Chandra et al., 2012), P_2O_5-based room temperature dehydration of small molecules (Fang et al., 2012), electrochemical methods (Li et al., 2010a,b,c,d; Bao et al., 2011), and chemical breakdown of carbon fiber (Peng et al., 2012), graphene (Zhuo et al., 2012), or graphite (Hens et al., 2012). However, there are four distinct limitations of the currently available FCNs. First, most of the synthetic methods produce weakly fluorescent FCNs with a quantum yield of less than 1%. Second, no methods are currently available for the large-scale synthesis of high-quality FCNs. Although some methods report milligram-scale FCNs with quantum yields of 5–60%, these methods require a sophisticated high-energy radiation-based synthesis followed by surface functionalization and size separation (Wang et al., 2010a,b). Third, although many methods report blue- and green-emitting FCNs, only a few methods report yellow- and red-emitting FCNs. In addition, yellow- and red-emitting FCNs are generally mixed with blue- and green-emitting FCNs and must be isolated through specialized size separation methods (Wang et al., 2010a,b; Liu et al., 2007a,b; Li et al., 2010a,b,c,d; Hens et al., 2012). Fourth, although the functionalization of nanoprobes is essential to the enhancement of the labeling specificity, such strategies are poorly developed for FCNs (Li et al., 2010a,b,c,d; Yu et al., 2005; Ray et al., 2009; Selvi et al., 2008; Bourlinos et al., 2008a,b; Fang et al., 2012; Peng et al., 2012), and in most cases, bare FCNs have been used for bioimaging and other applications (Yu et al., 2005; Ray et al., 2009; Selvi et al., 2008; Fang et al., 2012; Peng et al., 2012). A few synthesis processes are described in the following section.

1.2.1.1 Synthesis of Carbon Particle From Oxidation of Burning Candle Soot

Liu et al. (2007a,b) and Ray et al. (2009) used the burning candle soot for the synthesis of CNP for bioimaging and other device applications. Ray et al. (2009) showed the different steps in the preparation of CNP from soot as shown in Scheme 1.1. In this process carbon soot was collected from a burning candle putting aluminum foil on the top of burning candle. Then carbon soot

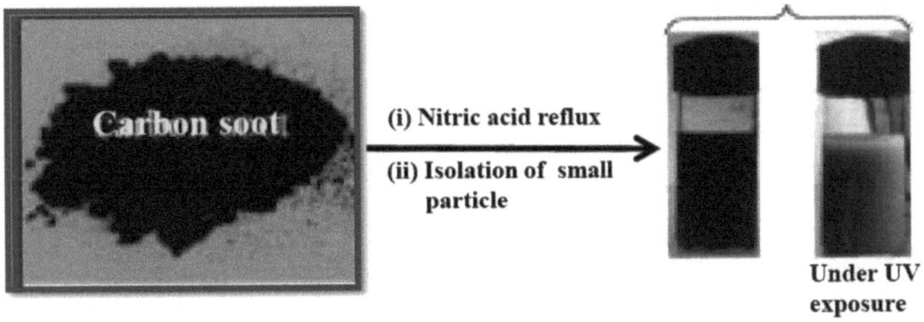

SCHEME 1.1
Steps in the preparation of fluorescent carbon nanoparticles from soot (Ray et al., 2009). *Copyright by the American Chemical Society.*

(25 mg) was mixed with 15 mL of 5 M nitric acid in a 25 mL three-necked flask. It was then refluxed at 100°C for 12 h with magnetic stirring. After that, the black solution was cooled and centrifuged at 3000 rpm for 10 min to separate out unreacted carbon soot. The light brownish-yellow supernatant was collected, which shows green fluorescence under ultraviolet (UV) exposure. The aqueous supernatant was mixed with acetone (water/acetone volume ratio was 1:3) and centrifuged at 14,000 rpm for 10 min. The black precipitate was collected and dissolved in 5–10 mL of water. The colorless and nonfluorescent supernatant was discarded. This step of purification separates excess nitric acid from the CNPs. This concentrated aqueous solution, having almost neutral pH, was taken for further use. The same synthesis technique was also performed for 6 h of reflux and 18 h of reflux. The supernatant obtained from the 6 h reflux was pale yellow and for the 18 h reflux, was dark yellow. The unreacted carbon soot was weighed, which was removed as precipitate, to find out the yield of soluble CNPs. The weight was ~22.5 mg for the 6 h reflux time (yield ~10%) and ~20 mg for the 12 and 18 h reflux times (yield ~20%). It shows that yield increases as reflux time increases from 6 to 12 h but, after that, no significant increases were observed. This solution has particles having sizes ranging from 20 to 350 nm and is called as-synthesized CPs (see Scheme 1.1).

For the separation of different particle size, solvent mixture with the combination of high-speed centrifuge-based separation process could be used. The aforementioned as-synthesized CPs are soluble in water, ethanol, and acetone but insoluble in chloroform. The aqueous and ethanoic particle solutions do not precipitate even at 16,000 rpm centrifugation. So it is better to choose a solvent mixture of water–ethanol–chloroform (single phase without any phase separation) for the size separation of particles where water–ethanol helps to solubilize the particles, but chloroform decreases their solubility. Next, a step-by-step separation is followed using different centrifugation speeds from 4000 to 16000 rpm.

In a typical process, the aqueous solution of CNPs was mixed with chloroform and ethanol, maintaining a water/chloroform/ethanol volume ratio of 1:1:3. Next, 2 mL of this solution was centrifuged at 4000 rpm for 10 min. The precipitate was collected and dissolved in 2 mL of freshwater. The supernatant was then again centrifuged at 5000 rpm for 10 min, and the precipitate was collected and redissolved in 2 mL of water. The same procedure was applied at 6000 and 8000 rpm. The supernatant obtained after 8000 rpm was not getting precipitated even at 16,000 rpm. This solution was collected and evaporated to dryness to remove ethanol and chloroform and finally dissolved in water. This solution has particle sizes of 2–6 nm and is named CNPs that are used for characterization and application. The yield of CNPs from CPs is ~50%, which means 25 mg of soot can produce ~2–3 mg of CNPs. In this calculation, it is assumed that carbon dioxide formation is negligible during nitric acid oxidation.

1.2.1.2 Synthesis of Fluorescence Carbon Nanoparticles From Carbohydrate Carbonization Method

The well-known carbohydrate carbonization method usually produces nonfluorescent CPs of greater than 10 nm size (Sun et al., 2004) or weakly fluorescent blue-green emitting CPs (Peng et al., 2009). But Bhunia et al. (2013) extensively varied the carbonization condition to finely tune the nucleation growth so that CNPs of less than 10 nm size can be prepared with narrow size distribution. Bhunia et al. (2013) followed two different approaches that have been used to control the nucleation growth as shown in Scheme 1.2. In the first approach (Scheme 1.2), different carbohydrate molecules have been used as carbon sources that dehydrate in different temperature. In the second approach, dehydrating agents and reaction temperature are varied to control the nucleation-growth kinetics. This synthetic condition is similar to well-established synthetic conditions for metal/metal oxide/semiconductor nanoparticles (Jana et al., 2008) and produces hydrophobic FCN with fatty amine capping. This approach is most suitable for synthesis of blue- and green-

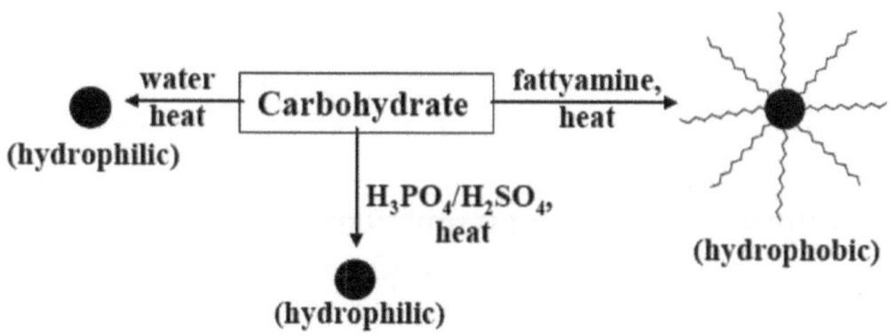

SCHEME 1.2

Different controlled carbonization approaches of carbohydrates in producing fluorescent carbon nanoparticles (Bhunia et al., 2013). *Copyright by Nature publishing group.*

emitting FCN. The best condition for yellow- and red-emitting FCN is concentrated phosphoric acid–based carbonization of carbohydrate at 80–90°C. Presented synthetic method restricts the particle size less than 10 nm and provides appropriate chemical composition of FCN for tunable emission with high fluorescence quantum yield. These FCNs have tunable emission (such as blue, green, yellow, and red) with quantum yield between 6% and 30% and have been synthesized in milligram to gram scale and transformed into various functional nanoprobes for different biological labeling and imaging applications (Bhunia et al., 2013).

In the aforementioned process, Bhunia et al. (2013) synthesized hydrophobic FCN_{blue} and FCN_{green} by mixing 50–200 mg of carbohydrate with 1–2 g of octadecylamine and octadecene and then heating the mixture at 70–300°C for 10–30 min. For example, FCN_{blue} was synthesized by mixing 150 mg of carbohydrate, 0.5 g of octadecylamine, and 10 mL of octadecene and heating the mixture at 80°C for 20 min. Similarly, FCN_{green} was synthesized by mixing 160 mg of carbohydrate and 1.6 g of octadecylamine and heating the mixture at 80°C for 10 min. Next, the reaction mixture was cooled to room temperature, and the dark brown solid was mixed with 5 mL of chloroform or acetone and was shaken vigorously. The soluble FCNs were extracted using organic solvent, and octadecylamine remained as an insoluble white solid. The organic solvent was then evaporated, and the brown, solid FCNs were solubilized in common organic solvents such as acetone, chloroform, toluene, or cyclohexane. However, these hydrophobic FCNs are partially soluble in methanol and ethanol and are insoluble in water. The hydrophobic FCNs were further purified by column chromatography. In this separation process, silica gel (60–120 mesh) was used as the stationary phase, and dichloromethane and methanol were used as the mobile phase. This process separates most of the remaining octadecylamine from the FCNs.

For the synthesis of hydrophilic FCN_{blue} and FCN_{green}, 0.2–1.6 g of carbohydrate is dissolved in 3–10 mL of water, and the pH was made acidic or basic using solutions of NaOH or HCl (Bhunia et al., 2013). After that the solution was heated at ~70–90°C for 20–30 min. Alternatively, 1–4 mL of concentrated carbohydrate solution (350 mg mL^{-1}) was mixed with 5–20 mL of concentrated sulfuric acid and heated to 70–90°C for 20–60 min. For example, FCN_{blue} was synthesized by dissolving 1.5 g of carbohydrate in 5 mL of water, the pH was made acidic by the addition of HCl, and the solution was heated at 90°C for 60 min. Similarly, FCN_{green} was synthesized by dissolving 0.7 g of carbohydrate in 5 mL of water, the pH was made basic by the addition of NaOH solution, and the solution was heated at 90°C for 20 min. FCN_{green} was also synthesized by mixing 1–4 mL of concentrated carbohydrate solution (350 mg mL^{-1}) with 5–20 mL of concentrated sulfuric acid and then heating the solution at 70–90°C for 5–20 min. The yellow/brown solution was neutralized by the addition of NaOH solution and was then dialyzed to remove the excess salts and reagents.

Hydrophilic FCN_{yellow} and FCN_{red} were synthesized by mixing 1–4 mL of aqueous carbohydrate solution (200–400 mg mL^{-1}) with 2.5–10 mL of concentrated phosphoric acid and heating the solution at ~80–90°C for 15–60 min (Bhunia et al., 2013). For example, 700 mg of carbohydrate was dissolved in 2 mL of water, mixed with 5 mL of concentrated phosphoric acid, and heated at 90°C for 15 min to obtain FCN_{yellow}. Similarly, 200 mg of carbohydrate and 0.5 mL of PEG-diamine were dissolved in 2 mL of concentrated phosphoric acid and heated at 90°C for 6 h to obtain FCN_{red}. The synthesized yellow/brown solution was neutralized with NaOH. The resulting precipitate was repeatedly washed or the resulting solution was dialyzed to remove excess salt. Different synthetic methods along with the quantum yield of FCNs are summarized in Table 1.1.

1.2.1.3 Functionalization of Fluorescence Carbon Nanoparticle

Water-soluble FCN having primary amine on their surface has been used by Bhunia et al. (2013) for covalent functionalization with affinity molecules. Hydrophobic FCN that is water insoluble and do not have such amine functionality has been transformed into water-soluble and primary amine–terminated FCN using polymer coating. In a typical procedure of polymer coating, 1 mL chloroform solution of hydrophobic FCN (~10 mg mL^{-1}) was mixed with 0.5 mL chloroform solution of polymaleicanhydride-1-octadecene solution (80 mg dissolved in 0.5 mL chloroform) and sonicated for 5 min. Next, chloroform solution of O,O′-bis(2-aminopropyl) polypropylene glycol-block-polyethylene glycol-block-polypropylene glycol (PEG-diamine) (15 mg PEG-diamine dissolved in 0.5 mL chloroform) was mixed to FCN solution. After overnight reaction, the whole solution was evaporated to dryness and 2–3 mL aqueous Na_2CO_3 solution was added and kept at room temperature for 1–2 h. FCN was completely dissolved in water, which has been used for functionalization. Functionalization of primary amine–terminated FCN has been achieved by Bhunia et al. (2013) using well-known bioconjugation chemistry. Folic acid functionalization of FCN was achieved by using N-hydroxysuccinimide-folate–based approach. Typically, the aqueous solution of FCN was prepared in borate buffer of pH 9 and mixed with dimethyl-formamide solution of N-hydroxysuccinimide-folate. After overnight reaction, the solution was dialyzed using dialysis membrane (MWCF 12–14 kDa) to remove excess reagents.

TAT peptide (CHHHHHHHHHGRKKRRQRRR, MW 2871, 95% purity) functionalization was achieved using well-known 4-maleimido-butyric acid N-hydroxysuccinimide ester reagent. The maleimido group of this reagent reacts with thiol group of cysteine-terminated TAT peptide, and N-hydroxysuccinimide group reacts with the primary amine group of FCN.

Table 1.1 Different Synthetic Methods Along With Quantum Yield (QY) for Fluorescent Carbon Nanoparticles (Bhunia et al., 2013)

Method	Emission (QY)	Size (nm)	Limitations	Refs.
High energy radiation of diamond	Red, NIR (–)	>100	Large size, one color, difficult for large-scale synthesis	Gruber et al. (1997); Yu et al. (2005)
Laser ablation of graphite and electric discharge of carbon	Blue to red (up to 60%)	<10	Surface passivation step and size separation is necessary, yellow/red-emitting particles have very weak emission, difficult to make in milligram scale	Wang et al. (2010a,b); Sun et al. (2006)
Oxidation of candle soot	Blue to red (<3%)	<10	Difficulty of size separation, difficult to make in milligram scale, yellow/red-emitting particles have weak emission	Liu et al. (2007a,b); Ray et al. (2009)
Carbonization of carbohydrate	Blue, green (up to 10%)	<10	Difficult to make yellow/red-emitting particle, mixed with all emission colors	Selvi et al. (2008); Peng et al. (2009); Zhang et al. (2010a,b,c); Yang et al. (2011); Yang et al. (2012)
Carbonization of small molecules	Blue, green (5–50%)	<10	Yellow/red-emitting particles are difficult to make, mixed with all emission colors	Bourlinos et al. (2008a,b); Wang et al. (2010a,b)
Carbonization of polymer	Blue to red (15% for blue)	<10	Low quantum yield for yellow/red emission, mixed with all emission colors	Liu et al. (2009a,b)
Electrochemical	Blue to yellow (upto 12%)	1–5	Difficult to make in large scale, red-emitting particles are difficult to make	Li et al. (2010a,b,c,d); Bao et al. (2011)
Chemical breakdown of carbon fiber/graphene/graphite	Blue, green (0.5–28%)	<100	Difficult to make yellow/red-emitting particle	Cao et al. (2013); Peng et al. (2009); Hens et al. (2012)

This results in covalent linking between FCN and TAT peptide. Typically, the aqueous solution of FCN was mixed with dimethylformamide solution of conjugation reagent and after 30 min TAT peptide solution was added to it. After overnight reaction, the solution was dialyzed using dialysis membrane (MWCF 12−14 kDa) to remove excess reagents and unbound peptide.

However, the aforementioned process for the synthesis of CNPs has tunable emission and are considered to be the next generation green nanomaterials (Baker et al., 2010; Wang et al., 2010a,b; Li et al., 2010a,b,c,d; Cao et al., 2013; Krysmann et al., 2012; Yu et al., 2012; Guo et al., 2012; Zhu et al., 2012). These nanomaterials are promising alternatives to fluorescent semiconductor nanocrystals (Medintz et al., 2005; Pelley et al., 2009). CNPs are potential optical detection probes (Zhu et al., 2012), bioimaging probes (Li et al., 2010a,b,c,d), light-emitting diode materials (Guo et al., 2012), and efficient visible light-active photocatalysts (Yu et al., 2012).

1.2.2 Synthesis and Functionalization of Carbon Nanotubes

CNTs consist exclusively of carbon atoms arranged in a series of condensed benzene rings rolled up into a tubular structure. This novel artificial nanomaterial belongs to the family of fullerenes, the third allotropic form of carbon. Based on the number of layers, structures of CNTs are classified into two types: SWCNTs and MWCNTs. SWCNTs consist of a single graphene cylinder with diameter varying between 0.4 and 2 nm and usually occur as hexagonal close-packed bundles. MWCNTs consist of two to several coaxial cylinders, each made of a single graphene sheet surrounding a hollow core. The outer diameter of MWCNTs ranges from 2 to 100 nm, while the inner diameter is in the range of 1−3 nm, and their length is 0.2 μm to several micrometers (Hirlekar et al., 2009; Bekyarova et al., 2005). From a chemical reactivity point of view, CNT can be differentiated into two zones: the tips and the sidewalls. An important factor that controls these unique properties comes from a variation of tubule structures that are caused by the rolling up of the graphene sheet into a tube. There are three distinct ways for the molecule to do the rolling, depending on its direction: armchair, zigzag, and chiral. Detailed explanations of CNT structures can be found in several reviews (Hirlekar et al., 2009; Liao et al., 2011; Liu et al., 2009a,b; Digge et al., 2012).

Three main techniques generally used for SWCNT and MWCNT production are arc-discharge method (using arc vaporization of two carbon rods), laser ablation method (using graphite), and CVD (using hydrocarbon sources: CO, methane, ethylene, acetylene). After preparation, CNTs are submitted to purification by acid refluxing, surfactant-aided sonication, or air oxidation procedure to eliminate impurities such as amorphous carbon, fullerenes, and transition metals introduced as catalysts during the synthesis (Singh et al., 2012; Usui et al., 2012; Digge et al., 2012).

However, pristine CNTs are not soluble in aqueous solutions because they have highly hydrophobic surfaces. Surface functionalization is required to solubilize CNTs and to render biocompatibility and low toxicity for their medical applications (Liu et al., 2009a,b). The functionalization procedure of CNTs can be divided into two main approaches, depending on the nature of the biomolecule linked to the CNT, that is, covalent attachment (chemical bond formation) and noncovalent attachment (physioadsorption) (Zhang et al., 2010a,b,c; Yang et al., 2007).

The covalent functionalization of CNTs is generally obtained by oxidation with strong acids (HNO_3). During the process, carboxyl (−COOH) groups are formed at the open sides (tips) and at the defects on the sidewalls of SWCNT or MWCNT, then, further covalent conjugation with amino acid. For the creation of −COOH on the sidewalls of CNTs, nitrene cycloaddition, arylation using diazonium salts, or 1,3-dipolar cycloadditions are usually employed (Zhang et al., 2010a,b,c; Kateb et al., 2010; Liu et al., 2007a,b; Yang et al., 2007). Schematic illustration of covalent functionalization of CNTs is summarized in Fig. 1.1 (He et al., 2014).

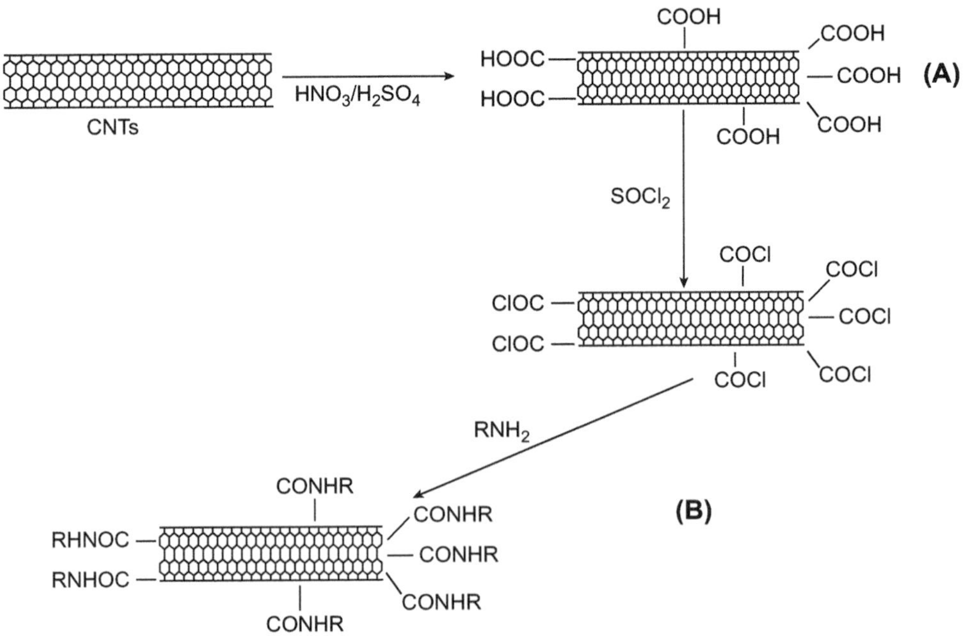

FIGURE 1.1
Covalent functionalization of carbon nanotubes: (A) oxidation reaction by strong acid and (B) further attaching hydrophilic molecules by amidation reactions. *Copyright permission from He, H., et al., 2014. Carbon nanotubes: applications in pharmacy and medicine. Bio Med. Res. Int. 2013, 578290:1–12.*

The noncovalent functionalization of CNTs can be carried out by coating CNTs with amphiphilic surfactant molecules or polymers (polyethylene glycol). The large aromatic (π-electrons) hydrophobic surface of CNTs makes them ideal partners for noncovalent interactions with suitable complementary molecules and macro-biomolecules (DNA). These interactions can take place both on the inside and outside of CNTs. However, macromolecules cannot be linked on their inside (Zhang et al., 2010a,b,c; Digge et al., 2012; Yang et al., 2007). Schematic noncovalent functionalization of CNTs is illustrated in Fig. 1.2 (He et al., 2014).

After functionalization, CNTs become hydrophilic and are ready to be linked with drugs or biomolecules (genes, DNA, proteins, enzymes, biosensors, etc.) for their delivery into the target cells or organs. Detailed techniques of functionalization could be found in the literatures (Usui et al., 2012; Zhang et al., 2010a,b,c; Liu et al., 2007a,b; Chen et al., 2011a,b; Jiang et al., 2012a,b; Xiao et al., 2012; Digge et al., 2012).

1.2.3 Synthesis of Graphene and Graphene Oxide

Chemical structure of graphene, GO, r-GO and G–N composites are shown in Fig. 1.3 (Mondal and Jana, 2014). High-quality graphene is generally synthesized by micromechanical exfoliation on silicon substrate or CVD on transition metal surfaces, but these methods have low production efficiency. In addition, insufficient functional groups generated on graphene surfaces limit preparation

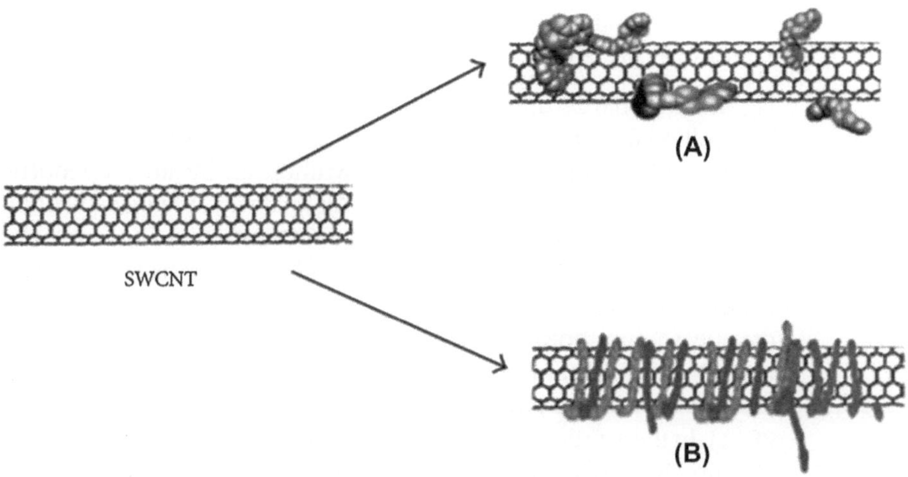

FIGURE 1.2

Noncovalent functionalization of carbon nanotubes with (A) surfactants such as protein adsorption and (B) polymers such as DNA wrapping. *SWCNT*, single-walled carbon nanotubes. *Copyright permission from He, H., et al., 2014. Carbon nanotubes: applications in pharmacy and medicine. Bio Med. Res. Int. 2013, 578290:1–12.*

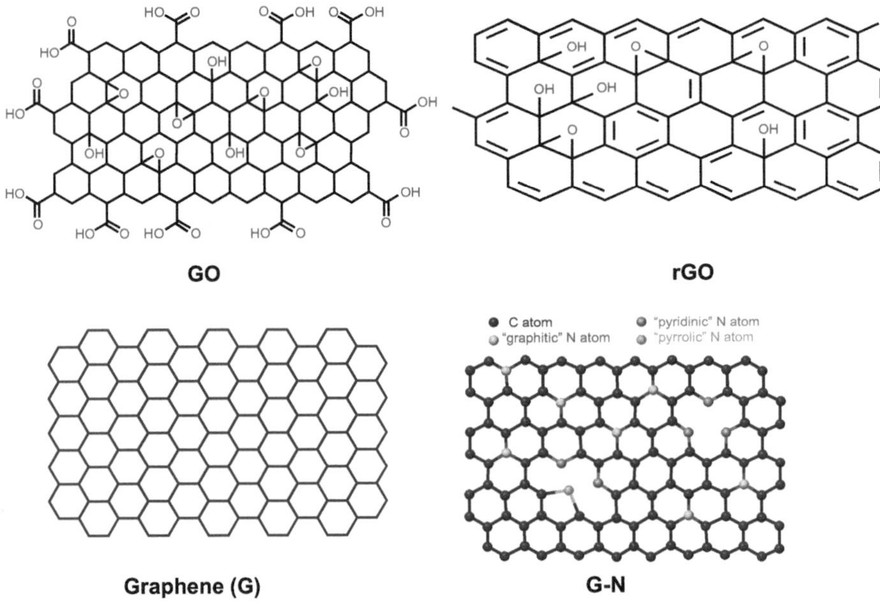

FIGURE 1.3
Chemical structure of graphene oxide (GO), reduced graphene oxide (r-GO), graphene, and graphene–nanoparticle (G–N) composites. *Copyright permission from Mondal et al. (2013).*

of graphene-based nanocomposites (Novoselov et al., 2004; Sarma et al., 2011; Compton et al., 2010; Mkhoyan et al., 2009). Thus colloidal GO, prepared by Hummer's method or modified Hummer's method, is commonly used for large-scale synthesis of graphene-based composites (Hummers et al., 1958; Marcano et al., 2010).

Fig. 1.4 summarize different synthetic approaches (Mondal and Jana, 2014). At first colloidal GO is synthesized from graphite flakes by modified Hummer's method. G–N composites are then synthesized by in situ reduction of metal ions in the presence of colloidal GO. Alternatively, preformed colloidal solution of nanoparticles is mixed with colloidal GO and then GO is reduced to prepare G–N composites. Different surfactants and polymers have been used to improve the dispersion of G–N composites. In addition different reducing agents (e.g., hydrazine, sodium borohydride, ascorbic acid, ethylene glycol, ammonia, alkali solution) and different reaction conditions (temperature, reaction time) have been used to control the property of G–N composites (Qiu et al., 2011; Yen et al., 2011). In these processes metal ions or prefunctionalized metal nanoparticles attach with GO surfaces through electrostatic, covalent, or weak interaction to form nanocomposites (Ren et al., 2011; Liu et al., 2012). Next, reduction of metal ion and/or GO is performed to produce G–N composites (Jiang et al., 2014; Kan and Wang, 2013). The GO-based composites are mainly used in sensing,

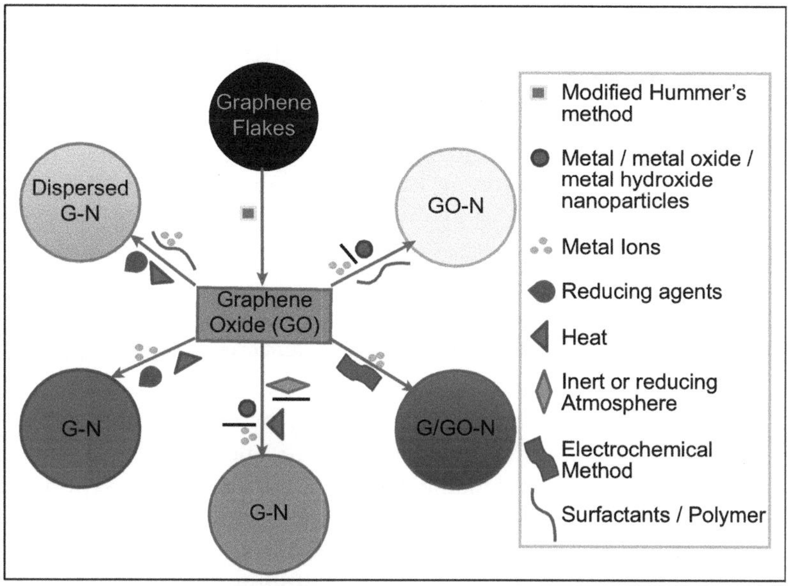

FIGURE 1.4

Schematic illustrations of synthetic strategies for graphene/graphene oxide (GO)—based composites with different nanoparticles. *Copyright permission from Mondal et al. (2013).*

photocatalysis, and water purification applications, whereas graphene-based composites are used for all types of applications.

1.2.3.1 Synthesis of Graphene Oxide by Hummer's Method

In a standard process, GOs are prepared by modified Hummer's method (Saha et al., 2010). In a beaker, 200 mg graphite powder, 100 mg sodium nitrate, and ~5 mL concentrated H_2SO_4 were mixed and cooled to 0°C. Then the solution was kept under vigorous stirring. Next, 600 mg $KMnO_4$ was added to this solution in a stepwise manner so that the temperature was <20°C during these $KMnO_4$ addition steps. After the complete addition of $KMnO_4$, the temperature of the solution was slowly raised to 35°C and kept in this condition for 30 min. A brownish gray paste was formed. Next, ~10 mL water was added to the whole solution and the solution turned brownish yellow. The temperature of the solution was increased to 98°C during the addition of water and this temperature was maintained for 15 min. The whole solution was then mixed with 28 mL of warm water followed by addition of 500 mL 3% H_2O_2 that reduces the residual permanganate. The light yellow particles were washed thoroughly with warm water seven to eight times. The solid was air dried and dissolved in ~20 mL distilled water by 15 min sonication. Then it was centrifuged at 3000 rpm for 30 min. The supernatant was collected as GO solution and used for polyacrylate coating.

1.2.3.2 Synthesis of Polymer-Coated Graphene Oxide and Graphene

In a standard method (Saha et al., 2010), four sets of 2 mL microcentrifuge tubes were used. In one of them, 100 μL of N-(3-aminopropyl) methacrylamide hydrochloride solution (1.8 mg dissolved in 1 mL water) was inserted. In a similar way, 100 μL of polyethylene glycol methacrylate (360 mL dissolved in 1 mL water), 100 μL of bis[2-(methacryloyloxy)-ethyl] phosphate (50 μL dissolved in 1 mL of H_2O), and 100 μL of 3-sulfopropyl methacrylate potassium salt (245 mg dissolved in 1 mL water) were inserted separately in three other micro-centrifuge tubes. Then 0.5 mL of Igepal and 1.4 mL of cyclohexane were added to each micro-centrifuge tube and the solutions were shaken vigorously to make them optically clear. Next, monomeric reverse micelle solutions of the four micro-centrifuge tubes were added to a three-neck flask containing 2 mL of cyclohexane. Then 500 μL of concentrated aqueous GO solution was added to the reaction mixture and 100 μL of organic base (N,N,N, N-tetramethylethylenediamine) was added to the reaction mixture. The whole solution was optically clear and was kept under magnetically stirred conditions and purged with nitrogen for ~15 min to make the reaction mixture O_2 free. After that ammonium persulfate solution (3 mg dissolved in 100 μL of H_2O) was injected to the reaction mixture as the initiator. The reaction was continued for 1 h in N_2 atmosphere and then was quenched by adding a small amount of ethanol. Ethanol addition precipitates the polymer-coated GO, which was washed repeatedly with $CHCl_3$ and ethanol. Next, they were dissolved in doubly distilled water. To remove the unbound polymer, this solution was then centrifuged at 16,000 rpm for 10 min and the precipitated particles were redispersed in freshwater. This solution was used for further applications.

In another process (Bhunia et al., 2011), 10 mg didodecyl dimethyl ammonium bromide was added to 5 mL of as prepared GO and sonicated for 7 min. After that pH was adjusted to 8 by adding ~75 mg of Na_2HPO_4 followed by 5 min sonication. Next, 3 mL of $CHCl_3$ was added, stirred for few seconds, and kept untouched for phase separation. Colorless $CHCl_3$ phase becomes brown leaving the colorless aqueous phase, suggesting that GO was transferred from aqueous to organic phase. The bottom organic phase was collected and mixed with $CHCl_3$ solution of polymaleic anhydride-alt-1-octadecene (40 mg dissolved in 0.5 mL $CHCl_3$) and sonicated for 5 min. Next, a $CHCl_3$ solution of diamine [O,O′-bis(2-aminopropyl) polypropylene glycol-block-polyethylene glycol block-polypropylene glycol] (60 μL dissolved in 500 μL $CHCl_3$) was added in two steps. First, ~100 μL of diamine solution was added and sonicated for 1 min. The rest of the diamine solution was added after 20 min. After overnight, the $CHCl_3$ was evaporated and a brownish precipitate was mixed with 3 mL of water and 5 mg of Na_2CO_3. The whole mixture was sonicated for 5–10 min and left overnight. Optically clear brown solution was obtained, which was used as stock solution.

The polymer-coated GO was further purified from free polymer using acetone-induced precipitation of GO. Typically, 1.0 mL of the polymer-coated GO solution is mixed with 1.0 mL of acetone and centrifuged at 7000 rpm for 1 min. A brown precipitate was formed, leaving the colorless supernatant. The brown precipitate was collected and dissolved in 200 μL of H_2O. Next, 600 μL acetone was added and again centrifuged at 10,000 rpm for 2 min. Brown precipitate was collected, dissolved in 2.5 mL of H_2O, and used for further experiments.

Polymer-coated GO was converted into polymer-coated G *via* hydrazine-based reduction. Typically, 40 μL of hydrazine hydrate was added to the 2.5 mL solution of polymer-coated GO and heated to70–80°C for 1 h. A deep black solution appeared. The solution was dialyzed for overnight to remove excess hydrazine.

1.2.3.3 Functionalization of Graphene and Graphene Oxide

In the functionalization process (Saha et al., 2010), polymer-coated GO/G solution was taken in carbonate buffer solution and mixed with 1:1 molar mixture of glucosamine and glutaraldehyde solution. After 1 h, borohydride solution was added to reduce the imine bond formed between amine and aldehyde. The solution was then dialyzed to remove free reagents. The polyacrylate coating method and functionalization are shown in Scheme 1.3. Colloidal GO

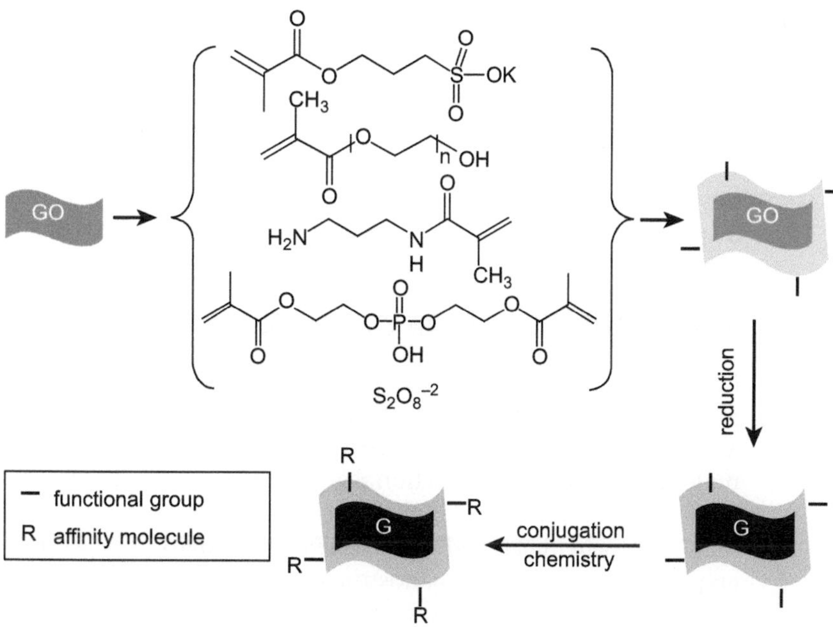

SCHEME 1.3

Polyacrylate coating step for colloidal graphene oxide (GO) followed by their conversion into graphene (G) and subsequent functionalization (Saha et al., 2010). *Copyright by the American Chemical Society.*

was prepared first using Hummer's method as discussed in previous section. Next, it was coated with polyacrylate in reverse micelle medium. Typically, concentrated aqueous solution of GO and acrylate monomers are dissolved in Igepal–cyclohexane reverse micelle to produce an optically clear solution, and then polymerization was initiated in oxygen-free atmosphere using persulfate initiator. After the polymerization was over, the particles were precipitated using ethanol addition, washed with chloroform and ethanol, and finally dispersed in water. Use of reverse micelle medium has the advantage of solubilizing hydrophilic GO and acryl monomers so that polymerization starts from a homogeneous solution and polymer capping can be very efficient (Basiruddin et al., 2010).

Different acrylate monomers; e.g., N-(3-aminopropyl) methacrylamide, 3-sulfopropyl methacrylate, polyethylene glycol methacrylate, and acrylic acid; different cross-linkers, e.g., bis[2-(methacryloyloxy)-ethyl] phosphate and methylenebisacrylamide; and their various combinations were tested. These monomers and cross-linkers were selected for testing because of their good water solubility, better capping property, and reported better performance in producing water-soluble nanoparticle (Wei et al., 2009; Saha et al., 2009; Basiruddin et al., 2010). Cross-linkers were used in 5–10 mol% to minimize the particle–particle linkage during coating. It was found that 3-sulfopropyl methacrylate and its combination with polyethylene glycol methacrylate and N-(3-aminopropyl) methacrylamide work well in producing colloidally stable GO. However, polyethylene glycol methacrylate, N-(3-aminopropyl) methacrylamide, acrylic acid, or their combinations do not produce stable GO solution. Both cross-linkers in 5–10 mol% work well, but bis[2-(methacryloyloxy)-ethyl] phosphate was used because of its lower toxicity. Polymer-coated GO is water soluble and it can be converted into water-soluble colloidal graphene (G) *via* hydrazine reduction.

1.3 PROPERTIES OF CARBON NANOMATERIALS

CNPs, CNTs, graphene, and GOs are considered as a new class of CNMs with properties that differ significantly from other forms of carbon such as graphite and diamond. The ability to custom syntheses of these carbon nanostructure materials with attached functional groups or to assemble analogues has opened up new avenues to design high surface area catalyst supports and materials with high photochemical and electrochemical activity. Unlike the conventional graphite phase, carbon nanostructures possess metallic or semiconductor properties that can induce catalysis by participating directly in the charge transfer process. Furthermore, the electrochemical properties of these materials facilitate modulation of their charge transfer properties and aid in the design of catalysts for hydrogenation, sensors, fuel cells, and biomedical applications.

The fundamental properties of these materials are listed and discussed in the next section.

1.3.1 Properties of Carbon Nanoparticles

1.3.1.1 Physical and Structural Properties

The nature of fluorescence spectra strictly depends on different size/types of particles with different colors. So identification/separation of CNP/FCN from mixture of particles is very important. Liu et al. (2007a,b) separated CNP/FCN from the mixture of nanoparticles of different size by the gel electrophoresis technique. Ray et al. (2009) performed the size separation of particles to isolate CNP/FCN of different fluorescent property. In size separation process, Ray et al. (2009) performed in a solvent mixture with the combination of high-speed centrifuge-based separation. It is identified that mixture of water–ethanol–chloroform as a single-phase solvent is effective for size separation of CPs, where water–ethanol helps to solubilize the CPs, but chloroform decreases their solubility. A step-by-step separation was followed using different centrifugation speed from 4000 to 16,000 rpm and found that the smallest particle does not precipitate even at 16,000 rpm that have highest fluorescence. Bhunia et al. (2013) used a variety of carbohydrates and different nucleation-growth conditions to obtain FCN having tunable emission and with high quantum yield. Fig. 1.5 shows the four types of FCN with different emission colors. Although all samples show excitation-dependent emission spectra, each sample has most intense emissions at certain excitation. Sizes and different physical properties of different FCNs are tabulated within Fig. 1.5.

Transmission electron microscopy (TEM) study of CPs obtained from candle soot by Ray et al. (2009) shows that as-synthesized CPs have broad size distribution from 20 to 350 nm, but FCNs have small and narrow particle size distribution from 2 to 6 nm as shown in Fig. 1.6A–B. Both the CP and FCN have good graphitization and the interlayer spacing between graphitic sheets $d_{(002)} = 0.33$ nm as obtained from high-resolution TEM as shown in the inset Fig. 1.6B, which is very close to that of the ideal graphite. A similar type of CNP (1.5–2.5 nm) is reported following an aqueous route with the help of silica sphere as a carrier (Liu et al., 2009a,b). This type of size-dependent fluorescence QY is observed for CPs produced *via* laser ablation technique (Cao et al., 2013) and thermal decomposition (Batalov et al., 2009; Zyubin et al., 2009) and particles obtained from candle soot (Liu et al., 2007a,b). Ray et al. (2009) observed that the CP and FCN obtained from burning candle soot has a very strong tendency of aggregation during TEM grid preparation or SEM slide preparation as shown in Fig. 1.6A. The aggregation is so high that it has difficulty in finding significant amount of isolated small FCNs under TEM either from as-synthesized CP solution or from FCN solution. A similar type of aggregation is observed by Iijima (1991) where small CPs are found to aggregate into

FCN	Molecular weight/ Particle size	Elemental composition (C:H:N:O)	$Emission_{max}$ (excitation)	Molecular excitation coefficient of excitation	Fluoresce quantum yield
FCN_{Blue}	400-2200Da	65:6:8:21	440 nm (370 nm)	2×10^3	6-30%
FCN_{Green}	2500-14000Da	75:10:5:10	500 nm (400 nm)	5×10^4	14%
FCN_{Yello}	1-4 nm	50:15:2:33	560 nm (425 nm)	4×10^3	12%
FCN_{Red}	4-10 nm	75:10:5:10	500 nm (400 nm)	7×10^5	7%

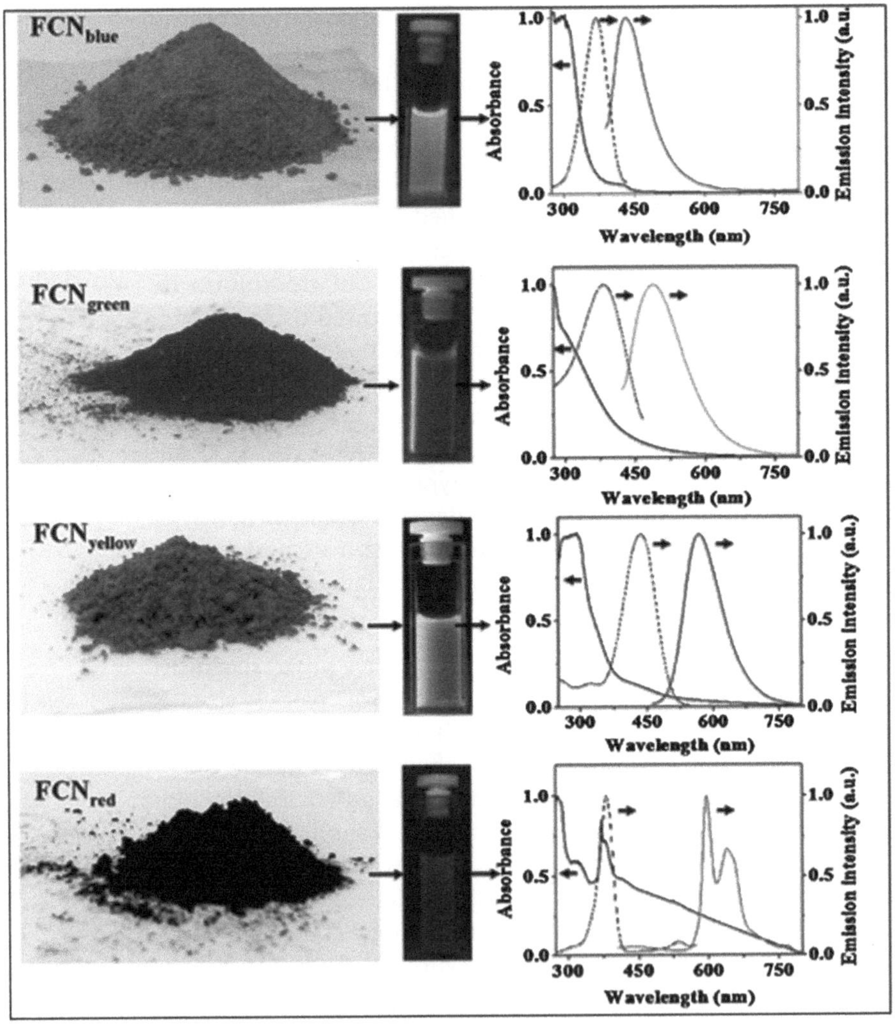

FIGURE 1.5

Digital image of gram-scale solid fluorescent carbon nanoparticle (FCN) samples, digital images of their solutions under respective excitations, and their absorption (*solid lines*), excitation (*dotted lines*), and emission (*colored lines*) spectra. Emission spectra have been measured by exciting at 370 nm for FCN_{blue}, by exciting at 400 nm for FCN_{green}, by exciting at 425 nm for FCN_{yellow}, and by exciting at 385 nm for FCN_{red}. All excitation spectra are recorded in respective emission maxima. Also different properties of these FCNs are tabulated (Bhunia et al., 2013). *Copyright by the Nature publishing group.*

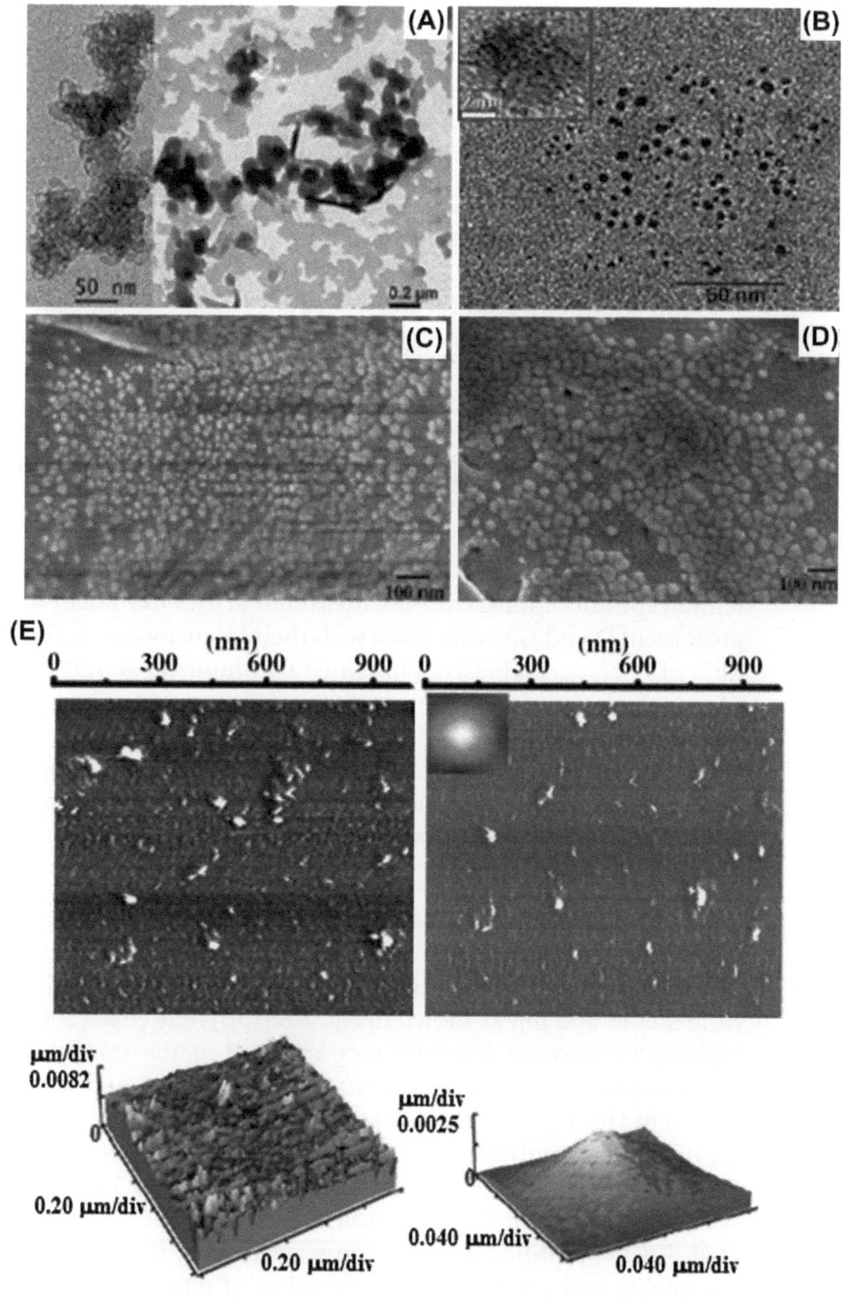

FIGURE 1.6

(A) Transmission electron microscopy (TEM) images of as-synthesized carbon particle showing broad size/shape distribution as well as extensive particle agglomeration. (B) TEM images of small-size FCN with high-resolution image of one particle at the inset. (C) Scanning electron microscopy image of single drop of FCN solution onto Si-substrate. (D) Scanning electron microscopy image of FCN solution on Si-substrate but after having three successive drops. (E) AFM image of FCN taken at different position and magnification (Ray et al., 2009). *Copyright by the American Chemical Society.*

approximately 80-nm-size nanohorn structures. Ray et al. (2009) found that the multiple-drop samples show larger particles than the single drop. It was estimated that the particle size of CNPs as 12–15 nm for the single-drop sample (Fig. 1.6C), but size increases when the sample was prepared from three drops (Fig. 1.6D). Fig. 1.6E shows typical AFM image of FCN structures.

Fig. 1.7A shows the Raman spectra of the FCN and soot, respectively, obtained from Ray et al. (2009). Spectrum of FCN shows high photoluminescence background compared to soot. The two signature peaks for carbon, i.e., D band and G band, are clearly seen for FCNs and soot where D band corresponds to the disordered structure of crystalline sp^2 cluster and G band corresponds to the in-plane stretching vibration mode E_{2g} of single crystal graphite. The intensity ratio (I_D/I_G), which is often used to correlate the structural purity to the graphite, also indicates that FCN is composed of mainly nanocrystalline graphite (Xia et al., 2004). The size of the nanocrystalline graphite obtained from the relation deduced by Ferrari et al. (2000) was calculated as 2.2 nm. Fig. 1.7B shows the EPR spectra of FCN measured at 25°C in comparison to a control CNP sample having poor fluorescence, whereas Fig. 1.7C shows Raman spectra of different FCN measured at 785 nm laser excitation showing prominent D and G bands along with their intensity ratios. Fig. 1.7D shows the XRD of FCN_{green} and FCN_{red} obtained by Bhunia et al. (2013) from oxidation of carbohydrates, showing peak at $2\theta \sim 20°$ corresponding to amorphous CPs and the hump at $2\theta \sim 40°$ corresponding to (100) plane of particles, and Fig. 1.7E shows ^{13}C NMR spectrum of FCN_{green} indicating the presence of both sp^3 (corresponding δ values of 8–80 ppm) and sp^2 (corresponding δ values of 90–180 ppm) carbon atoms.

1.3.1.2 Optical Luminescence Properties

It has been shown by Ray et al. (2009) that the fluorescence spectra of FCN are obtained from the burning of candle soot at different excitation energy ranging from 325 to 600 nm as shown in Fig. 1.8 (top). Ray et al. (2009) found that the highest fluorescence intensity was achieved at the excitation wavelength of 450 nm that shows maximum emission at 520 nm and is green emission as shown in Fig. 1.8. As a comparison, UV–visible absorption spectra of aqueous solution of as-synthesized CP and smallest-size FCN of 2–6 nm are shown in Fig. 1.8, where CP shows clear absorption near 325 nm. Fig. 1.9 (bottom) shows the different FCN photoluminescence spectra obtained by Bhunia et al. (2013) at different excitation wavelengths. Ray et al. (2009) also tested the quality and yield of CP as a function of reflux time and found that 12 h is the optimum time. If the reflux time is less, the yield is low, and for longer reflux time, the yield does not increase appreciably (Ray et al., 2009). Ray et al. (2009) also observed that the intensity of fluorescence of various phosphate buffer solution of FCN does not change appreciably on solution pH

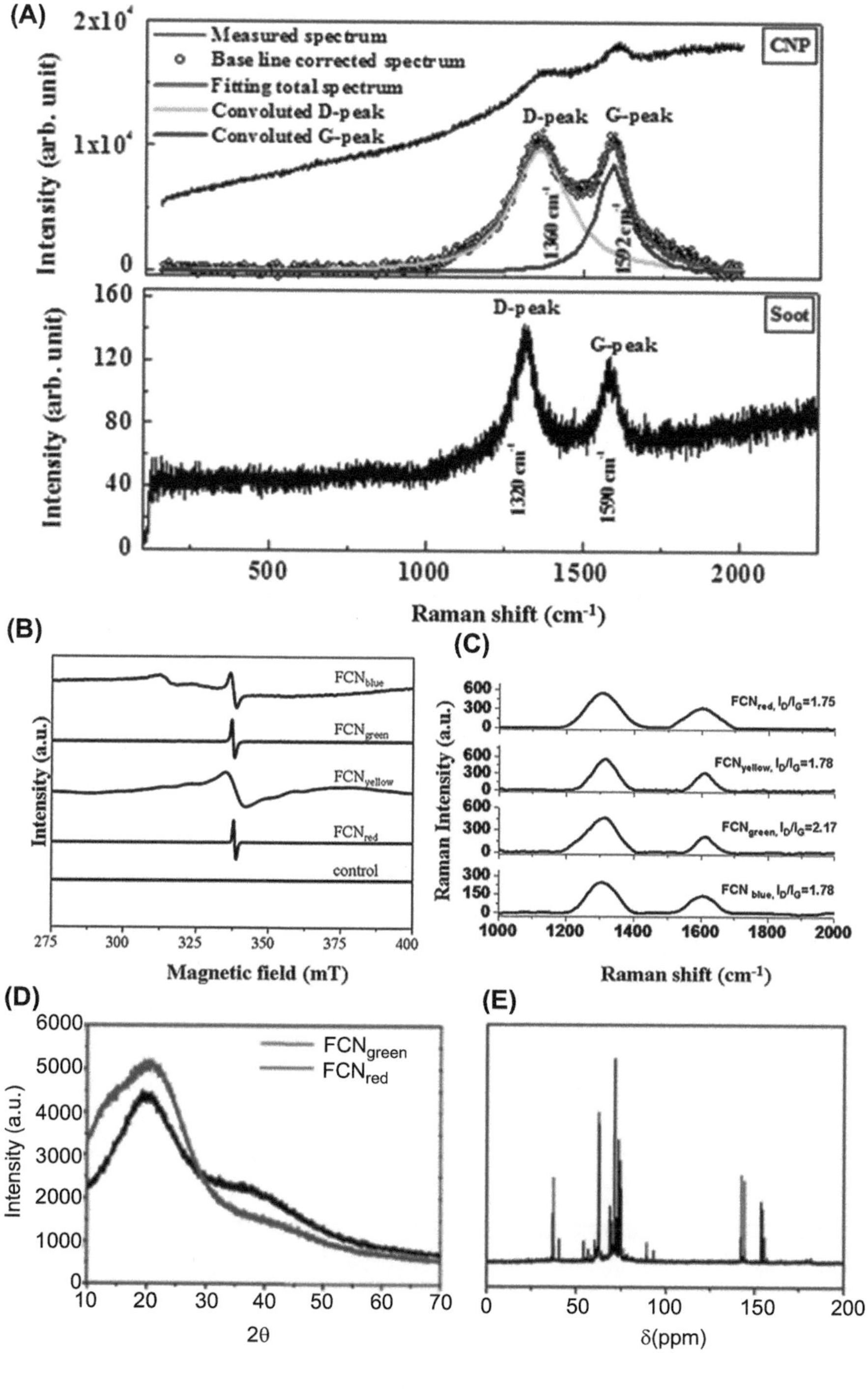

FIGURE 1.7

(A) Raman spectra of the FCN and raw powder of soot (Ray et al., 2009); (B) EPR spectra of FCN measured at 25°C in comparison to a control CNP sample having poor fluorescence; (C) Raman spectra of fluorescent CNPs measured at 785 nm laser excitation showing prominent D and G bands along with their intensity ratio; (D) XRD of FCN_{green} and FCN_{red}, showing peak at $2\theta \sim 20°$ corresponding to amorphous carbon particles and the hump at $2\theta \sim 40°$ corresponding to (100) plane of particles; and (E) ^{13}C NMR spectrum of FCN_{green} indicating the presence of both sp^3 (corresponding δ values of 8–80 ppm) and sp^2 (corresponding δ values of 90–180 ppm) carbon atoms (Bhunia et al., 2013).

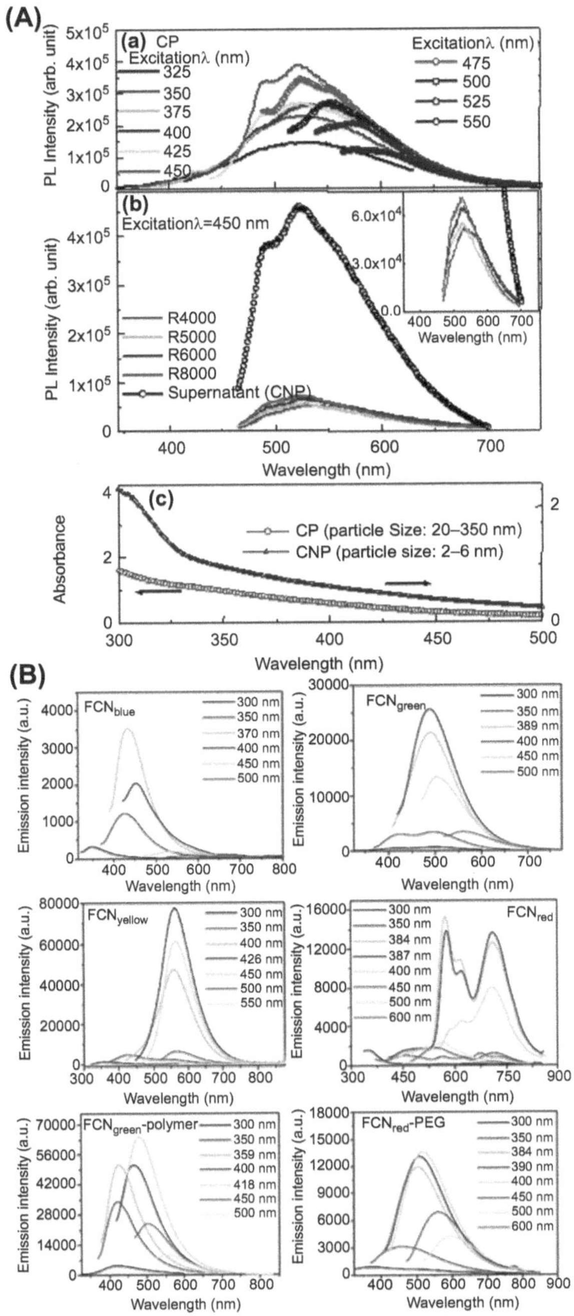

FIGURE 1.8

(Top) Fluorescence spectra of (A) as-synthesized carbon particle (CP), (B) different size CPs obtained at different centrifugation (intense spectrum for smallest size of 2–6 nm), and (C) UV–visible absorption spectra of aqueous solution of as-synthesized CP and the smallest size FCN of 2–6 nm (Ray et al., 2009). (Bottom) Emission spectra of different FCN (Bhunia et al., 2013).

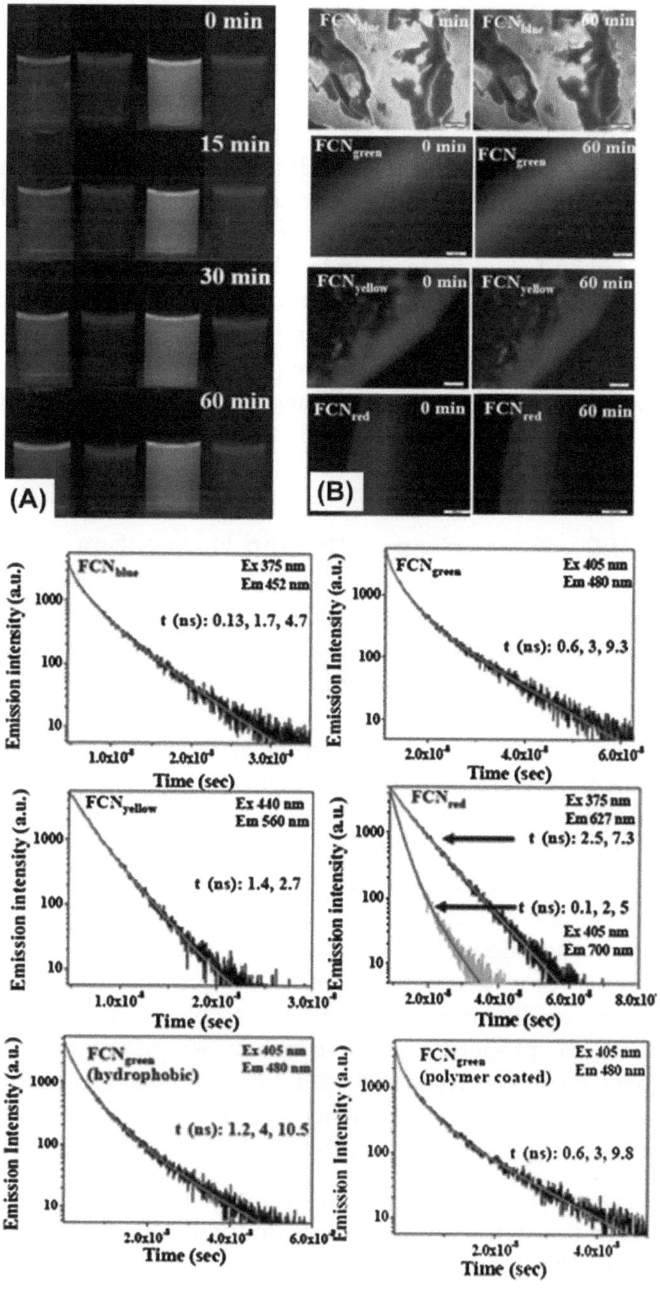

FIGURE 1.9

(Top) Photo-stability studies of (A) different FCN solutions and (B) different FCN films (Bhunia et al., 2013). (Bottom) Fluorescence lifetime decay spectra of solutions of different FCN (Bhunia et al., 2013). *Copyright by the Nature publishing group.*

from 7 to 9. To extend the application potential of FCN as fluorescent label, Bhunia et al. (2013) have investigated the fluorescence stability under different conditions as shown in Fig. 1.9 (top). It shows the fluorescence of FCN solutions under different pH conditions, showing that fluorescence is stable in a wider range of pH and under continuous excitations. In this case, the films of FCNs are prepared by depositing respective aqueous solutions on the glass slides and then imaged under fluorescence mode with blue/green excitation before and after 60 min excitation with 365 nm UV lamp. Fig. 1.9 (bottom) shows lifetime study of FCN and shows that the value is in the range of nanosecond timescale. In this figure (Fig. 1.9 bottom), black/green lines represent experimental data and red/blue lines correspond to fitted data. Experimental excitation/emission wavelength and resultant lifetime values are shown inside each graph.

1.3.2 Properties of Carbon Nanotubes

CNTs are considered to have several unique chemical, optical, electrical, and structural properties that make them attractive as drug delivery and biosensing platforms for the treatment of various diseases (Bianco et al., 2005) and the noninvasive monitoring of blood levels and other chemical properties of the human body, respectively (Clendenin et al., 2007). They are chemically very stable, mechanically very strong and conduct electricity. So they open up new perspectives for various applications, such as nanotransistors in circuits, as artificial muscles, and different biological and medical applications. CNTs can be metallic or semiconducting depending on their structure. This is due to the symmetry and unique electronic structure of graphene. Thus, some nanotubes have conductivities higher than those of copper, while others behave more like silicon. The nanotube structure usually has an effect on the electrical properties of graphene owing to its electronic structure that is quite unique. There are, however, a number of exceptions in the general rule regarding the electrical properties of these nanotubes. Some superconductivity properties have also been registered with some of these nanotubes, and although there are several other experiments that disagree with this assertion, these reports continue to create a stir among experts.

Because of to their nanoscale dimensions, electron transport in CNTs will take place through quantum effects and will only propagate along the axis of the tube. These electrical and structural properties best serve CNTs as far as biosensing is concerned because current changes in the CNTs can signify specific biological entities they are designed to detect. The fact that CNTs are small (nanometer scale) allows them to deliver smaller doses of drugs to specific disease cells in the body thus reducing side effects and harm to healthy cells unlike conventional drugs, while improving disease cell targeting efficiency (Harris, 2009). CNTs have been also observed to have enhanced solubility when functionalized with lipids, which would make their movement through the human body easier and would also reduce the risk of blockage of vital body organ pathways. As far as optical properties are concerned, CNTs have been shown to exhibit strong optical absorbance in certain spectral windows such as NIR (near-infrared) light, and when functionalized with tumor cell, specific binding entities have allowed the selective destruction of disease (e.g., cancer) cells with NIR in drug delivery applications.

The thermal properties of CNTs are quite unique too because these nanotubes are mostly expected to have good conduction. Ballistic conduction is actually a property that is shown by the nanotubes as they go about their thermal conduction. Another vital issue in the properties of these nanotubes is the issue of toxicity. When dealing with nanotechnology, many experts have raised several questions about the toxicity of the nanotubes. This study and research is still in and there is quite a lot criticism still going on. It is expected that over the years the results will help to give experts a little more understanding of how these nanotubes deal with the question of toxicity.

1.3.3 Properties of Graphene and Graphene Oxide

Because of the presence of oxygen functionalities, GO can easily disperse in organic solvents, water, and different matrixes. This is a major benefit when combining the material with polymer or ceramic matrixes to enhance their mechanical and electrical properties. With respect to electrical conductivity, GO functions as an electrical insulator, because of the disturbance of its sp^2 bonding networks. It is important to reduce the GO so as to recover the honeycomb hexagonal lattice of graphene, to restore electrical conductivity. It was observed that after a large number of oxygen groups have been removed, it is not easy to disperse the r-GO, because this material tends to produce aggregates. The properties of graphene can be changed by the functionalization of GO. The chemically altered graphenes obtained by this method could possibly be used in several applications. Depending on the intended application, the GO can be functionalized in a number of ways. One way to ensure that the chemically altered graphenes disperse easily in organic solvents is to use amines through organic covalent functionalization, for instance. This makes the material better suited to production of biodevices and optoelectronics, and for use in drug delivery. It has also been shown that the possibility to attach fullerene-functionalized secondary amines and porphyrin-functionalized primary amines to GO platelets, to enhance the nonlinear optical performance of the material. GO could potentially be used as an intermediary in the production of single-layer or few-layer graphene sheets. To achieve this, an oxidization and reduction process should be developed that can isolate carbon layers and separate them without changing their structure.

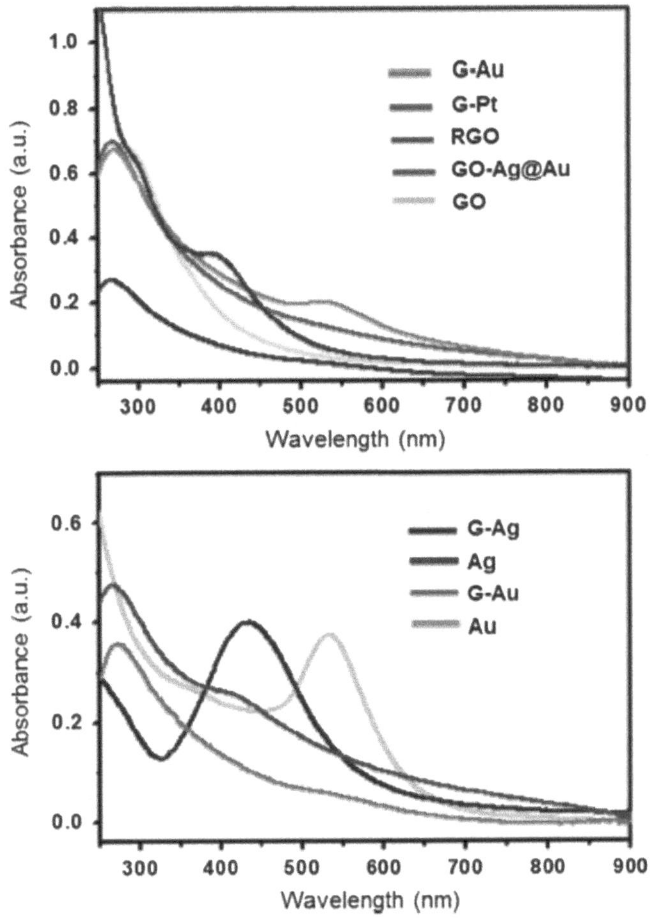

FIGURE 1.10

UV—visible absorption spectra of graphene oxide (GO), reduced graphene oxide (r-GO), and different G—N composites made of silver nanoparticle (G-Ag), gold nanoparticle (G-Au), and platinum nanoparticle (G-Pt). *Copyright permission from Mondal et al. (2013).*

It is observed that the UV—visible absorption spectrum of GO clearly represents the n-π* transition at around 300 nm, whereas π-π* transition is not observed because of lower percentage of conjugated double bonds (Fig. 1.10). The chemical reduction of GO by hydrazine leads to increased percentage of conjugated double bonds and appearance of π-π* transition band at 260 nm. Optical property of G—N composites shows combined property of both components (Fig. 1.10). For example, G—N composites with plasmonic silver or gold nanoparticles show respective plasmonic band along with graphene band. However, plasmonic band is extensively damped by graphene. The plasmonic bands for silver and gold nanoparticles are observed at ~400 nm and ~530 nm, respectively, along with the π-π* transition band at ~260 nm and n-π* transition band at ~300 nm. In the case of platinum and other nanoparticle-based composites, Plasmon band is absent and only graphene band is observed.

Raman spectroscopy is a nondestructive tool for the analysis of order and disorder of the crystallized carbon samples (Wang et al., 2008; Malarda et al., 2009). In pure form, graphene has ordered structure showing intense graphitic bands (G) at 1600 cm^{-1}, which arise from the scattering of the E_{1g} phonon of the sp^2-hybridized carbon (Fig. 1.11). As the disorder arises on graphene surfaces, a new defective band (D) at 1300 cm^{-1} arises from a breathing mode of point photons of A1g symmetry. Such defects can arise due to chemical oxidation, doping, and thermal annealing and result in the disruption of sp^2 domain of graphene surfaces. The disordering of the graphene-based material can be elucidated by the ratio of the intensity of defective bands to graphitic bands (I_D/I_G). For example, the GO and r-GO have the defective structure of graphene, which represents the D band at 1300–1350 cm^{-1} and the G band at 1600 cm^{-1} and the I_D/I_G ratio is >1. This indicates prominent defective nature of graphitic structure. Incorporation of nanoparticles on the graphene surfaces further increases the disordering of graphitic structure and thus the I_D/I_G ratio becomes in the range of 1.9–2.1 (Corral et al., 2011; Subrahmanyam et al., 2010; Liu et al., 2014).

Characteristic reflection from both components can be observed from XRD of G–N composites (Fig. 1.12). The r-GO has two broad reflections of graphitic

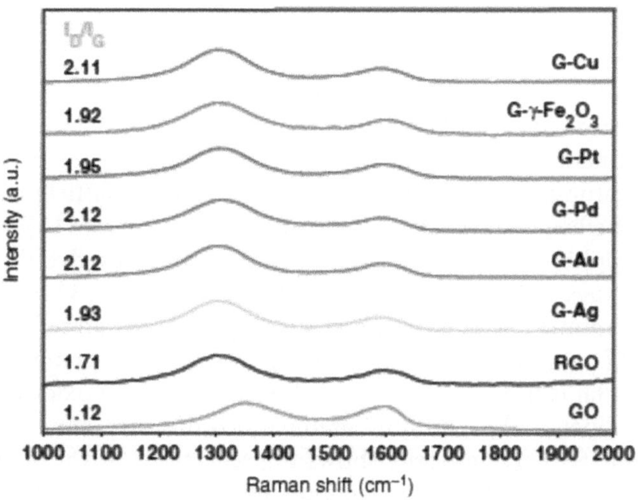

FIGURE 1.11

Characteristic Raman spectra of different G–N composites made of silver nanoparticle (G-Ag), gold nanoparticle (G-Au), platinum nanoparticle (G-Pt), palladium nanoparticle (G-Pd), copper nanoparticle (G-Cu), and iron oxide nanoparticle (G-γ-Fe_2O_3), showing the G-band and D-bands of graphene. *Copyright permission from Mondal et al. (2013).*

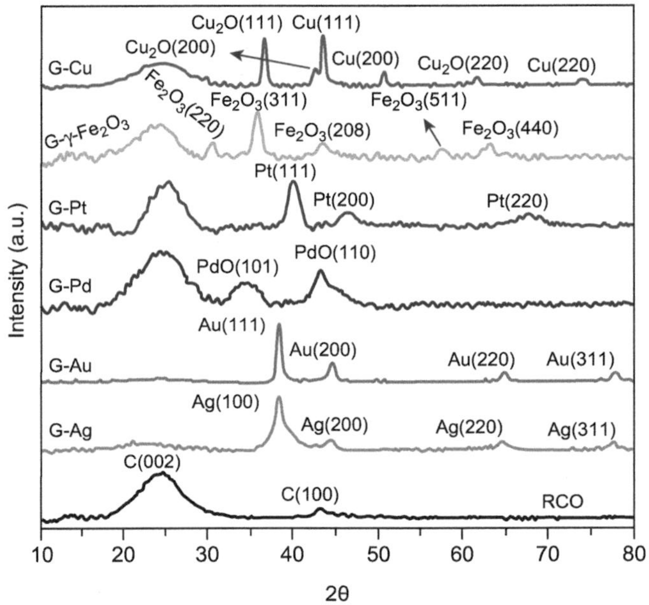

FIGURE 1.12

Characteristic XRD spectra of different G—N composites made of silver nanoparticle (G-Ag), gold nanoparticle (G-Au), platinum-nanoparticle (G-Pt), palladium nanoparticle (G-Pd), copper nanoparticle(G-Cu), and iron oxide nanoparticle (G-γ-Fe_2O_3), showing the reflections from nanoparticle and carbon. *Copyright permission from Mondal et al. (2013).*

planes of carbon (002) and carbon (100). The number of layers and *d* spacing between the planes can be derived from Scherer equation, and as the number of stacking layers increases, the reflection from crystalline carbon becomes sharper. G—N composites show respective reflections from nanoparticle components along with reflection for graphene component (Fig. 1.12). For example, G—N composites with silver nanoparticle show reflections because of silver planes of (100), (200), (220), and (311) as well as reflection of graphitic (002) planes. The G—Au composites give the signature of Au metal planes reflection at (111), (200), (220), and (311) along with reflection of graphitic plane. G-Pt composites show the broad reflections of platinum plane at (111), (200), and (220), along with graphite plane at (002). When less noble metal is used, the reflection of respective metal oxide dominates. For example, copper- and palladium-based nanocomposites show reflection of metal oxides. The G—Cu composite shows the reflections of Cu metal at (111), (200), and (220) and reflections of copper oxide at (200), (111), and (220). G—Pd composites show the reflections of palladium oxide at (101) and (110) along with the signature of graphite. The composite nature of G-γ-Fe_2O_3 is well elucidated from the appearance of reflection because of γ-Fe_2O_3 and graphene.

1.4 CONCLUSION AND PERSPECTIVES OF CARBON NANOMATERIALS

Fluorescent CNP of 1–10 nm in diameter could be obtained using different oxidation and carbonization processes from soot particles and carbohydrates. In case of oxidation process, the surface oxidation and subsequent nitrogen and oxygen doping afforded light-emitting property of CP. It is to be noted that the light emitted by these CPs depends on the wavelength of light used for excitation. Different sizes of CPs are isolated and it is found that emission quantum yield is size-dependent, i.e., smaller the size, better is their photoluminescence efficiency. The fluorescence property of these particles is useful for cell imaging application. These fluorescent CNPs enter into cell without any further functionalization, and fluorescence property of the particle can be used to track their position in cell using conventional fluorescence microscope. The discovery of fluorescent CNPs will no doubt lead to more research in this filed, particularly these particles have the potential in biomedical application where cadmium-based quantum dot shows toxic effects. In case of CPs obtained from carbonization process of carbohydrates, highly fluorescent CNPs with tunable visible emission from blue to red are synthesized in gram scale and transformed into various biological labels. Small size, stable emission, and low toxicity of these nanoprobes keep them in advantageous position over widely used semiconductor nanocrystals. Future work should focus in utilizing the full potential on these nanomaterials, particularly in biomedical science.

Although significant progress has been made in this area, further development is needed with particular focus on following areas. First, the current status of understanding of the origin of fluorescence is poor and multiple mechanisms have been proposed. This lack of clear understanding limits further advancement of synthetic methods. For example, if the doping of heteroatom is the main reason for high fluorescence quantum yield, then synthetic condition would focus in that direction. This understanding would also help to develop high-quality red/NIR-emitting FCN. Second, functionalization of FCN is not yet advanced and only few functional FCNs have been reported. This area should use the knowledge advanced in the area of graphene functionalization and work together with the scientist of organic chemistry background. Third, more detailed study should be conducted on FCN-induced in vitro and in vivo cytotoxicity. This would provide more understanding on long-term effect of FCN to our environment and to human consumption. Fourth, more in vivo studies should be conducted to exploit the advantages of FCN. These should include targeting different organ and brain targeting. With these advancements, it is expected that this research area would create new excitements in next 5–10 years.

References

Baker, S.N., Baker, G.A., 2010. Luminescent carbon nanodots: emergent nanolights. Angew. Chem. Int. Ed. 49, 6726–6744.

Balandin, A.A., 2011. Thermal properties of graphene and nanostructured carbon materials. Nat. Mat. 10, 569–581.

Bao, L., et al., 2011. Electrochemical tuning of luminescent carbon nanodots: from preparation to luminescence mechanism. Adv. Mater. 23, 5801–5806.

Basiruddin, S., et al., 2010. Functionalized gold nanorod solution via reverse micelle based polyacrylate coating. Langmuir 26, 7475–7481.

Batalov, A., et al., 2009. Low temperature studies of the excited-state structure of negatively charged nitrogen-vacancy color centers in diamond. Phys. Rev. Lett. 102, 195506:1–4.

Becerril, H.A., et al., 2008. Evaluation of solution-processed reduced graphene oxide films as transparent conductors. ACS Nano 2, 463–470.

Bekyarova, E., et al., 2005. Applications of carbon nanotubes in biotechnology and biomedicine. J. Biomed. Nanotech 1 (1), 3–17.

Bhunia, S.K., et al., 2011. Peptide-functionalized colloidal graphene via interdigited bilayer coating and fluorescence turn-on detection of enzyme. ACS Appl. Mat. Interf. 3, 3335–3341.

Bhunia, S.K., et al., 2013. Carbon nanoparticle-based fluorescent bioimaging probes. Sci. Rep. 3, 1473:1–7.

Bianco, A., et al., 2005. Applications of carbon nanotubes in drug delivery. Current Opinion Chem. Biol. 9 (6), 674–679.

Bourlinos, A.B., et al., 2008a. Photoluminescent carbogenic dots. Chem. Mater. 20, 4539–4541.

Bourlinos, A.B., et al., 2008b. Surface functionalized carbogenic quantum dots. Small 4, 455–458.

Cahalan, M.D., et al., 2002. Two-photon tissue imaging: seeing the immune system in a fresh light. J. Nat. Rev. Immunol. 2, 872–880.

Cao, L., et al., 2013. Photoluminescence properties of grapheme versus other carbon nanomaterials. Acc. Chem. Res. 46, 171–180.

Chandra, S., et al., 2012. Tuning of photoluminescence on different surface functionalized carbon quantum dots. RSC Adv. 2, 3602–3606.

Chen, D., et al., 2010. Graphene-based materials in electrochemistry. Chem. Soc. Rev. 39, 3157–3180.

Chen, Z., et al., 2011a. Comparative study on contents of oxygen-containing groups on multiwalled carbon nanotubes functionalized by three kinds of acid oxidative methods. Chin. J. Anal. Chem. 39 (5), 718–722.

Chen, Z., et al., 2011b. Adsorption behaviour of epirubic in hydrochloride on carboxylated carbon nanotubes. Int. J. Pharm. 405 (1–2), 153–161.

Chen, D., et al., 2012. Graphene oxide: preparation, functionalization, and electrochemical applications. Chem. Rev. 112, 6027–6053.

Clendenin, J., et al., 2007. An aligned carbon nanotube biosensor for DNA detection. Proc. 2nd. IEEE Conf. Nanotechnol. 1028–1033.

Compton, O.C., Nguyen, S.T., 2010. Graphene oxide, highly reduced graphene oxide, and graphene: versatile building blocks for carbon based materials. Small 6, 711–723.

Corral, I.L., et al., 2011. DFT study of hydrogen adsorption on palladium decorated graphene. J. Phys. Chem. C 115, 4315–4323.

Debgupta, J., et al., 2014. Photophysical and photoconductivity properties of thiol functionalized graphene-CdSe QD composites. RSC Adv. 4, 13788–13795.

Dey, R.S., et al., 2012. A rapid room temperature chemical route for the synthesis of graphene: metal-mediated reduction of graphene oxide. Chem. Commun. 48, 1787–1789.

Digge, M.S., et al., 2012. Applications of carbon nanotubes in drug delivery: a review. Int. J. Pharm. Tech. Res. 4 (2), 839–847.

Dreyer, D.R., et al., 2011. Reduction of graphite oxide using alcohols. J. Mater. Chem. 21, 3443–3447.

Dreyer, D.R., 2010. The chemistry of graphene oxide. Chem. Soc. Rev. 39, 228–240.

Dua, V., et al., 2010. All-organic vapour sensor using inkjet-printed reduced graphene oxide. Angew. Chem. 122, 2200–2203.

El-Sheikh, A.H., Sweileh, J.A., 2011. Recent applications of carbon nanotubes in solid phase extraction and pre-concentration: a review. Jordan J. Chem. 6 (1), 1–16.

Fan, Z.J., et al., 2011. Facile synthesis of graphene nano sheets via Fe reduction of exfoliated graphite oxide. ACS Nano 5, 191–198.

Fang, Y., et al., 2012. Easy synthesis and imaging applications of cross-linked green fluorescent hollow carbon nanoparticles. ACS Nano 6, 400–409.

Fei, H., et al., 2014. Preparation of carbon-coated iron oxide nanoparticles dispersed on graphene sheets and applications as advanced anode materials for lithium-ion batteries. Nano Res. 7, 502–507.

Ferrari, A.C., et al., 2000. Interpretation of Raman spectra of disordered and amorphous carbon. J. Phys. Rev. B 61, 14095–14107.

Fu, C.C., et al., 2007. Characterization and application of single fluorescent nanodiamonds as cellular biomarkers. Proc. Natl. Acad. Sci. USA 104, 727–732.

Galano, A., 2010. Carbon nanotubes: promising agents against free radicals. Nanoscale 2 (3), 373–380.

Gao, L., et al., 2007. Carbon dots for multiphoton bio-imaging. J. Am. Chem. Soc. 129, 11318–11319.

Gao, J., et al., 2010a. Environment-friendly method to produce graphene that employs vitamin C and amino acid. Chem. Mater. 22, 2213–2218.

Gao, X., et al., 2010b. Hydrazine and thermal reduction of graphene oxide: reaction mechanisms, product structures, and reaction design. J. Phys. Chem. C114, 832–842.

Geim, A.K., Novoselov, K.S., 2007. The rise of graphene. Nat. Mat. 6, 183–191.

Geim, A.K., 2009. Graphene: status and prospects. Science 324, 1530–1534.

Glinka, Y.D., et al., 1999. Multiphoton-excited luminescence from diamond nanoparticles. J. Phys. Chem. B 103, 4251–4263.

Gruber, A., et al., 1997. Scanning confocal optical microscopy and magnetic resonance on single defect centers. Science 276, 2012–2014.

Guo, H.L., et al., 2009. A green approach to the synthesis of graphene nanosheets. ACS Nano 3, 2653–2659.

Guo, X., et al., 2012. Facile access to versatile fluorescent carbon dots toward light emitting diodes. Chem. Commun. 48, 2692–2694.

Guo, S., Zhang, S., Sun, S., 2013. Tuning nanoparticle catalysis for the oxygen reduction reaction. Angew. Chem. Int. Ed. 52, 8526–8544.

Harris, P.J.F., 2009. Carbon Nanotube Science: Synthesis, Properties and Applications. Cambridge University Press, Cambridge.

He, H., et al., 2014. Carbon nanotubes: applications in pharmacy and medicine. Biol. Med. Res. Int. 2013, 578290:1–12.

Hens, S.C., et al., 2012. Photoluminescent nanostructures from graphite oxidation. J. Phys. Chem. C 116, 20015–20022.

Hirlekar, R., et al., 2009. Carbon nanotubes and its applications: a review. Asian J. Pharm. Clin. Res. 2 (4), 17–27.

Hou, J., et al., 2011. Graphene based electrochemical energy conversion and storage: fuel cells, supercapacitors and lithium ion batteries. Phys. Chem. Chem. Phys. 13, 15384–15402.

Huang, L.C.L., Chang, H.C., 2004. Adsorption and immobilization of cytochrome *c* on nanodiamonds. Langmuir 20, 5879–5884.

Huang, C., et al., 2011. A graphene oxide/haemoglobin composite hydrogel for enzymatic catalysis in organic solvents. Chem. Comm. 47, 4962–4964.

Huang, C., et al., 2012. Graphene based catalysts. Energy Environ. Sci. 5, 8848–8868.

Hummers Jr., W.S., Offeman, R.E., 1958. Preparation of graphitic oxide. J. Am. Chem. Soc. 80, 1339.

Iijima, S., 1991. Helical microtubes of graphitic carbon. Nature 354, 56–58.

Jana, N.R., Ying, J.Y., 2008. Synthesis of functionalized Au nanoparticles for protein detection. Adv. Mater. 20, 430–434.

Jiang, L., et al., 2012a. Adsorption behavior of pazufloxacinmesilate on amino-functionalized carbon nanotubes. J. Nanosci. Nanotechnol. 12 (1–9), 7271–7279.

Jiang, L., et al., 2012b. Preparation methods of amino-functionalized carbon nanotubes and their application in pharmaceutical field. Prog. Pharm. Sci. 36 (9), 400–405.

Jiang, L., et al., 2014. Facile and novel electrochemical preparation of a graphene-transition metal oxide nanocomposite for ultrasensitive electrochemical sensing of acetaminophen and phenacetin. Nanoscale 6, 207–214.

Jung, H.S., et al., 2014. Nano graphene oxide-hyaluronic acid conjugate for target specific cancer drug delivery. RSC Adv. 4, 14197–14200.

Kamat, P.V., 2010. Graphene-based nano-architectures. Anchoring semiconductor and metal nanoparticles on a two-dimensional carbon support. J. Phys. Chem. Lett. 1, 520–527.

Kan, J., Wang, Y., 2013. Large and fast reversible Li-ion storages in Fe_2O_3-graphene sheet-on-sheet sandwich-like nanocomposites. Sci. Rep. 3, 3502:1–10.

Kateb, B., et al., 2010. Multi-walled carbon nanotube (MWCNT) synthesis, preparation, labelling, and functionalization. Meth. Mol. Biol. 651, 307–317.

Kleijn, S.E.F., et al., 2014. Electrochemistry of nanoparticles. Angew. Chem. Int. Ed. 53, 3558–3586.

Kong, X., et al., 2005a. Polylysine-coated diamond nanocrystals for MALDI-TOF mass analysis of DNA oligonucleotides. Anal. Chem. 77, 4273–4277.

Kong, X.L., et al., 2005b. High-affinity capture of proteins by diamond nanoparticles for mass spectrometric analysis. Anal. Chem. 77, 259–265.

Kroto, H.W., et al., 1985. C_{60}: buckminster fullerene. Nature 318 (6042), 162–163.

Krysmann, M.J., et al., 2012. Formation mechanism of carbogenic nanoparticles with dual photoluminescence emission. J. Am. Chem. Soc. 134, 747–750.

Kumar, P.V., et al., 2014. Scalable enhancement of graphene oxide properties by thermally driven phase transformation. Nat. Chem. 6, 151–158.

Lee, C., et al., 2008. Measurement of the elastic properties and intrinsic strength of monolayer graphene. Science 321, 385–388.

Li, H., et al., 2010a. Water-soluble fluorescent carbon quantum dots and photo-catalyst design. Angew. Chem. Int. Ed. 49, 4430–4434.

Li, Q., et al., 2010b. Photo-luminescent carbon dots as biocompatible nano-probes for targeting cancer cells in vitro. J. Phys. Chem. C 114, 12062–12068.

Li, S., et al., 2010c. Fluorescence study on the interactions between carbon nanotubes and bovine serum albumin. Spectrosc. Spectral Anal. 30 (10), 2689–2692.

Li, Y., et al., 2010d. Catalytic performance of Pt nanoparticles on reduced graphene oxide for methanol electro-oxidation. Carbon 48, 1124–1130.

Li, H., et al., 2011a. One-step ultrasonic synthesis of water-soluble carbon nanoparticles with excellent photo-luminescent properties. Carbon 49, 605–609.

Li, Q., et al., 2011b. Highly efficient visible-light-driven photocatalytic hydrogen production of CdS-cluster-decorated graphene nano sheets. J. Am. Chem. Soc. 133, 10878–10884.

Li, L., et al., 2013. Interaction of carboxylated single-walled carbon nanotubes with bovine serum albumin. Spectroc. Acta A 105, 45–51.

Li, S., 2008. Applications of carbon nanotubes in drug and gene delivery. Prog. Chem. 20 (11), 1798–1803.

Liao, H., et al., 2011. Applications of carbon nanotubes in biomedical studies. Methods Mol. Biol. 726, 223–241.

Lim, T.-S., et al., 2009. Fluorescence enhancement and lifetime modification of single nano-diamonds near a nano-crystalline silver surface. Phys. Chem. Chem. Phys. 11, 1508–1514.

Liu, H., et al., 2007a. Fluorescent carbon nanoparticles derived from candle soot. Angew. Chem. Int. Ed. 46, 6473–6475.

Liu, Z., et al., 2007b. Supramolecular chemistry on water-soluble carbon nanotubes for drug loading and delivery. ACS Nano 1 (1), 50–56.

Liu, R., et al., 2009a. An aqueous route to multicolour photo-luminescent carbon dots using silica spheres as carriers. Angew. Chem. Int. Ed. 48 (25), 4598–4601.

Liu, Z., et al., 2009b. Carbon nanotubes in biology and medicine: in vitro and in vivo detection, imaging and drug delivery. Nano Res. 2 (2), 85–120.

Liu, L., et al., 2012. Graphene oxide enwrapped Ag_3PO_4 composite: towards a highly efficient and stable visible-light induced photo-catalyst for water purification. Catal. Sci. Technol. 2, 2525–2532.

Liu, R., et al., 2014. Metal nanoparticle/carbon quantum dot composite as a photo-catalyst for high-efficiency cyclohexane oxidation. ACS Catal. 4, 328–336.

Loryuenyong, V., et al., 2013. Preparation and characterization of reduced graphene oxide sheets via water-based exfoliation and reduction methods. Adv. Mater. Sci. Eng. 2013, 923403:1–5.

Machadoab, B.F., Serp, P., 2012. Graphene-based materials for catalysis. Catal. Sci. Technol. 2, 54–75.

Malarda, L.M., M, et al., 2009. Raman spectroscopy in graphene. Phys. Rep. 473, 51–88.

Manimaran, M., Jana, N.R., 2007. Detection of protein molecules by surface-enhanced raman spectroscopy-based immunoassay using 2–5 nm gold nanoparticle lables. J. Raman Spectrosc. 38, 1326–1331.

Marcano, D.C., et al., 2010. Improved synthesis of graphene oxide. ACS Nano 4, 4806–4814.

Medintz, I.L., et al., 2005. Quantum dot bio-conjugates for imaging, labelling and sensing. Nat. Mater. 4, 435–446.

Mikhailov, S., 2011. In: Handbook of Physics and Applications of Graphene-Experiments, vol. 5. In Tech, ISBN 978-953-307-217-3.

Mills, A., et al., 1993. Water purification by semiconductor photocatalysis. Chem. Soc. Rev. 22, 417–425.

Mkhoyan, K.A., et al., 2009. Atomic and electronic structure of graphene-oxide. Nano Lett. 9, 1058–1063.

Mochalin, V.N., Gogotsi, Y., 2009. Wet chemistry route to hydrophobic blue fluorescent nanodiamond. J. Am. Chem. Soc. 131, 4594–4595.

Mohammed, A.M., et al., 2014. Application of gold nanoparticles for electrochemical DNA biosensor. J. Nanomat. 2014, 683460:1–7.

Mondal, P., et al., 2013. Enhanced catalytic performance by copper nanoparticle–graphene based composite. RSC Adv 3, 5615–5623.

Mondal, A., Jana, N.R., 2014. Graphene-nanoparticle composites and their applications in energy. Environ. Biomed. Sci. Rev. Nanosci. Nanotechnol. 3, 177–192.

Muszynski, R., et al., 2008. Decorating graphene sheets with gold nanoparticles. J. Phys. Chem. C112, 5263–5266.

Nethravathi, C., Rajamathi, M., 2008. Chemically modified graphene sheets produced by the solvo-thermal reduction of colloidal dispersions of graphite oxide. Carbon 46, 1994–1998.

Neugart, F., et al., 2007. Dynamics of diamond nanoparticles in solution and cells. Nano Lett. 7, 3588–3591.

Novoselov, K.S., et al., 2004. Electric field effect in atomically thin carbon films. Science 306, 666–669.

Pan, D., et al., 2010. Observation of pH-, solvent-, spin-, and excitation-dependent blue photoluminescence from carbon nanoparticles. Chem. Commun. 46, 3681–3683.

Park, S., Ruoff, R.S., 2009. Chemical methods for the production of graphenes. Nat. Nanotech 4, 217–224.

Pei, S., Cheng, H.M., 2011. The reduction of graphene oxide. Carbon 50, 3210–3228.

Pelley, J.L., et al., 2009. State of academic knowledge on toxicity and biological fate of quantum dots. Toxicol. Sci. 112, 276–296.

Peng, H., Travas-Sejdic, J., 2009. Simple aqueous solution route to luminescent carbogenic dots from carbohydrates. Chem. Mater. 21, 5563–5565.

Peng, J., et al., 2012. Graphene quantum dots derived from carbon fibers. Nano Lett. 12, 844–849.

Qiao, Y., Li, C.M., 2011. Nanostructured catalysts in fuel cells. J. Mater. Chem. 21, 4027–4036.

Qiu, J.D., et al., 2011. Controllable deposition of platinum nanoparticles on graphene as an electrocatalyst for direct methanol fuel cells. J. Phys. Chem. C115, 15639–15645.

Rao, C.N.R., et al., 2009. Graphene: the new two-dimensional nanomaterial. Angew. Chem. Int. Ed. 48, 7752–7777.

Ray, S.C., et al., 2009. Fluorescent carbon nanoparticles: synthesis, characterization, and bioimaging application. J. Phys. Chem. C 113, 18546–18551.

Ren, W., et al., 2011. A binary functional substrate for enrichment and ultrasensitive SERS spectroscopic detection of folic acid using graphene oxide/Ag nanoparticle hybrids. ACS Nano 5, 6425–6433.

Rosenand, Y., Elman, N.M., 2009. Carbon nanotubes in drug delivery: focus on infectious diseases. Expert Opin. Drug Delivery 6 (5), 517–530.

Saha, A., et al., 2009. Functionalized plasmonic-fluorescent nanoparticles for imaging and detection. J. Phys. Chem. C 113, 18492–18498.

Saha, A., et al., 2010. Functionalized graphene and graphene oxide solution *via* polyacrylate coating. Nanoscale 2, 2777–2782.

Sarina, S., et al., 2013. Enhancing catalytic performance of palladium in gold and palladium alloy nanoparticles for organic synthesis reactions through visible light irradiation at ambient temperatures. J. Am. Chem. Soc. 135, 5793–5801.

Sarma, S.D., Adam, S., Hwang, E.H., Rossi, E., 2011. Electronic transport in two-dimensional graphene. Rev. Mod. Phys. 83, 407–470.

Seger, B., Kamat, P.V., 2009. Electro-catalytically active graphene platinum nanocomposites. Role of 2-D carbon support in PEM fuel cells. J. Phys. Chem. C 113, 7990–7995.

Selvi, B.R., et al., 2008. Intrinsically fluorescent carbon nano spheres as a nuclear targeting vector: delivery of membrane-impermeable molecule to modulate gene expression in vivo. Nano Lett. 8, 3182–3188.

Shena, Y.F., et al., 2009. Preparation and application of magnetic Fe_3O_4 nanoparticles for wastewater purification. Sep. Purific. Tech. 68, 312–319.

Singh, B.G.P., et al., 2012. Carbon nanotubes. A novel drug delivery system. Int. J. Res. Pharm. Chem. 2 (2), 523–532.

Some, S., et al., 2013. High-quality reduced graphene oxide by a dual-function chemical reduction and healing process. Sci. Rep. 3, 1929:1–5.

Stankovich, S., et al., 2006. Graphene-based composite materials. Nature 442, 282–286.

Stoller, M.D., et al., 2008. Graphene based ultra-capacitors. Nano Lett. 8, 3498–3502.

Subrahmanyam, K.S., et al., 2010. A study of graphene decorated with metal nanoparticles. Chem. Phys. Lett. 497, 70–74.

Sun, X., et al., 2004. Colloidal carbon spheres and their core/shell structures with noble metal nanoparticles. Angew. Chem. Int. Ed 43, 597–601.

Sun, Y.P., et al., 2006. Quantum-sized carbon dots for bright and colorful photoluminescence. J. Am. Chem. Soc. 128, 7756–7757.

Thomas, H.R., et al., 2013. Deoxygenation of graphene oxide: reduction or cleaning? Chem. Mater. 25, 3580–3588.

Tran, P.A., Zhang, L., Webster, T.J., 2009. Carbon nanofibers and carbon nanotubes in regenerative medicine. Adv. Drug Delivery Rev. 61, 1097–1114.

Upadhyay, R.K., et al., 2014. Role of graphene/metal oxide composites as photo-catalysts, adsorbents and disinfectants in water treatment: a review. RSC Adv. 4, 3823–3851.

Ushizawa, K., et al., 2002. Covalent immobilization of DNA on diamond and its verification by diffuse reflectance infrared spectroscopy. Chem. Phys. Lett. 351, 105–108.

Usui, Y., et al., 2012. Carbon nanotubes innovate on medical technology. Med. Chem. 2 (1), 1–6.

Veetil, J., Ye, K., 2009. Tailored carbon nanotubes for tissue engineering applications. Biotechnol. Prog. 25, 709–721.

Wang, G., et al., 2008. Facile synthesis and characterization of graphene nano sheets. J. Phys. Chem. C 112, 8192–8195.

Wang, F., et al., 2010a. One-step synthesis of highly luminescent carbon dots in non-coordinating solvents. Chem. Mater. 22, 4528–4530.

Wang, X., et al., 2010b. Bandgap-like strong fluorescence in functionalized carbon nanoparticles. Angew. Chem. Int. Ed. 49, 5310–5314.

Wang, X., et al., 2011. Microwave assisted one-step green synthesis of cell-permeable multicolour photo-luminescent carbon dots without surface passivation reagents. J. Mater. Chem. 21, 2445–2450.

Wang, P., et al., 2012. Plasmonic photo-catalysts: harvesting visible light with noble metal nanoparticles. Phys. Chem. Chem. Phys. 14, 9813–9825.

Wee, T.-L., 2007. Two-photon excited fluorescence of nitrogen-vacancy centers in proton-irradiated type Ib diamond. J. Phys. Chem. A 111, 9379–9386.

Wei, Y., et al., 2009. Surface coating directed cellular delivery of TAT-functionalized quantum dots. Bio-Conjugate Chem. 20, 1752–1758.

Xia, Y., et al., 2004. Synthesis of Ordered Mesoporous Carbon and Nitrogen-Doped Carbon Materials with Graphitic Pore Walls via a Simple Chemical Vapor Deposition Method. Adv. Mater 16, 1553–1558.

Xiao, D., et al., 2012. Magnetic carbon nanotubes: synthesis by a simple solvo-thermal process and application in magnetic targeted drug delivery system. J. Nanopart. Res. 14, 984:1–12.

Xiao, D., et al., 2013. Development of novel molecularly imprinted magnetic solid-phase extract ion materials based on magnetic carbon nanotubes and their application for the determination of gatifloxacin in serum samples coupled with high performance liquid chromatography. J. Chromat. A 1274, 44–53.

Xiaoqiang, A., Jimmy, C.Y., 2011. Graphene-based photocatalytic composites. RSC Adv. 1, 1426–1434.

Xu, C., et al., 2013. Graphene oxide-TiO_2 composite filtration membranes and their potential application for water purification. Carbon 62, 465–471.

Yang, W., et al., 2007. Carbon nanotubes for biological and biomedical applications. Nanotech 18, 412001:1–12.

Yang, Z.-C., Li, X., Wang, J., 2011. Intrinsically fluorescent nitrogen-containing carbon nanoparticles synthesized by a hydrothermal process. Carbon 49, 5207–5212.

Yang, Y., et al., 2012. One-step synthesis of amino-functionalized fluorescent carbon nanoparticles by hydrothermal carbonization of chitosan. Chem. Commun. 48, 380–382.

Yang, M.Q., Zhang, N., Xu, Y.J., 2013. Synthesis of fullerene-, carbon nanotube-, and graphene-TiO_2 nanocomposite photo-catalysts for selective oxidation: a comparative study. ACS Appl. Mater. Interfaces 5, 1156–1164.

Yangab, M.Q., Xu, Y.J., 2013. Selective photo-redox using graphene-based composite photocatalysts. Phys. Chem. Chem. Phys. 15, 19102–19118.

Yen, M.Y., et al., 2011. Platinum nanoparticles/graphene composite catalyst as a novel composite counter electrode for high performance dye-sensitized solar cells. J. Mater. Chem. 21, 12880–12888.

Yogeswaran, U., et al., 2007. Nanocomposite of functionalized multiwall carbon nanotubes with nafion, nano platinum, and nano gold biosensing film for simultaneous determination of ascorbic acid, epinephrine, and uric acid. Anal. Biochem. 365, 122–131.

Yu, B.Y., Kwak, S.-Y., 2012. Carbon quantum dots embedded with mesoporous hematite nano spheres as efficient visible light-active photocatalysts. J. Mater. Chem. 22, 8345–8353.

Yu, G., et al., 2011. Enhancing the supercapacitor performance of graphene/MnO2 nanostructured electrodes by conductive wrapping. Nano Lett. 11, 4438–4442.

Yu, S.J., et al., 2005. Bright fluorescent nano-diamonds: no photo-bleaching and low cytotoxicity. J. Am. Chem. Soc. 127, 17604–17605.

Yu, J., 2011. Preparation of hydroxypropyl-β-cyclodextrin cross-linked multi-walled carbon nanotubes and their application in enantioseparation of clenbuterol. Chin. J. Chem. 29 (5), 893–897.

Zha, J., et al., 2011. Studies on the interaction of gatifloxacin with bovine serum albumin in the presence of carbon nanotubes by fluorescence spectroscopy. Spectrosc. Spectral Anal. 31 (1), 149–153.

Zhang, S., Pan, N., 2013. Supercapacitor performance of crumpled and planar graphene materials produced by hydrogen gas reduction of graphene oxide. J. Mater. Chem. A 1, 7957–7962.

Zhang, B., et al., 2010a. A novel one-step approach to synthesize fluorescent carbon nanoparticles. Eur. J. Inorg. Chem. 26, 4411–4414.

Zhang, J., et al., 2010b. Controlled synthesis of green and blue luminescent carbon nanoparticles with high yields by the carbonization of sucrose. New J. Chem. 34, 591–593.

Zhang, Y., et al., 2010c. Functionalized carbon nanotubes for potential medicinal applications. Drug Disc. Today 15 (11–12), 428–435.

Zhang, W., et al., 2011. The application of carbon nanotubes in target drug delivery systems for cancer therapies. Nano. Res. Lett. 6, 555–577.

Zhao, Q.-L., et al., 2008. Facile preparation of low cytotoxicity fluorescent carbon nanocrystals by electrooxidation of graphite. Chem. Commun. 5116–5118.

Zhi, L.J., Müllen, K., 2008. A bottom-up approach from molecular nanographenes to unconventional carbon materials. J. Mater. Chem. 18, 1472–1484.

Zhou, J., et al., 2007. An electrochemical avenue to blue luminescent nanocrystals from multiwalled carbon nanotubes (MWCNTs). J. Am. Chem. Soc. 129, 744–745.

Zhu, A., et al., 2012. Carbon-dot-based dual-emission nano-hybrid produces a ratio metric fluorescent sensor for in vivo imaging of cellular copper ions. Angew. Chem. Int. Ed. 51, 7185–7189.

Zhuo, S., et al., 2012. Up-conversion and down-conversion fluorescent graphene quantum dots: ultrasonic preparation and photo-catalysis. ACS Nano 6, 1059–1064.

Zyubin, A.S., et al., 2009. Quantum chemical modeling of photoabsorption properties of two- and three-nitrogen vacancy point defects in diamond. J. Phys. Chem. C 113, 10432–10440.

Application of Carbon-Based Nanomaterials for Removal of Biologically Toxic Materials

2.1 INTRODUCTION

Toxicity is the degree to which a substance can damage an organism. Toxicity can refer to the effect on a whole organism, such as an animal, bacterium, or plant, as well as the effect on a substructure of the organism, such as a cell (cytotoxicity) or an organ such as the liver (hepatotoxicity). Toxicity of materials are dose dependent; even water can lead to water intoxication when taken in too high a dose, whereas for even a very toxic substance such as snake venom, there is a dose below which there is no detectable toxic effect. Toxicity is species specific, making cross-species analysis problematic. Newer paradigms and metrics are evolving to bypass animal testing, while maintaining the concept of toxicity endpoints. A wide variety of organic pollutants are responsible for environmental and water pollution (Benotti et al., 2009). These pollutants cause various environmental and health-related issues worldwide.

A class of organic pollutants known as endocrine disruptors can interfere with hormone systems, resulting in reproductive problems, cancer, hypospadias, miscarriages, endometriosis, and infertility (Boas et al., 2012a,b). Thus, efficient removal/degradation of organic pollutants is an important issue in environmental pollution monitoring and even for water purification (Zhang et al., 2006). The development of cost-effective and stable materials and methods for providing freshwater in adequate amounts is the need of the water industry. Several methods exist for their separation and removal, and among them, adsorption-based separation is the most widely used because of its simplicity and general applicability (Zhang et al., 2006; Basile et al., 2011; Sui et al., 2011). Traditional water/wastewater treatment technologies remain ineffective for providing adequate safe water because of the increasing demand of water coupled with stringent health guidelines and emerging contaminants. Nanotechnology-based multifunctional and highly efficient processes are providing affordable solutions to water/wastewater treatments that do not rely on large infrastructures or centralized systems. However, separation efficiency varies depending on the chemical and physical nature of sorbents and pollutants (Zhang et al., 2006; Basile et al., 2011; Sui et al., 2011). Carbon

CONTENTS

Carbon Nanomaterials for Biological and Medical Applications. http://dx.doi.org/10.1016/B978-0-323-47906-6.00002-3

nanomaterials are very useful in developing more efficient and cheaper nanostructured and reactive membranes for water/wastewater treatment and desalination (Srivastava et al., 2004). Carbon nanomaterials offer opportunities to control the cost of desalination and increase their energy efficiency, and among these are carbon nanoparticles (CNPs), carbon nanotubes (CNTs) (Holt et al., 2006; Fornasiero et al., 2008), and graphene/graphene oxides (GOs) (Sint et al., 2008: Jiang et al., 2009; Bai et al., 2010).

The aim of the present study in this chapter is to review the possible applications of the carbon nanomaterials/fibers for the removal of pollutants from biologically toxic materials. This chapter will briefly overview the availability and practice of different carbon nanomaterials such as CNPs, CNTs, graphene, and GO/reduced graphene oxide (rGO) for the removal of biologically toxic materials. Finally, recommendations are made based on the current practices of carbon nanomaterial applications for the removal of those toxic materials.

2.2 APPLICATION FOR THE REMOVAL OF BIOLOGICALLY TOXIC MATERIALS

2.2.1 Removal of Biologically Toxic Materials Using Carbon Nanoparticles

Determining toxicity of CNPs is quite complicated. The surface coating, the presence of metal catalysts and/or graphite, the exposure to UV radiation, the dispersion properties, the tendency to deposit as aggregates, because of high van der Waals forces, influence greatly the behavior of these particles. Furthermore, species differences complicate research because some species, such as rats, are more sensitive to particles, making it difficult to extrapolate the results to humans. However, some of examples are discussed in the next section.

2.2.1.1 Removal of Toxic Materials by Carbon Nanoparticles by the Process of Sorption

In general, nanoparticles (NPs) can be used as sorbent materials in two configurations: chemically bonded, through a covalent bond to microparticles, and/or directly used as raw materials. When these materials are used as sorbents, the interaction of the analyst can be produced directly on the NP surface or in the interstices of the aggregate. The noncovalent interaction established between the analyst and the NPs or the nanostructured materials includes ionic interactions (dipole–dipole), hydrogen bonds, dispersion forces, $\pi-\pi$ stacking, dative bonds, and the hydrophobic effect. The presence of functionalized NPs or supramolecular aggregates allows the possibility of incorporating one or more of these interactions. However, nanomaterials, being used as sorbents for removing heavy metal ions in wastewater, should satisfy the following criteria:

1. The nanosorbents themselves should be nontoxic.
2. The sorbents present relatively high sorption capacities and selectivity to the low concentration of pollutants.
3. The adsorbed pollutant could be removed from the surface of the nanoadsorbent easily.
4. The sorbents could be infinitely recycled.

Carbon-based nanomaterials are used widely in the removal of heavy metals in 2 decades, because of their nontoxicity and high sorption capacities. Activated carbon (AC) is used firstly as sorbents, but it is difficult to remove heavy metals at parts per billion levels. Then, with the development of nanotechnology, CNTs, fullerenes, and graphene are synthesized and used as nanosorbents. So far, a variety of nanomaterials such as CNTs (Dai, 2002), carbon-based materials such as nanodiamonds (Raty et al., 2005), fullerenes (Jiang et al., 2005), nanoonions (Roy et al., 2003), graphene/graphite nanosheets (Li et al., 2009), and polymeric sorbents have been used for the removal of toxic materials, and the results indicate that these nanomaterials show high adsorption capacity. The sorptive capacity of conventional carbonaceous sorbents is limited by the density of the surface active sites, the activation energy of sorptive bonds, the slow kinetics and nonequilibrium of sorption in heterogeneous systems, and the mass transfer rate to the sorbent surface. The large dimensions of traditional sorbents also limit their transport through low-porosity environments and complicate the applications in subsurface remediation. Carbonaceous nanosorbents with their high surface-area-to-volume ratio, controlled pore size distribution, and surface chemistry overcome many of these intrinsic limitations. Sorption studies using carbon-based nanomaterials report rapid equilibrium rates, high adsorption capacity, effectiveness over a broad pH range, and consistency with Brunauer–Emmett–Teller, Langmuir, or Freundlich isotherms (Yang et al., 2006a,b). Direct sorption of organic contaminants to the nanomaterial surface is driven by the same fundamental hydrophobic, dispersion, and weak dipolar forces to determine sorption energies in conventional systems (Allen-King et al., 2002). The higher equilibrium rates of carbonaceous nanosorbents over AC are attributed to π electron polarizability or π–π electron donor–acceptor interactions within aromatic sorbents, reduced heterogeneity of adsorption energies, and the absence of pore diffusion as an intermediate mechanism in adsorption (Peng et al., 2003). This conclusion is reinforced by Yang et al. (2006a,b) comparing a variety of carbonaceous nanosorbents. Another advantage to carbonaceous nanosorbents is the virtual absence of hysteresis between adsorption and desorption isotherms for liquids and gases under atmospheric pressure. Enhanced atmospheric pressure relevant to gas adsorption in hydrogen storage applications may restore hysteresis in the system by reducing the energy barrier to fill nonwetting CNT pores and the intraparticle region of the nanoaggregates (Punyamurtula et al., 2007). While rapid equilibrium rates and high sorbent capacity are powerful attributes of carbonaceous nanosorbents, their true revolutionary potential lies in the diverse pathways for tailored manipulations of their surface chemistry. Tailoring the dominant physical and chemical adsorption forces via selective functionalization yields carbonaceous nanomaterials that complement the existing suite of relatively unspecific conventional sorbents. Functionalized nanosorbents may provide an optimized approach for targeting micropollutants, removing contaminants (Yan et al., 2006). CNTs functionalized with hydrophilic −OH and −COOH groups exhibited superior sorption of low molecular weight and polar compounds (Lu et al., 2005). In contrast to the relatively nonspecific, hydrophobic sorption mechanisms describing organic sorption, inorganic sorption to carbonaceous nanostructures is characterized by specific complexation reactions. Surface functional group density rather than total surface area becomes the primary determinant of inorganic sorption capacity. Metal speciation or competing complexation reactions render sorption capacity sensitive to changes in pH (Wang et al., 2005). Despite the synthesis costs, the cost-effectiveness of carbon-based nanomaterials versus traditional AC, the environmental applications of nanomaterials sorption capacity being not limited to the removal or remediation of contaminants, but to controlled delivery of fertilizers and pesticides as well. NPs, particularly in soils, can be divided into natural and anthropogenic particles, which are further separated into carbon-containing and inorganic NPs. Natural and anthropogenic combustion processes that take place both in stationary and in mobile sources emit a variety of particles. From these particles the soot fraction of the black carbon corresponds to the nano-sized bacterial cellulose (BC). Soot as a product of recondensation processes during incomplete combustion of fossil and renewable fuels is emitted into the atmosphere from where it is deposited onto soils and water bodies. Although fullerenes are considered as engineered NPs, they are also natural particles (fullerenes). The majority are believed to have formed from polycyclic aromatic hydrocarbons (PAH) during metamorphosis at high temperatures and in the presence of elemental sulfur (Heymann et al., 2003). Dissolved organic carbon and organic colloids in the submicron size range have been recognized as distinct organic phase to which pollutants are partitioning. By their shear abundance, such sorbents may attenuate the truly dissolved exposures of organic pollutants. Soot is an important adsorbent for organic compound, the nonlinear adsorption of organic compounds onto BC completely dominating total sorption at low aqueous concentrations in soils and sediments (Cornelissen et al., 2005). The efficient sorption to BC pulls toxic PAHs, polychlorinated biphenyls, dioxins, polybrominated diphenyl ethers, and pesticides into sediments and soils (Koelmans et al., 2006); the presence of BC explaining the sorption of organic compounds into soils and sediments is much higher in comparison with the absorption into the organic matter alone.

Fullerenes have also been tested for the adsorption of organic compounds, adsorption depending to a great extent on the dispersion state of the C60 (Cheng et al., 2005) incomplete dispersion leading to significant adsorption–desorption hysteresis (Yang and Xing, 2007). Ballesteros et al. (2000) found that fullerenes are not very good sorbents for organic compounds, but they are efficient for the removal of organometallic contaminants.

2.2.2 Removal of Biologically Toxic Materials Using Carbon Nanotubes

The unique strength of CNTs in adsorption of biological contaminants is mainly because of their structural and functional properties. The microbial cytotoxicity of CNTs has a partial influence on the concentration of bacteria. CNTs form aggregated pores because of the entanglement of tens and hundreds of individual tubes that are adhered to each other as a result of van der Waals forces of attraction (Agnihotri et al., 2005; Donaldson et al., 2006; Yan et al., 2008). These aggregated pores have the dimensions of a mesopore or higher (Benny et al., 2008; Chen et al., 2006; Lee et al., 2005; Li et al., 2003a,b,c; Liao et al., 2008; Liu et al., 2006; Mauter and Elimelech, 2008; Niu et al., 2007; Yang et al., 2006a,b) and are able to provide large external surface areas that can immobilize large biological contaminants.

2.2.2.1 Removal of Bacterial Pathogens Using Carbon Nanotubes

Biological contaminants can be classified into three categories, namely microorganisms, natural organic matter (NOM), and biological toxins. Microbial contaminants include human pathogens and free-living microbes (Escobar et al., 2001; Majsterek et al., 2004; Berry et al., 2006; Dugan and Williams, 2006; Srinivasan et al., 2008). The removal of cyanobacterial toxins is an issue in conventional water treatment systems (Huang et al., 2007; Al Momani et al., 2008). Many adsorbents including AC have reasonably good removal efficiencies and again a number of factors influence the removal process (Pendleton et al., 2001; Lee et al., 2004; Yan et al., 2006; Wang et al., 2007).

Contamination from bacteria, protozoans, and viruses is possible in both groundwater and surface water. Standard chlorine-based disinfection produces harmful by-products. Chlorine dioxide is expensive and results in the production of hazardous substances such as chlorite and chlorate in the manufacturing process. Ozone, on the other hand, has no residual effects but produces unknown organic reaction products. For UV disinfection, longer exposure time is required for effectiveness and also there is no residual effect.

Despite advances in disinfection technology, outbreaks from waterborne infections are still occurring. So advanced disinfection technologies must, at least, eliminate the emerging pathogens, in addition to their suitability for large-scale adsorption.

There are many different types of nanomaterials such as Ag, titanium, and zinc capable of disinfecting waterborne disease–causing microbes. Because of their charged structures, they possess antibacterial properties. TiO_2 photocatalysts and metallic and metal-oxide NPs are among the most promising nanomaterials with antimicrobial properties. The efficiency of metal ions in water disinfection has been highlighted by many researchers (Jain and Pradeep, 2005). Other adsorbents including AC have reasonably good removal efficiencies, and a number of factors influence the removal process (Pendleton et al., 2001; Lee et al., 2004; Yan et al., 2006; Wang et al., 2007).

However, the CNTs are proved to be very effective in removing *bacterial pathogens*. CNTs (one of nanosorbents), which have been used for the removal of biological impurities, have received special attention for their excellent capabilities of removing biological contaminants from water (Savage and Diallo, 2005). CNTs possess antimicrobial characteristics against a wide range of microorganisms including bacteria such as *Escherichia coli* and *Salmonella* (Zhu et al., 2006; Deng et al., 2008a,b; Nepal et al., 2008; Upadhyayula et al., 2008a,b,c; Akasaka and Watari, 2009) and viruses (Brady-Estevez et al., 2008; Mostafavi et al., 2009). The adsorption of cyanobacterial toxins on CNTs is also higher when compared with carbon-based adsorbents mainly because of large specific surface area, external diameter of CNTs, large composition of mesoporous volume, and so forth (Long and Yang, 2001; Yan et al., 2004; Ye et al., 2007; de Albuquerque et al., 2008). Researchers have attributed the antimicrobial effects of CNTs to their unique physical, cytotoxic, and surface functionalizing properties (Upadhyayula et al., 2009); their fibrous shape (Kang et al., 2007, 2008a,b,c); the size and length of the tubes; and number of layers (single- or multiwalled) (Li et al., 2008a,b; Wick et al., 2007). The mechanisms of killing bacteria by CNTs are also due to the production of oxidative stress, disturbances to cell membrane, and so forth (Vecits et al., 2010). Although single-walled CNTs (SWCNTs) are more detrimental against microorganisms than multiwalled CNTs (MWCNTs, Kang et al., 2008a,b,c), dispersity of CNTs is a more important parameter than length (Arias and Yang, 2009). Many researchers have observed an extremely high adsorption rate of bacteria by SWCNTs, in addition to their high sorption capacities by many researchers (Jia et al., 2005; Donaldson et al., 2006; Dumortier et al., 2006; Endo et al., 2006; Nuzzo, 2006; Niu et al., 2007; Kar et al., 2008; Jugan et al., 2009; Liu et al., 2009). Filtration membranes containing radially aligned CNTs are very effective in removing both bacteria and viruses in very short time because of size exclusion and depth filtration (Brady-Estevez et al., 2010; Vecitis et al.,

2011; Rahaman et al., 2012; Ahmed et al., 2012; Tiraferri et al., 2011) and thus enable such filters to be used as cost-effective and point-of-use water disinfection devices. CNTs can also reduce membrane biofouling, and a nanocomposite membrane of SWCNTs and polyvinyl-N-carbazole showed high inactivation of bacteria upon direct contact (Ahmed et al., 2012). Another example of controlling the biofouling in thin film nanocomposite membranes is the SWCNTs covalently bonded to thin film composite membrane surface, which have exhibited moderate antibacterial properties (Tiraferri et al., 2011).

2.2.2.2 Removal of Dissolved Salts From Water Using Carbon Nanotubes

Removal of dissolved salts (desalination) is considered an important alternative process for obtaining freshwater. Membrane-based desalination processes cover most of the desalination capability out of which only reverse osmosis (RO) accounts for 41% (Humplik et al., 2011). Parameters that control the desalination cost include maximizing the flux of water through membrane to minimize the fouling. Recent developments in membrane technology have resulted in energy efficiency in RO plants (Fritzmann et al., 2007). Nanofiltration has also been evaluated for desalinating seawater (Mohsen et al., 2003). Nanomaterials are very useful in developing more efficient and cheaper nanostructured and reactive membranes for water/wastewater treatment and desalination such as CNT filters (Srivastava et al., 2004). Nanomaterials offer opportunities to control the cost of desalination and increase its energy efficiency, and among these are CNTs (Holt et al., 2006; Fornasiero et al., 2008) and graphene (Sint et al., 2008; Jiang et al., 2009; Bai et al., 2010). The controlled synthesis of both the length and diameters of CNTs has enabled them to be used in RO membranes to achieve high water fluxes (Pint et al., 2008; Arjmandi et al., 2009; Song and Corry, 2009).

By grafting functional groups, such as carboxyl, at the opening of CNTs, membranes have better selective rejection of some components, but this has resulted in reduced permeability rendering CNTs incapable for desalination (Hoek and Ghosh, 2009; Nednoor et al., 2005; Mauter and Elimelech, 2008). Hinds (2012) concluded a uniform CNT diameter of less than 0.8 nm for high salt rejection. Nanocomposite membranes may serve as ideal membranes for desalination, but a basic understanding of transport mechanism along with proper pore size selection by keeping the uniformity is required for economically feasible and commercially acceptable desalination membranes. The effects of real seawater feed on the efficiency of different nanomaterials need to be investigated in terms of long-term operation and maintenance of membrane performance.

2.2.2.3 Removal of Organic Contaminants Using Carbon Nanotubes

NOM constitutes a diverse group of hydrophobic (humic and fulvic acids) and hydrophilic organic compounds, and it contributes significantly toward water contamination (Vecitis et al., 2010; Kang et al., 2008a,b,c; Arias and Yang, 2009; Jia et al., 2005; Donaldson et al., 2006; Dumortier et al., 2006; Endo et al., 2006). A variety of carbon-based adsorbents have been used for the removal of NOM from raw water, and several factors affect this sorption of NOM (Nuzzo, 2006; Niu et al., 2007; Kar et al., 2008; Jugan et al., 2009).

Among the different types of nanomaterials, nanosorbents such as CNTs, polymeric materials (e.g., dendrimers), and zeolites have exceptional adsorption properties and are applied for the removal of organics from water/wastewater (Savage and Diallo, 2005). CNTs have received special attention because of their exceptional water treatment capabilities, and adsorption of organics on CNTs is widely studied and extensively reviewed (Rao et al., 2007). The removal of NOM by CNTs is higher when compared with carbon-based adsorbents mainly because of large surface areas and other factors (Saleh et al., 2008; Yan et al., 2008). CNTs are also effective in removing polycyclic aromatic organic compounds (Hedderman et al., 2006; Yang et al., 2006a,b; Gotovac et al., 2007) and atrazine (ATZ, Yan et al., 2008). Nanoporous AC fibers, prepared by electrospinning of CNTs, showed much higher organic sorption equilibrium constants for benzene, toluene, xylene, and ethylbenzene than granular AC (Mangun et al., 2001). The sorption of 1,2-dichlorobenzene by CNTs has been shown to be effective in one study (Peng et al., 2003). Compared with AC, both SWCNTs and MWCNTs displayed higher adsorption capacities for trihalomethanes (Lu et al., 2005, 2006a,b,c). MWCNTs have been used as sorbents for chlorophenols, herbicides, dichlorodiphenyltrichloroethanes (DDTs), and so forth (Cai et al., 2005; Zhou et al., 2006a,b, 2007). Finally, novel polymers with functionalized CNTs showed effective removal of organic pollutants from water (Salipira et al., 2007).

2.2.2.4 Removal of Heavy Metals and Ions Using Carbon Nanotubes

Different types of nanomaterials have been introduced for removal of heavy metals from water/wastewater such as nanosorbents including CNTs, zeolites, and dendrimers and they have exceptional adsorption properties (Amin et al., 2014; Savage and Diallo, 2005). The ability of CNTs to adsorb heavy metals is reviewed by many researchers (Rao et al., 2007) such as Cd^{2+} (Li et al., 2003a,b,c), Cr^{3+} (Di et al., 2006), Pb^{2+} (Li et al., 2005), and Zn^{2+} (Lu et al., 2006a,b,c) and metalloids such as arsenic (As) compounds (Peng et al., 2005). Composites of CNTs with Fe and cerium oxide (CeO_2) have also been reported to remove heavy metal ions in few studies (Lu et al., 2005, 2006a,b,c; Salipira et al., 2007). Cerium oxide NPs supported on CNTs are used effectively to adsorb arsenic (Di et al., 2006). Fast adsorption kinetics of CNTs is mainly due to the highly accessible adsorption sites and the short intraparticle diffusion distance (Rao et al., 2007). Some of other removals of pollutants by the CNTs are listed in Table 2.1.

Table 2.1 Removal of Pollutants Using Carbon Nanotubes (CNTs)

Type of Nanoparticle	Removal Target	Initial Concentration	Dose of NPs	Equilibrium Concentration	Adsorption Capacity (mg/g)	pH	References
CNTs	Pb(II)	10 mg/L	0.05 g/L	–	11.2	5.1	Li et al. (2002)
CNTs	Pb(II)	–	–	–	34	5	Li et al. (2003a,b,c)
CNTs	Trihalomethanes ($CHCl_3$)	–	–	–	2.7	3.0 to 7.0	Lu et al. (2005)
SWCNTs	Zn(II)	10–80 mg/L	–	–	437	–	Lu et al. (2006a,b,c)
MWCNTs	Zn(II)	10–80 mg/L	–	–	32.7	–	Lu et al. (2006a,b,c)

MWCNT, *multiwalled carbon nanotube;* NP, *nanoparticle;* SWCNT, *single-walled carbon nanotube.*

2.2.2.5 Removal of Pesticides Using Carbon Nanotubes

The application of pesticides in agricultural practices can increase yields and improve the quality and quantity of products; however, both soils and water sources can become contaminated with these compounds. The removal of residual pesticides from drinking water is of particular importance with respect to human health. Chen et al. (2011a,b) reported the removal of two herbicides (Fig. 2.1) from contaminated water using MWCNTs. The adsorption of diuron onto as-prepared and oxidized MWCNTs was studied by Deng et al. (2012), with the oxidation treatment increasing the surface area and pore volume and resulting in an increased adsorption. SWCNTs have been demonstrated

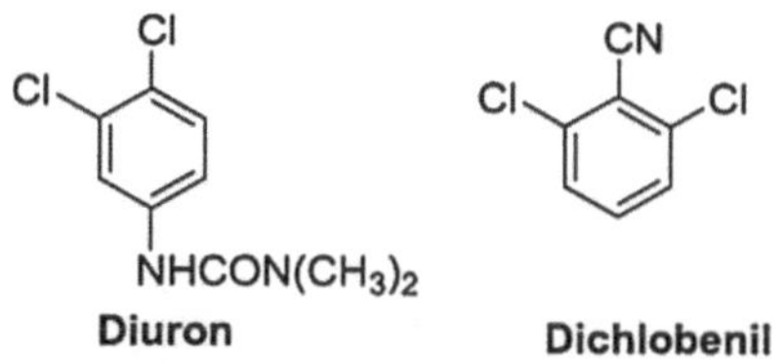

FIGURE 2.1

Chemical structures of diuron and dichlobenil. *Copyright permission from Chen, L., et al., 2011a. Complexation of microcystins and nodularin by cyclodextrins in aqueous solution, a potential removal strategy. Environ. Sci. Technol. 45, 2293–2300; Chen, G.C., et al., 2011b. Adsorption of diuron and dichlobenil on multiwalled carbon nanotubes as affected by lead. J. Hazard Mater. 188, 156–163.*

to have a higher adsorption capacity for 4-chloro-2-methylphenoxyacetic acid, a phenoxy acid herbicide, than three kinds of MWCNTs (with average outer diameters of 15, 30, and 50 nm) and several nanoscale metal oxides (Al_2O_3, TiO_2, and ZnO); the adsorption kinetics usually followed pseudo–second-order kinetics, with the adsorption process being spontaneous and exothermic (De Martino et al., 2012). The adsorption of ATZ from aqueous solution by granular AC and CNTs also occurred via the same kinetic and thermodynamic processes (Rambabu et al., 2012). Overall, the use of CNTs for pesticide removal seems to be less studied than for other organic contaminants. Considering the widespread use of pesticides, more efforts should be directed to investigate the removal of pesticide residues from drinking water and wastewater.

2.2.2.6 Removal of Pharmaceuticals Using Carbon Nanotubes

In light of their environmental persistence, bioaccumulation, and potential toxicity, all currently used human or animal drugs can exert potential negative effects on the environment. For example, steroid hormones, including estradiol, estrone (Fig. 2.2), ethinylestradiol, etc., are known carcinogens, and both natural and synthetic steroid hormones have been found in water systems at nanogram per liter levels. Olaquindox (Fig. 2.2) is a well-known food additive that is highly phototoxic, mutagenic, genotoxic, and carcinogenic. Tetracycline (Fig. 2.2) is one of the most widely used antibiotics in the world but has serious side effects on human health and potential negative effects on the environment when it accumulates in water systems (Yu et al.,

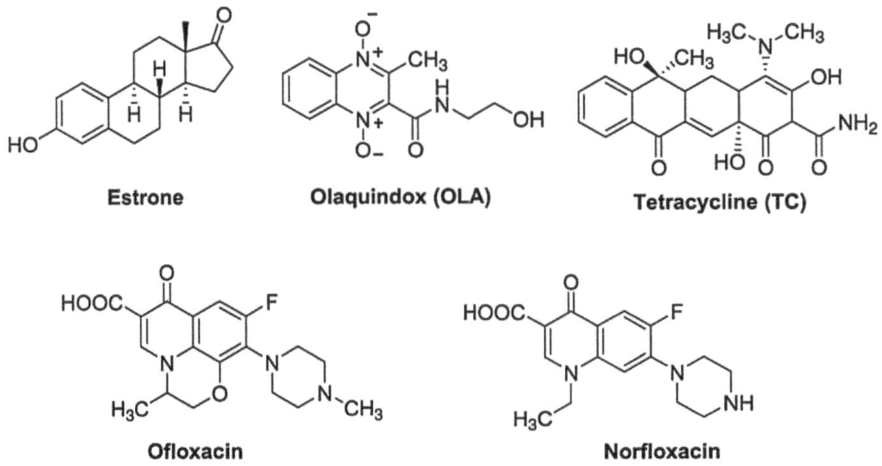

FIGURE 2.2

Chemical structures of pharmaceuticals/drugs targeted for removal by carbon nanotubes. *Copyright permission from Yu et al., 2012. Aqueous adsorption and removal of organic contaminants by carbon nanotubes. Sci. Total Environ. 482, 483, 241–251.*

2012). Thus, effectively removing pharmaceuticals/drugs from contaminated water is of great importance and has attracted significant attention; however, only a few recent reports on their removal by adsorption have been published. As a result of their highly specific surface area and large micropore volume, CNTs are considered superior adsorbents for the removal of pharmaceuticals (Heo et al., 2011; Joseph et al., 2011).

In particular, the removal of endocrine-disrupting compounds (EDCs) from drinking water has attracted considerable attention because of their interference with human reproductive systems by blocking or mimicking the activity of natural hormones. Joseph et al. (2011) investigated the adsorption of EDC from artificial seawater, brackish water, or the combination thereof using SWCNTs and reported a higher removal efficiency for 17α-ethinylestradiol (95–98%) than for bisphenol A (BPA; 75–80%). As a result of the excellent adsorption capacity of SWCNTs, a strong linear correlation between the retention and adsorption of BPA and 17β-estradiol in ultrafiltration (UF) membrane systems blended with SWCNTs (SWCNT–UF) has also been observed (Heo et al., 2012). The addition of SWCNTs to the UF process did not significantly exacerbate the permeate flux decline and total membrane resistance, which is important from a practical standpoint; however, the use of SWCNT–UF reduced the membrane resistance because the SWCNTs offered additional binding sites for the adsorbates (Heo et al., 2012).

To uncover the possible relationship between adsorbates and CNTs (Peng et al., 2012), three functionalized MWCNTs (hydroxylated, carboxylated, and graphitized MWCNTs), three *p*-MWCNTs with different diameters (15 nm, 30 nm, and 50 nm), and three functionalized SWCNTs (hydroxylated, carboxylated, and purified SWCNTs) were used to remove ofloxacin and norfloxacin (Fig. 2.2) from aqueous solutions (Peng et al., 2012). The results indicated that site-specific adsorption might be less important and that the hydrophobic effect contributes more to the adsorption effect. Adsorption of the antibiotics onto the CNTs was described as a structurally controlled process, which is different from that reported for hyper–cross-linked resin MN-202, animated polystyrene resin MN-150, or macroporous resin XAD-4 (Yang et al., 2012).

Importantly, the effective reuse of CNTs will affect their utility in long-term applications. A systematic investigation into the removal of two representative sulfonamide antibiotics demonstrated that small amounts of sulfamethoxazole (3%) and sulfapyridine (9%) were irreversibly bonded to a CNT/sand porous medium during fixed bed regeneration, thus reducing the column capacity (Tian et al., 2013). However, Shi et al. (2013) reported on a new magnetic mesoporous carbon nanocomposite (Fe_3O_4/C) for the removal of ciprofloxacin that retained over 85% of its adsorption capacity even after 10 recycle runs. A newly developed CNTs/Al_2O_3 hybrid adsorbent demonstrated high maximum sorption capacities of 157.4 and 106.5 μmol g^{-1} for diclofenac sodium and carbamazepine, respectively; more importantly, the adsorbent could be thermally regenerated at 400°C in air (Wei et al., 2013).

2.2.3 Removal of Biologically Toxic Materials Using Graphene/Graphene Oxide/Reduced Graphene Oxide

2.2.3.1 Removal of Endocrine-Disrupting Compounds From water

The contamination of water by EDCs, such as BPA, ATZ, 1-naphthol, dibutyl phthalate, DDT, and estrone, is a serious worldwide problem (Falconer et al., 2006; Benotti et al., 2009). Human exposure to EDCs leads to the inhibition of the biological activity of natural hormones through interactions with hormonal receptors and interferes with transport and metabolic processes of natural hormones (Schug et al., 2011; Balabanič et al., 2011; Boas et al., 2012a,b). This activity results in various diseases, such as reproductive problems, cancer, hypospadias, miscarriages, endometriosis, and infertility (Schug et al., 2011; Balabanič et al., 2011; Boas et al., 2012a,b). Although there are various processes by which EDC is released into the environment, the main source of contamination is from industrial wastewater. Thus, the efficient separation of EDCs from contaminated water is becoming an important issue in environmental pollution and water purification.

Currently, several methods exist for the separation of EDCs from wastewater (Ohko et al., 2001; Zhang et al., 2006; Choi et al., 2006; El-Naas et al., 2009; Lofgreen et al., 2011; Griffete et al., 2012; Sreeprasad et al., 2012; Zhao et al., 2012), including membrane filtration (Zhang et al., 2006), bacterial degradation (El-Naas et al., 2009), photocatalytic degradation (Ohko et al., 2001), molecular-imprinted approaches (Lofgreen et al., 2011; Griffete et al., 2012), adsorption-based approaches (Sreeprasad et al., 2012; Nakanishi et al., 2002; Yu et al., 2009; Kuo, 2009; Hu et al., 2010; Sui et al., 2011; Zaib et al., 2012; Chandra et al., 2010; Zhao et al., 2011a,b; Madadrang et al., 2012; Yang et al., 2012; Xu et al., 2012; Wu et al., 2012a,b,c), and other methods (Choi et al., 2006) (Table 2.2). Of these, the molecular-imprinted method seems to be the most promising, owing to its high separation selectivity. However, this method requires individually optimized materials for each type of EDC, and the separation efficiencies of the reported methods have generally been poor (Lofgreen et al., 2011; Griffete et al., 2012). The photocatalytic degradation method is less efficient, particularly at lower EDC concentrations (Ohko et al., 2001). Adsorption-based separation is the most widely used, owing to its simplicity and general applicability for a wide variety of EDCs. For this purpose, various carbonaceous material–based sorbents have been reported, including AC (Nakanishi et al., 2002; Yu et al., 2009), CNTs (Kuo, 2009; Hu et al., 2010; Zaib et al., 2012), porous carbon (Sui et al., 2011), GO (Yang et al., 2012; Xu et al., 2012), and graphene (Chandra et al., 2010; Zhao

Table 2.2 Summary of Different Methods for the Removal of Endocrine-Disrupting Compounds

Materials	Method of Separation	Maximum Removal Capacity (Milligram of Pollutant Removed/Gram of Sorbent)				References
		Bisphenol A	Atrazine	1-Naphthol	Dibutyl phthalate	
Graphene–iron oxide	Adsorption	360	200	680	200	Sinha et al. (2013)[a]
Graphene	Adsorption	182	–	–	–	Xu et al. (2012)
Graphene–iron oxide at 303K, 343K and graphene oxide–iron oxide 303K	Adsorption	–	–	243, 588, 228	–	Yang et al. (2012)
Sulfonated graphene	Adsorption	–	–	347	–	Zhao et al. (2011a,b)
Graphene magnetic nanocomposite	Adsorption	–	–	–	–	Wu et al. (2012a,b,c)
Graphene-coated fiber	Adsorption	–	–	–	–	Wu et al. (2012a,b,c)
Graphene-coated magnetic nanoparticles	Adsorption	–	–	–	–	Zhao et al. (2011a,b)
MWCNTs–iron oxide CD, MWCNTs iron–oxide, and oxidized MWCNTs	Adsorption	–	–	57, 47, 90	–	Sui et al. (2011)
SWNT and t-SWNT at 280, 295, and 315K	Adsorption	19, 15, 14 and 8, 6, 5	–	–	–	Zaib et al. (2012)
CNT and modified CNT	Adsorption	61, 70	–	–	–	Hu et al. (2010)
Porous carbon	Adsorption	186	–	–	–	Sui et al. (2011)
Magnetic NP–polymer	Molecular imprinted	4	–	–	–	Griffete et al. (2012)
Membrane	Membrane filtration	–	–	–	–	Zhang et al. (2006)
TiO_2	Photodegradation	–	–	–	–	Ohko et al. (2001)

[a]*The synthesis of graphene-γ-Fe_2O_3 composite materials and the approach for the separation of EDCs are* shown in *Scheme 2.1. First, hydrophobic γ-Fe_2O_3 nanoparticles (7–8 nm) were synthesized by high-temperature thermal degradation and then converted into primary amine–terminated water-soluble g-Fe_2O_3 particles through silica coating (silica shell thickness: 2–5 nm) (Jana et al., 2007). These resultant particles had a cationic surface charge and a hydrodynamic diameter of 10–15 nm (Jana et al., 2007).*

et al., 2011a,b; Madadrang et al., 2012; Xu et al., 2012; Wu et al., 2012a,b,c). However, their separation efficiencies vary depending on the chemical and physical nature of the sorbent and the chemical nature of the EDC. In addition, their separation efficiencies are generally poor, and the separations of very few EDCs have been reported (see Table 2.2). Ideally, a sorbent material should have a high surface area, fast adsorption rate, and high adsorption capacity, and they should be nontoxic, cheap, easily removable, and, preferably, reusable. Considering all of these requirements, graphene seems to be the most promising material, owing to its high surface area (about 2000–3000 $m^2\ g^{-1}$), which is suitable for the adsorption of organic molecules (Englert et al., 2011; Zhu et al., 2011; Dreyer et al., 2010). However, the potential application of graphene sheets is severely limited, owing to their inherent aggregation properties (Englert et al., 2011; Zhu et al., 2011; Dreyer et al., 2010) and the requirement of appropriate functionalization (Zhao et al., 2011a,b; Yang et al., 2012) or alternative strategies (Chandra et al., 2010) for the effective utilization of their exposed surface.

Herein we discuss the process of removal of various EDCs using graphene-based composites with γ-Fe_2O_3 NPs (graphene-γ-Fe_2O_3) (Sinha et al., 2013). The advantage of this composite material is that the γ-Fe_2O_3 NPs partially inhibit the graphene–graphene interactions, thereby allowing graphene to act as a dispersible sorbent without compromising much of its surface area. The dispersibility of the composite material allows the efficient adsorption of EDCs onto an exposed graphene surface, and the external magnet–induced rapid aggregation of composite offers the facile separation of any adsorbed EDCs.

The removal procedure involved the addition of solid composite materials into solutions of EDCs at different concentrations (5–30 $mg\ L^{-1}$), followed by stirring overnight. Next, the composite materials were separated by using a magnet, the supernatant water was collected, and the presence of any remaining EDC was estimated by spectrometry. Fig. 2.3 shows representative results of the removal efficiency of various EDCs by using these graphene-γ-Fe_2O_3 composites (Sinha et al., 2013). The results show that the general removal efficiencies at all of the tested concentrations are >85% for all of the EDCs considered. In addition, these composite materials can be reused without significant loss of their removal efficiency as shown in Fig. 2.4. Sinha et al. (2013) also found that the removal efficiency of a composite is better when the loading of γ-Fe_2O_3 NPs in the composite is 10 wt%, rather than 25 or 0 wt% as shown in Fig. 2.4. One tentative explanation is that a higher loading of γ-Fe_2O_3 lowers the overall surface area of the composite, by occupying most of the exposed graphene surface, thereby decreasing the number of active adsorption sites for EDCs. In the absence of any γ-Fe_2O_3 NPs, graphene–graphene interactions dominate, thus lowering both the surface area and the

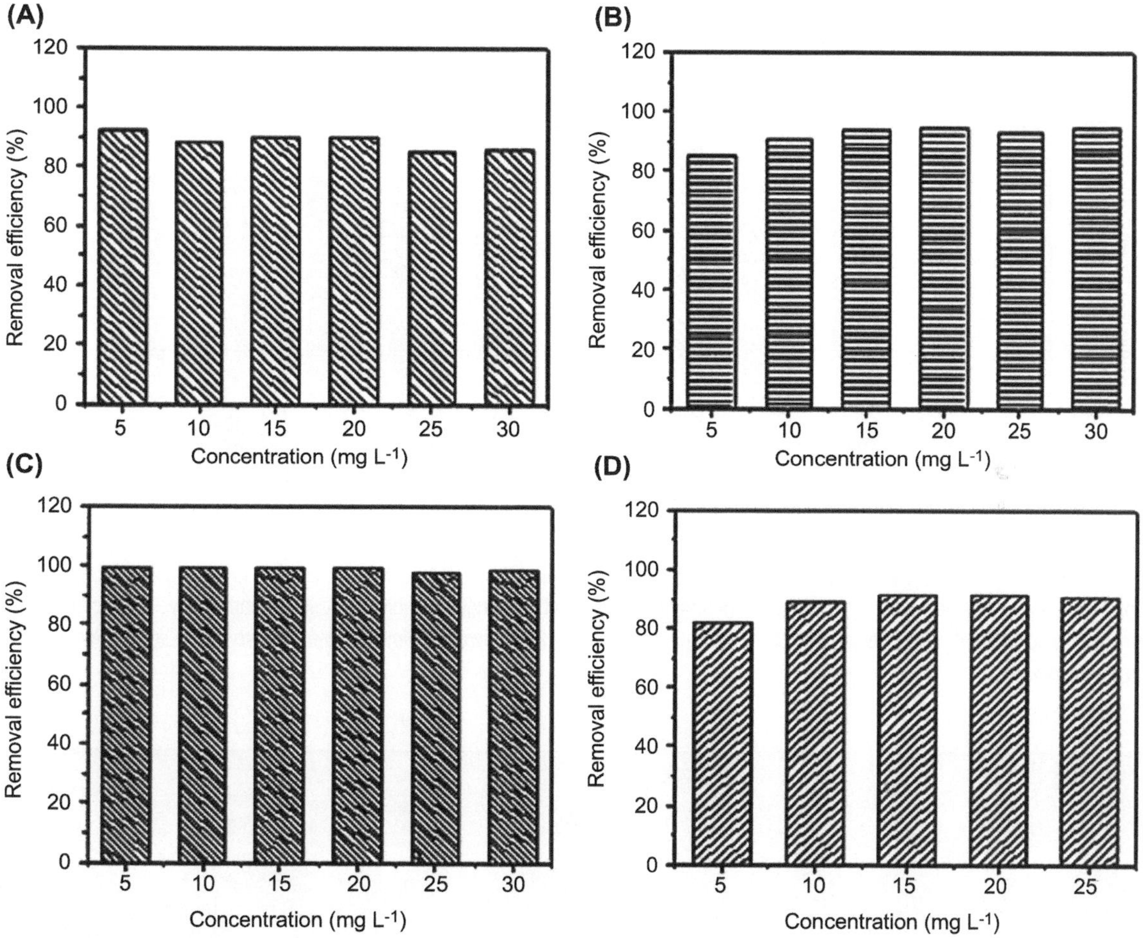

FIGURE 2.3

Removal efficiency of various endocrine-disrupting compounds using a graphene-γ-Fe_2O_3 composite with 10 wt.% γ-Fe_2O_3: (A) bisphenol A, (B) atrazine, (C) 1-naphthol, and (D) dibutyl phthalate. Experimental conditions: The graphene-γ-Fe_2O_3 composite material (1.5 mg) was added to the contaminated water solution (10 mL) and stirring was continued overnight at room temperature, before magnetic separation (Sinha et al., 2013). *Copyright Wiley-VCH Verlag GmbH & Co. KGaA. Reproduced with permission.*

number of EDC-binding sites, which lead to poor removal efficiency. This result suggests that the loading of NPs should be optimized for highest separation efficiency.

The separation efficiency of EDC removal by the graphene-γ-Fe_2O_3 composite was compared with that of AC (Fig. 2.5). For this comparison, Sui et al. (2011) used AC with a surface area (>500 $m^2\ g^{-1}$) higher than that of composite graphene-γ-Fe_2O_3 material. Sui et al. (2011) found that the removal efficiency of the composite was higher than that of AC, particularly when the EDC

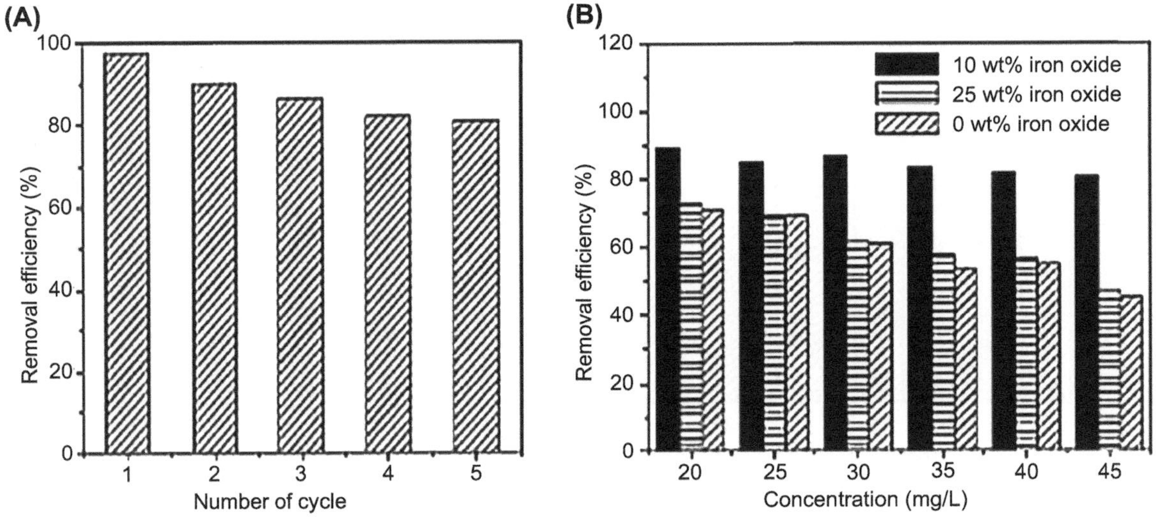

FIGURE 2.4

(A) Removal efficiency of recycled graphene-γ-Fe_2O_3 composite materials. 10 mL of l-naphthanol solution (50 mg/L) and 1.5 mg of graphene-γ-Fe_2O_3 materials were used for this experiment. (B) Removal efficiency of bisphenol A using graphene-γ-Fe_2O_3 composite materials having different weight percentage of γ-Fe_2O_3 (Sinha et al., 2013). *Copyright Wiley-VCH Verlag GmbH & Co. KGaA. Reproduced with permission.*

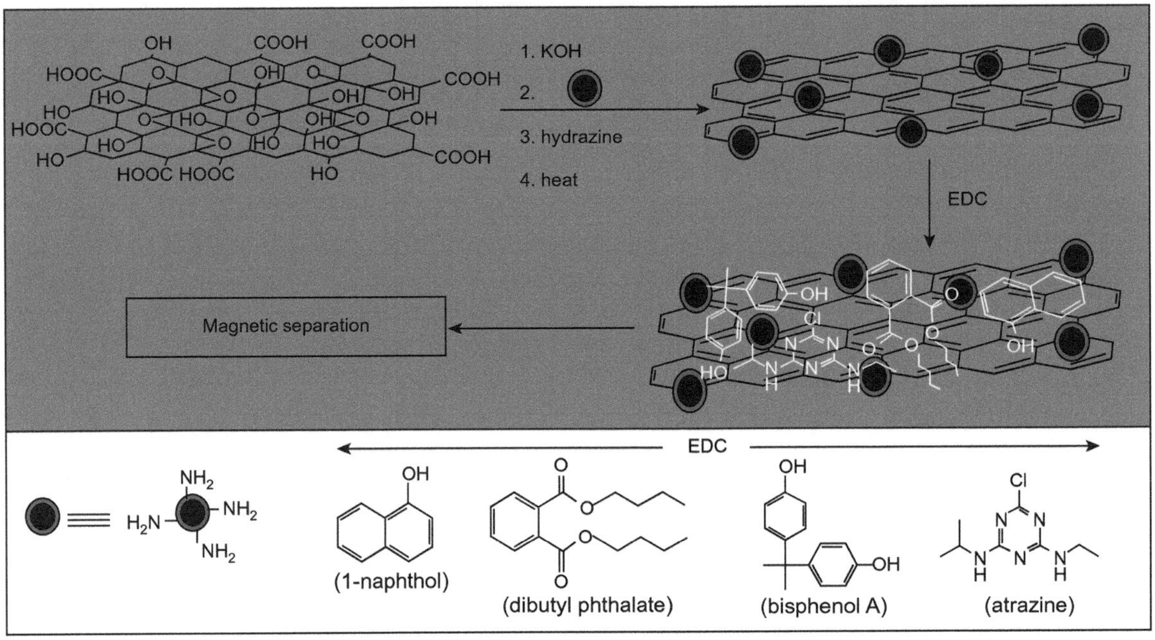

SCHEME 2.1

Synthesis of graphene-γ-Fe_2O_3 composites and endocrine-disrupting compound (EDC)-separation strategy (Sinha et al., 2013). *Copyright Wiley-VCH Verlag GmbH & Co. KGaA. Reproduced with permission.*

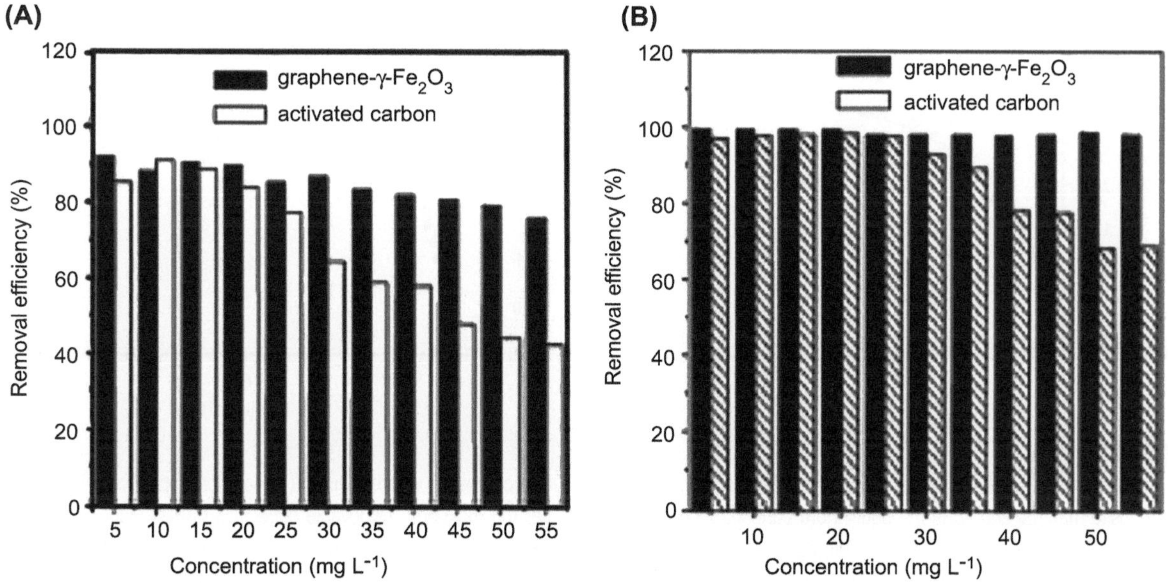

FIGURE 2.5

Comparison of the removal efficiency of bisphenol A (A) and 1-naphthol (B) by using activated carbon and a graphene-γ-Fe_2O_3 composite (10 wt% g-Fe_2O_3). Notably, the separation efficiency of activated carbon at higher concentrations of bisphenol A/1-naphthol was poor. Experimental conditions: The graphene-γ-Fe_2O_3 composite material (1.5 mg) was added to the contaminated water solution (10 mL) and stirring was continued overnight at room temperature, before magnetic separation (Sinha et al., 2013). *Copyright Wiley-VCH Verlag GmbH & Co. KGaA. Reproduced with permission.*

concentration was high. At lower EDC concentrations, the separation efficiencies were comparable. These results imply that the new composite offers higher adsorption capacity for EDC than AC. To further confirm this result, the maximum adsorption capacity for each EDC was determined by using their respective adsorption isotherms on the graphene-γ-Fe_2O_3 composite. In the cases of 1-naphthol and BPA, the Langmuir model was used for these measurements (Fig. 2.6). The measured maximum adsorption capacities for 1-naphthol and bisphenol A were 680 and 360 mg g^{-1}, respectively. However, poor solubility restricted the use of such an approach for ATZ and dibutyl phthalate, and, in these cases, the approximate maximum absorption capacity was determined from the removal capacity, which was measured at the highest ATZ/dibutyl phthalate concentration. The measured maximum adsorption capacities for ATZ and dibutyl phthalate were 200 and 200 mg g^{-1}, respectively. All of the measured maximum adsorption capacities are either comparable to or higher than previously reported values that are tabulated in Table 2.2 (Bhunia et al., 2014). Such high adsorption capacities can be explained by the formation of favorable $\pi-\pi$ and hydrophobic interactions between the EDCs and the extended graphene surface in the composite materials. The observed difference in adsorption capacities for different EDCs is due to their structural differences,

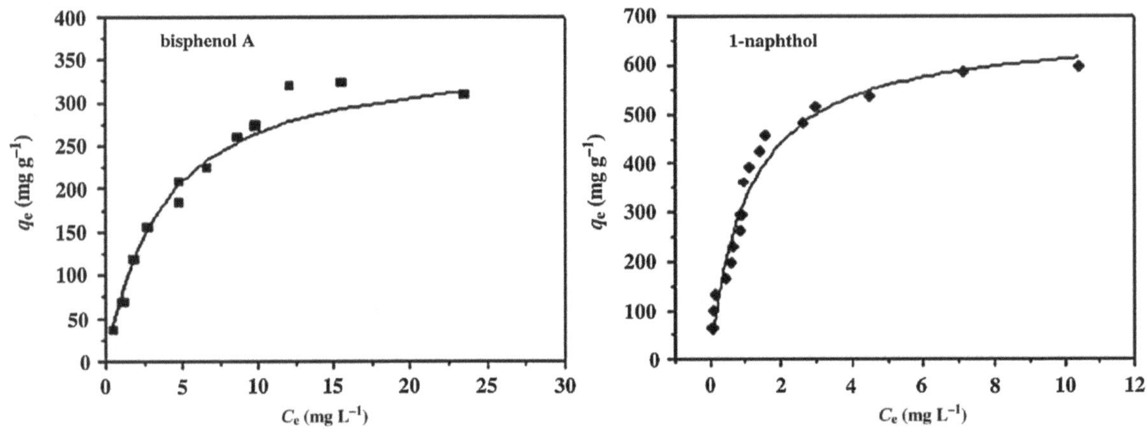

FIGURE 2.6
Adsorption isotherm of bisphenol A and 1-naphthol on a graphene-γ-Fe_2O_3 composite, where q_e is the adsorption capacity of the composite material and C_e is the equilibrium concentration of the adsorbate; the maximum removable capacities of bisphenol A and 1-naphthol were 360 mg g^{-1} and 680 mg g^{-1}, respectively, which were measured by the Langmuir model. Experimental conditions: $[C_{composite}] = 0.15$ g L^{-1}, $[C_{bisphenol\ A}]_{initial} = 5$–60 mg L^{-1}, $[C_{1\text{-}naphthol}]_{initial} = 5$–100 mg L^{-1}, T = room temperature (Sinha et al., 2013). *Copyright Wiley-VCH Verlag GmbH & Co. KGaA. Reproduced with permission.*

and more hydrophobic 1-naphthol is expected to form stronger interactions with graphene than BPA.

2.2.3.2 Removal of Colorless Endocrine-Disrupting Compounds by Photocatalytic Process

Sunlight-induced degradation of organic pollutants is one of the most attractive approaches for combating environmental pollution and wastewater treatment (Keane et al., 2014). A variety of photocatalysts have been developed toward this goal; efficient degradation of colorless organic pollutants by visible light is a challenging issue. TiO_2 is most frequently used because of its chemical stability, favorable optoelectronic properties, low toxicity, and low cost (Linsebigler et al., 1995). However, TiO_2 is photoactive in the UV region because of its higher band gap (3.2 eV), and thus, utilization of solar irradiation is not efficient (Linsebigler et al., 1995). A variety of new generation visible light photocatalysts are under development, and prominent examples include nitrogen-doped TiO_2 (Asahi et al., 2001), plasmonic NP–TiO_2 composite (Takai et al., 2011; Kochuveedu et al., 2012; Kale et al., 2014) plasmonic NP–quantum dot composite (Costi et al., 2008), and other nanocomposites (Hu et al., 2006; Wang et al., 2009, 2013; Liu et al., 2012a,b,c; Zhang et al., 2012, 2011a,b,c,d). In addition, attempts have also been made to design catalysts with efficient charge separation and storage from the photoexcited NP, and prominent examples include Ag–TiO_2 (Takai et al., 2011), graphene–TiO_2 (Zhang et al., 2010), and graphene–quantum dot (Kaniyankandy et al., 2012).

Graphene-based composites with plasmonic and semiconductor NPs offer unique advantage as a photocatalyst for the organic pollutants (Kamat, 2010). NPs and graphene can act as antennae for visible light. In addition, the highly conducting graphene surface offers high mobility of photo-generated charge carrier that in turn slows the recombination of photo-generated electron–hole pairs (Kamat, 2010). A hydrophobic graphene surface also offers the adsorption sites of organic molecules (Zhao et al., 2011a,b). All these factors result in the enhancement of photocatalytic activity (Kamat, 2010). Although there are few reports on graphene–NP composite as visible light photocatalyst, they are mostly demonstrated for dye degradation, and reports on the degradation of colorless organics are rare or inefficient (Hu et al., 2006; Wang et al., 2009, 2013; Zhang et al., 2011a,b,c,d, 2012; Liu et al., 2012a,b,c).

However, many organic pollutants are colorless and do not absorb visible light, and thus, visible light–induced degradation is a challenging issue (Benotti et al., 2009). In this section, the visible light photocatalyst for the degradation of colorless organic pollutants by the rGO-based composite with silver NP (rGO–Ag) that can act as an efficient degradation compound has been discussed. This process is developed by Bhunia et al. (2014) and a simple, low-cost, gram-scale synthesis method for rGO–Ag and used the composite for degradation of three well-known endocrine disruptors that are colorless (Scheme 2.2). Bhunia et al. (2014) found that photocatalytic efficiency by

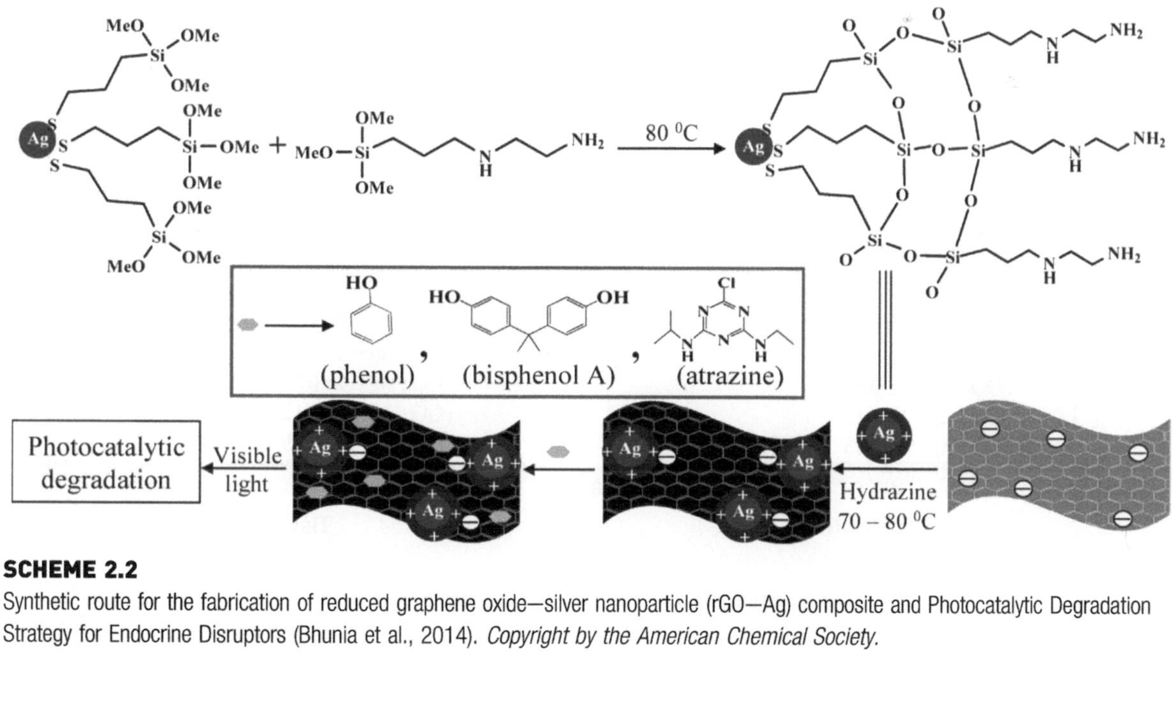

SCHEME 2.2

Synthetic route for the fabrication of reduced graphene oxide–silver nanoparticle (rGO–Ag) composite and Photocatalytic Degradation Strategy for Endocrine Disruptors (Bhunia et al., 2014). *Copyright by the American Chemical Society.*

rGO−Ag under visible light is significantly higher as compared to rGO or silver NP and depends on the optimum loading of Ag on rGO−Ag.

There are mainly three distinct novelties of the presented approach. First, rGO−Ag has been synthesized by simple electrostatic interaction between negatively charged GOs and silica-coated positively charged Ag NPs followed by the hydrazine-based reduction of GO. The synthetic approach can finely control the loading of Ag NPs, has been used for gram-scale synthesis of rGO−Ag, and can be easily adapted for even larger-scale synthesis. Second, the plasmonic property of Ag has been used for the efficient capturing of visible light and the photolysis of organic molecules that do not absorb any visible light. This approach can be used for the degradation of colorless organic pollutants. Third, it is found that photocatalytic efficiency depends on loading of Ag, and loading that is either too high or too low reduces the catalytic activity.

Three well-known endocrine disruptor compounds such as phenol, BPA, and ATZ have been selected for photocatalytic degradation study (Boas et al., 2012a,b). These three pollutants were selected because they are commonly found in wastewater with well-known toxic effect, and they do not absorb visible light (Boas et al., 2012a,b). Typically, an aqueous solution of endocrine disruptor compound is mixed with solid rGO−Ag and stirred in the dark for 2 h to reach the equilibrium adsorption state. Next, the mixture solution is irradiated with visible or UV light and a portion of the solution is used for quantitative investigation. Figs. 2.7 and 2.8 summarize the results of photocatalytic

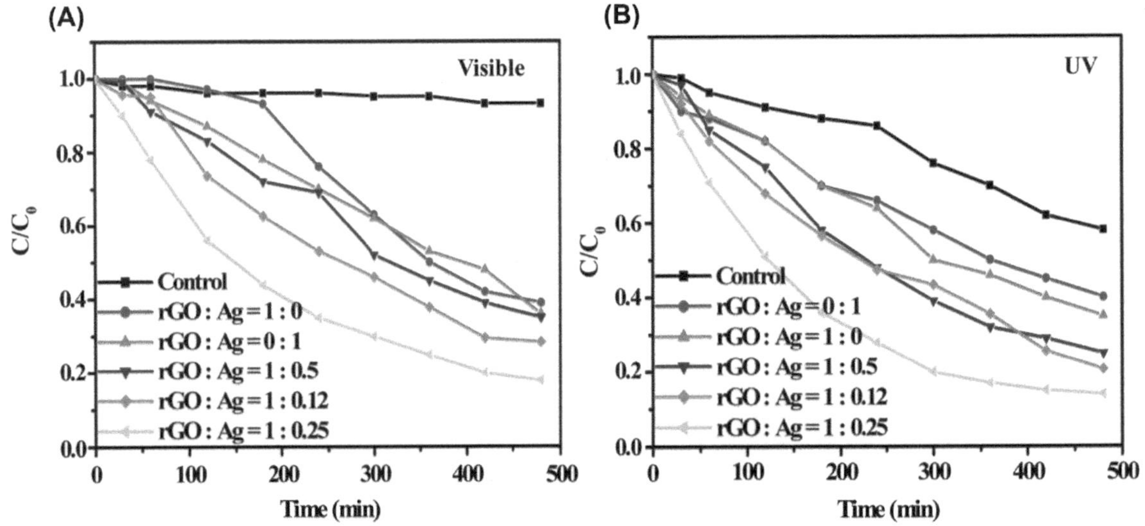

FIGURE 2.7
Photocatalytic degradation of phenol under (A) visible and (B) UV light irradiation using reduced graphene oxide−silver nanoparticle (rGO−Ag) composite with varying amounts of Ag showing that an rGO/Ag ratio of 1:0.25 displays higher photocatalytic activity than the other composites, only rGO nanoparticles, or only Ag nanoparticles. The control study indicates photolysis without any catalyst (Bhunia et al., 2014). *Copyright by the American Chemical Society.*

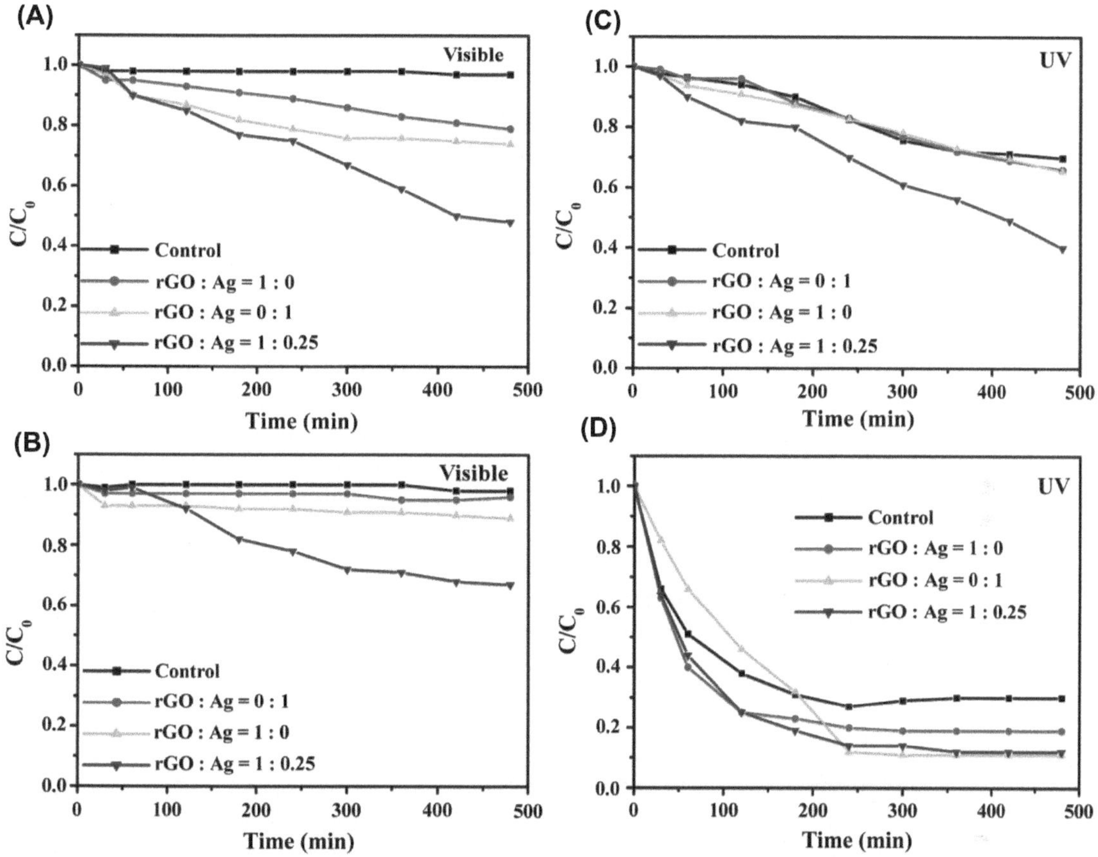

FIGURE 2.8

Reduced graphene oxide—silver nanoparticle (rGO—Ag) composite with rGO/Ag of 1:0.25 as photocatalyst for degradation of (A and C) bisphenol A and (B and D) atrazine under (A and B) visible and (C and D) UV light (Bhunia et al., 2014). *Copyright by the American Chemical Society.*

degradation under visible and UV light (Bhunia et al., 2014). The photocatalytic degradation result of phenol is shown in Fig. 2.7 (Bhunia et al., 2014). The initial concentration of phenol decreases with irradiation time, and the rate of photodegradation strongly depends on the nature of the light and the catalyst. In the absence of any catalyst, phenol does not degrade under visible light but slowly degrades under UV light. The presence of catalyst initiates the degradation of phenol under visible light and enhances the rate of degradation under UV light. Most importantly, the visible light–induced degradation of phenol by rGO—Ag is most effective compared to degradation by Ag NP or rGO. In addition, it is also observed that the rGO/Ag-based photocatalytic degradation rate is sensitive to Ag loading and is highest for rGO—Ag with a weight ratio of 1:0.25, but it is low for other compositions. A similar trend is observed for UV light–induced photocatalytic degradation of phenol, suggesting that rGO–Ag is also most effective under UV light. Photocatalytic degradation of BPA and ATZ shows that they do not degrade under visible light but degrade slowly under UV light (Fig. 2.8; Bhunia et al., 2014). However, the degradation under visible light is most enhanced by rGO–Ag compared to rGO and Ag NP.

One of the important aspects of the catalyst is that it should be reused with unaltered performance. The presented catalyst is easily separable for reuse after the reaction via decantation or low-speed centrifugation. The visible light photocatalytic performance of the rGO–Ag composite has been investigated by recycling it after the phenol degradation and then reusing it as catalyst. Results show that even after rGO–Ag is recycled five consecutive times, its photocatalytic activity does not significantly decrease as shown in Fig. 2.9(A). A TEM study of reused rGO–Ag shows that composite structure and Ag particle size remain intact as shown in Fig. 2.9(B). This fact indicates that rGO–Ag is stable enough for repeated use. This work shows that controlled loading of plasmonic NPs on a graphene surface is an important aspect for the effective utilization of visible light. Different rGO- or GO-based nanocomposites with reported photocatalytic activity are tabulated in Table 2.3 (Bhunia et al., 2014).

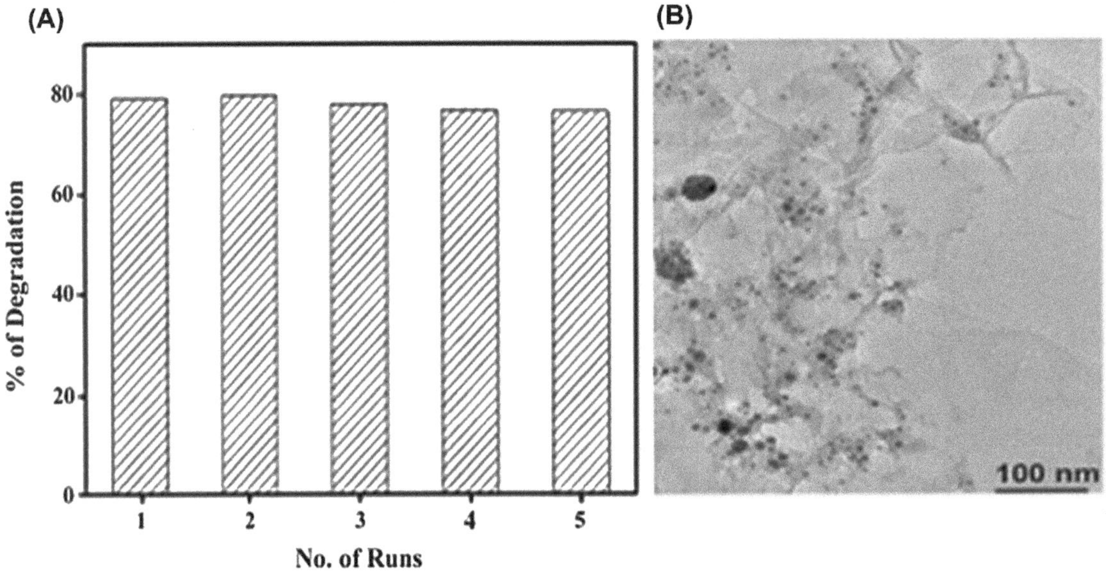

FIGURE 2.9

(A) Photocatalytic degradation of phenol using recycled reduced graphene oxide–silver nanoparticle (rGO–Ag) composite five consecutive times, showing that catalytic activity does not decrease with repeated use. (B) TEM image of the rGO–Ag composite after five consecutive times of use, showing that Ag nanoparticles are still well attached with rGO (Bhunia et al., 2014). *Copyright by the American Chemical Society.*

2.2.3.3 *Removal of Microcystin-LR by Cyclodextrin-Functionalized Magnetic Composite of Colloidal Graphene and Porous Silica*

Microcystins are cyclic heptapeptides produced by cyanobacteria (De-Figueiredo et al., 2004). Among the microcystins family, the microcystin-LR is known to be the most toxic (Dawson, 1998). As the cyanobacterial algal blooms in various water bodies are a worldwide issue, the contamination of drinking water by microcystins-LR and selective separation from contaminated water become critical issues for advanced water purification technology (De-Figueiredo et al.,

Table 2.3 Summary of Different Reduced Graphene Oxide (Graphene/rGO)- or Graphene Oxide (GO)-Based Nanocomposites With Reported Photocatalytic Activity

Composite Material	Particle Size	Wavelength of Light Used	Targeted Pollutant	Pollutant Concentration	References
rGO–Ag					Bhunia et al. (2014)[a]
Ag@AgCl/graphene	0.4–1.1 μm	$\lambda > 400$ nm	2,4-dichloro phenol, rhodamin B	10 mg/L	Zhang et al. (2011a,b,c,d)
$ZnFe_2O_4$/GR	40 nm	$\lambda > 420$ nm	*p*-chlorophenol	10 mg/L	Hou et al. (2013)
GO–Ag_3PO_4	Irregular morphology	$\lambda > 420$ nm	Bisphenol A, rhodamin B	20 mg/L, 10 mg/L	Wanga et al. (2014)
GO–Ag_3PO_4	500 nm	420 nm $< \lambda <$ 630 nm	Acid orange 7, phenol	50 mg/L, 10 mg/L	Liu et al. (2012a,b,c)
Titanate/GR	7–9 nm	$\lambda > 510$ nm	Phenol	80 mg/L	Hu et al. (2013)
Ag/AgBr/TiO_2		$\lambda = 254$ nm, $\lambda > 400$ nm	Volatile organic compound	250 mg/L	Zhang et al. (2011a,b,c,d)
Ag_3PO_4/rGO	80 nm	$\lambda > 420$ nm	Methyl orange, methylene blue	10 mg/L	Dong et al. (2013)
Ag-AgBr/TiO_2-rGO	TiO_2 = 17–18 nm, AgBr = 39 nm	Different color of LED	Penicillin G	10 mg/L	Wang et al. (2013)
TiO_2	7 nm	UV light	Bisphenol A	40 mg/L	Ohko et al. (2001)

[a]***Preparation of Reduced Graphene Oxide–Silver Nanoparticle (rGO–Ag) Composite***: *rGO–Ag is synthesized according to the method reported (Salam et al., 2014) with a finer adjustment of silver loading. Typically, three separate batches of 50 mL of GO aqueous solution were prepared with a concentration of GO of ~1 mg/mL and kept under stirring condition. Next, 25, 12, and 6 mL of aqueous silica-coated Ag nanoparticle solution (concentration ~1 mg/mL) were added dropwise to three different vials. The solutions were stirred continuously for 15 min, and then 0.5 mL hydrazine hydrate solution was added to each batch and heated at approximately 70–80°C for 2 h. As time progressed, the brown solution turned black, and a precipitate was formed. The black precipitate was washed with distilled water several times and dried. The dried composites were designated as rGO–Ag with rGO/Ag weight ratios of 1:0.5, 1:0.25, and 1:0.12, respectively. This approach was also used for larger-scale preparation of rGO–Ag composites. For example, 60 mL of aqueous silica-coated Ag nanoparticle solution (concentration ~2 mg/mL) was added dropwise to 250 mL of GO solution (concentration ~2 mg/mL) under stirring condition. Stirring was continued for 15 min, and then 5 mL of hydrazine hydrate solution was added and heated at approximately 70–80°C for 2 h. Finally, black precipitate of rGO–Ag (with rGO/Ag = 1:0.25) was washed with distilled water several times and dried with a yield of ~0.6 g (Bhunia et al., 2014).*

2004; Dawson, 1998; Hoegera et al., 2005; Schwarzenbach et al., 2006). Microcystins exist in several forms because of the variation of amino acids, and the presence of the hydrophobic Adda (3-amino-9-methoxy-2,6,8-trimethyl-10-phenyldeca-4,6-dienoic acid) moiety, which is the key component that binds with protein phosphatase to exhibit its toxicological activity (MacKintoshl et al., 1990; Runnegar et al., 1993). Microcystin-LR is reported to induce acute and chronic toxicity to animals and humans (De-Figueiredo et al., 2004; Dawson, 1998; Codd et al., 2005; Briand et al., 2003; Jasionek et al., 2010; Li et al., 2015). Exposure to microcystin-LR causes diarrhea, vomiting, piloerection, and weakness, and all these toxic effects lead to liver damage (De-Figueiredo et al., 2004; Dawson, 1998; Codd et al., 2005; Briand et al., 2003). In addition, it is shown that microcystin-LR promotes tumor growth (Matsushima et al., 1992; Christen et al., 2013). According to WHO guidelines, the permissible limit of microcystin-LR in drinking water is 1.0 μg/L. Thus, separation/removal of microcystin-LR from drinking water is a very challenging issue. Removal of microcystin-LR is challenging because of their high stability and resistance against chemical hydrolysis and oxidation (Song et al., 2006). Several traditional methods such as filtration, coagulation, flocculation, sedimentation, oxidation, and biological treatment have been investigated (Lawton et al., 1999). Conventional filtration, coagulation, flocculation, and sedimentation are effective for separation of cyanobacterial cells but ineffective for the separation of dissolved cyanotoxins (Teixeira et al., 2005; Yuana et al., 2002; Wu et al., 2011a,b). Advanced oxidation methods such as chlorination and ozonation can remove cyanotoxin but require high dosage, which causes other toxic side products (Rodriguez et al., 2007; Kronberget al., 1998). Biological treatments are a less efficient and time-consuming process (Bournea et al., 2006). Photocatalytic degradation of microcystin is shown to be effective but not cost-effective and tested only on the laboratory scale (Pelaez et al., 2013; Long et al., 2006; Liang et al., 2014). In contrast, adsorption-based separation (Wu et al., 2011a,b) is most effective for removal of toxic pollutants from water, and AC (Huang et al., 2007) is most widely used for this purpose. However, AC-based separation of microcystin is limited because of small micropores in carbon that cannot accommodate the large microcystin molecule. Recently, mesoporous carbon (Zhang et al., 2011a,b,c,d; Teng et al., 2013a,b), mesoporous silica (Teng et al., 2013a,b; Chen et al., 2009; Lu et al., 2011; Deng et al., 2008a,b), and bare GO (Pavagadhi et al., 2013) have been used for the separation of microcystins. Although separation efficiency can be largely improved, all these materials do not exhibit selective separation of microcystin-LR. Herein the functional nanocomposites for selective separation of microcystin-LR from contaminated water have been discussed. Sinha et al. (2015a,b) synthesized the magnetic composite of colloidal graphene and porous silica and then functionalized with cyclodextrin, which offers host–guest interaction with microcystin-LR. The high surface area of the nanocomposite offers high loading of cyclodextrin, the dispersible form of the nanocomposite offers easier accessibility of cyclodextrin to microcystin-LR, and the magnetic component offers easier separation of microcystin-LR. Sinha et al. (2015a,b) selected graphene and mesoporous silica as they are known to have high surface area and are widely used in biomedical (Li et al., 2012; Chung et al., 2013) and water purification (Sinha et al., 2012, 2013; Geng et al., 2012; Madadrang et al., 2012; Gao et al., 2011; Chandra et al., 2011; Ma et al., 2012; Gao et al., 2013; Yang et al., 2013; Zhao et al., 2011a,b) application, and their magnetic composites have been used for improved separation (Sinha et al., 2012, 2013; Geng et al., 2012; Zhao et al., 2014) and other applications (Ray et al., 2015; Machado et al., 2012). Cyclodextrin is selected for functionalization as it is known to selectively interact with a variety of small molecules via host–guest interaction (Sinha et al., 2015a,b; Del Valle, 2004; Bibby et al., 2003; Bhattarai et al., 2014; Guo et al., 2010; Mondal et al., 2012), and recently it has been shown to interact with microcystin-LR (Chen et al., 2011a,b). Cyclodextrin is a cyclic oligomer of glucose composed of six to eight glucose units and has a hydrophobic cavity at the center of its molecular arrangement (Del Valle, 2004). The sizes of the hydrophobic cavity increase from α- and β- to γ-cyclodextrin. It is shown that the Adda moiety of microcystin-LR is involved in the host–guest interaction with cyclodextrin, and γ-cyclodextrin offers the most stable inclusion complex as compared to β- or α-cyclodextrin (Chen et al., 2011a,b). Here it is discussed that the cyclodextrin-functionalized nanocomposite also shows a similar trend and offers enhanced performance in separation of microcystin-LR (Scheme 2.3).

In this process, a stock solution of microcystin-LR with the concentration of 500 μg/mL was prepared by dissolving in methanol–water (1:9 v/v) solution. In a separate vial stock, solutions of G-Fe_2O_3-CD and MMS-CD were prepared. Next, solution of G-Fe_2O_3-CD or MMS-CD was mixed with the solution of microcystin-LR with varied concentration. The typical concentrations of G-Fe_2O_3-CD and MMS-CD were 0.05 and 1.0 mg/mL, respectively, and the final volume of water was 1.0 mL. The whole solution was stirred for 2 h and then particles were separated by laboratory-based bar magnet. The supernatant solution was used for the estimation of the remaining microcystin-LR by UV–visible spectroscopy or HPLC. The HPLC-based analysis of microcystin-LR involved a C18 column and UV detector. Two solvent systems were prepared composed of 0.05% aqueous trifluoroacetic acid (solution A) and methanol (solution B). The elution gradient started at 70% of solution A that decreased to 30% after 12 min, and after 2 min of stability it was back to 70% of solution A within 15 min. The sample injection volume was 50 μL with the flow rate of 1.0 mL/min, and the wavelength was set to 239 nm for microcystin-LR estimation. Now, the separation (removal) of microcystin-LR by the G-Fe_2O_3-CD and MMS-Cd has been discussed. The main advantages of cyclodextrin-functionalized nanocomposites toward separation of microcystin-LR from

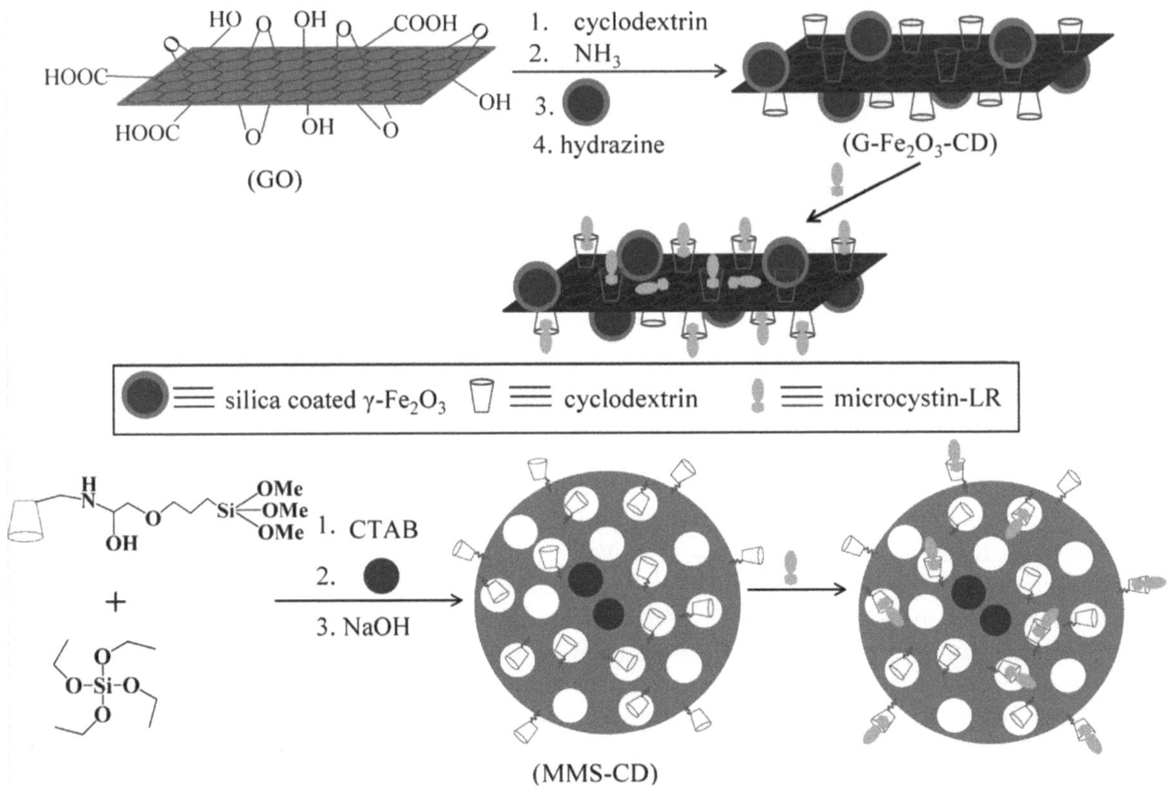

SCHEME 2.3

Synthesis Strategies for Cyclodextrin-Functionalized Magnetic Graphene Composite (G-Fe_2O_3-CD) and Cyclodextrin-Functionalized Magnetic Mesoporous Silica (MMS-CD).

Synthesis of Cyclodextrin-Functionalized Magnetic Graphene Composite (G-Fe_2O_3-CD): Silica-coated iron oxide NP solution with primary amine—terminated functional groups is synthesized using our previously reported method (Jana et al., 2007) and a stock solution was prepared with a concentration of 2 mg/mL. GO was prepared from natural graphite powder by modified Hummer's method (Hummers et al., 1958), and colloidal solution was prepared with a concentration of 1 mg/mL. In a separate vial, cyclodextrin solution was prepared with the concentration of 50 mg/mL. Next, 10 mL of GO solution was mixed with 10 mL of cyclodextrin solution followed by the addition of 200 μL of NH_3 solution (25 wt%) under stirring condition. After 30 min of stirring, 1 mL of silica-coated γ-Fe_2O_3 solution was added, and the whole mixture was stirred for another 30 min. Next, 200 μL of hydrazine solution (98%) was added, and the temperature of the solution was increased to 70—80°C and maintained for 4 h. The color of the solution gradually turns black along with the appearance of partial precipitation. Next, the reaction was stopped, and 0.5 mL of NaCl solution (~20 mg/mL) was added to precipitate the composite materials. The precipitate was washed several times with distilled water, and it was finally dispersed in water for further use.

Functionalization of G-Fe_2O_3 with α-, β-, or γ-cyclodextrin (Guo et al., 2010; Mondal et al., 2012) was achieved following the same procedure with the use of respective cyclodextrins. Dextran functionalized G-Fe_2O_3 was synthesized following the same procedure except that dextran was used instead of cyclodextrin. Nonfunctionalized G-Fe_2O_3 was synthesized by the afore-mentioned procedure without using any cyclodextrin or dextran.

Synthesis of Cyclodextrin-Functionalized Magnetic Mesoporous Silica (MMS-CD): At first 1 mL as-synthesized hydrophobic γ-Fe_2O_3 was purified by a well-known precipitation-redispersion method and dissolved in 1 mL chloroform. Next, 4 mL of water have been investigated by Sinha et al. (2015a,b) and are discussed as follows.

2.2.3.3.1 Separation of Microcystin-LR by Cyclodextrin-Functionalized Magnetic Graphene Composite and Cyclodextrin-Functionalized Magnetic Mesoporous Silica

Fig. 2.10 and Table 2.4 show the separation approach and removal efficiency of microcystin-LR by different nanocomposites (Sinha et al., 2015a,b). Typically, G-Fe_2O_3-γ-CD or MMS-CD nanocomposite is mixed with the aqueous solution of microcystin-LR and then the nanocomposite is removed by magnet. Next, the amount of remaining microcystin-LR in water has been estimated using HPLC or UV–visible spectroscopy. Digital images in Fig. 2.10(A, B) show the response of colloidal solution of nanocomposites by laboratory-based bar magnet that attracts them, leaving the clear bulk solution. This property of the material helps easier separation of adsorbed materials. Sinha et al. (2015a,b) determined the removal efficiency for G-Fe_2O_3-γ-CD and MMS-CD, keeping the fixed concentration of microcystin-LR (6 μg/mL), and then selected an optimum concentration of each material for microcystin-LR separation. The results indicate that MMS-CD is less efficient than G-Fe_2O_3-γ-CD, and so more MMS-CD is needed for separation of a similar amount of microcystin-LR separation. Thus, we have selected some optimum concentration of each material for microcystin-LR separation. Typical analysis results are shown in Fig. 2.9(C, D) (Sinha et al., 2015a,b). It is shown that absorbance and HPLC signal of microcystin-LR decrease remarkably after treatment with nanocomposites. Each separation experiment has been repeated three times, and average values are shown in Fig. 2.10(E, F) with ~10% error (Sinha et al., 2015a,b). The results summarize the separation efficiencies due to different functionalization. The results show that microcystin-LR can be separated without any functionalization, which is

aqueous CTAB solution (0.15 M) was added under stirring condition and heated to 50–60°C. After a few minutes γ-Fe_2O_3 NPs were transferred into the aqueous phase and chloroform was evaporated. In a separate vial, 200 mg of amine-functionalized γ-cyclodextrin was dissolved in 4 mL of dry dimethylformamide and mixed with 40 μL of (3-glycidyloxypropyl) trimethoxysilane and reacted overnight. Next, 4 mL of aqueous γ-Fe_2O_3 solution was diluted to 40 mL by adding water and mixed with 600 μL of NaOH (1 M) solution. Next, 2 mL of ethanol solution of tetraethoxysilane (240 μL of tetraethoxysilane was mixed with 8 mL of ethanol) and 1 mL of dimethylformamide solution of cyclodextrin-functionalized silane were added. Next, the whole solution was stirred for another 3 h, and particles were precipitated by adding excess ethanol. The particles were separated by centrifuging and washed with water and ethanol to remove unreacted reagents. Finally, the CTAB template was removed by the NH_4NO_3-based extraction method and stored for further use. Functionalization of MMS with α- and γ-cyclodextrin was achieved following the same procedure with the use of respective cyclodextrins. Nonfunctionalized MMS was synthesized by the afore-mentioned procedure without using cyclodextrin-functionalized silane. *Reproduced from Sinha, A. et al., 2015a. β-cyclodextrin functionalized magnetic mesoporous silica colloid for cholesterol separation. ACS Appl. Mater. Interfaces 7, 1340–1347; Sinha, A. et al., 2015b. Separation of microcystin-LR by cyclodextrin-functionalized magnetic composite of colloid graphene and porous silica. ACS Appl. Mater. Interfaces 7, 9911–9919. Copyright by the American Chemical Society.*

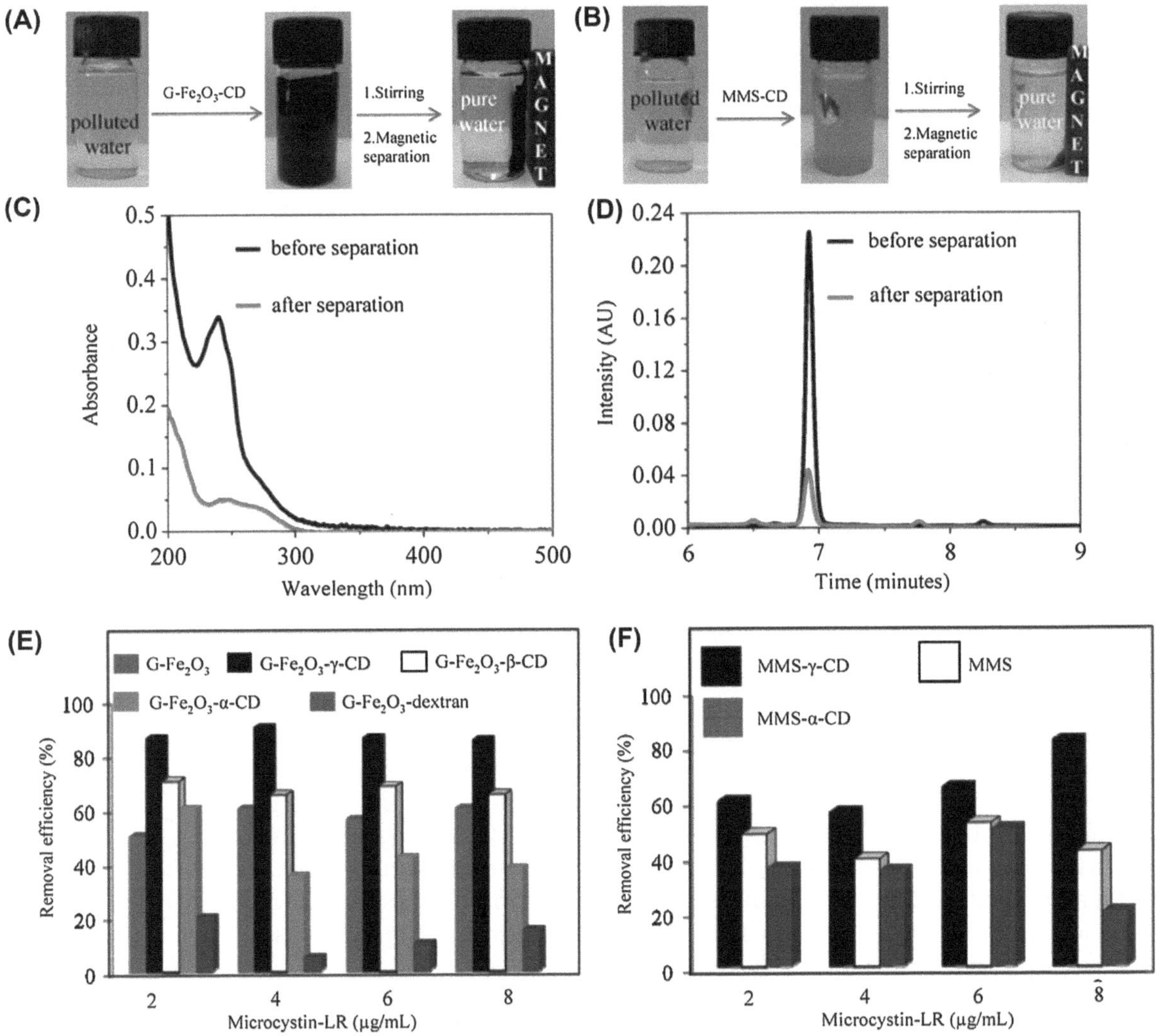

FIGURE 2.10

(A, B) Separation of microcystin-LR from water using G-Fe_2O_3-CD (A) and cyclodextrin-functionalized magnetic mesoporous silica (MMS-CD) (B). (C) Typical UV—visible absorption spectra of microcystin-LR solution before and after treatment with G-Fe_2O_3-CD. (D) HPLC chromatogram of microcystin-LR solution before and after treatment with MMS-CD. (E) Removal efficiency of microcystin-LR using G-Fe_2O_3-γ-CD and other functional G-Fe_2O_3. The concentration of nanocomposite has been kept at 0.05 mg/mL. (F) Removal efficiency of microcystin-LR using MMS-CD and other functional MMS. The concentration of nanocomposite has been kept at 1.0 mg/mL (Sinha et al., 2015a,b). *Copyright by the American Chemical Society.*

due to nonspecific adsorption with nanocomposites. However, separation performance is greatly enhanced after functionalization with cyclodextrin, and among all the cyclodextrins, the γ-cyclodextrin offers the best performance. Such best performance by γ-cyclodextrin over α- and β-cyclodextrin is

Table 2.4 Removal Capacity of Different Nanocomposites With Respect to Weight Percent of Cyclodextrin Present (Sinha et al., 2015a,b)

		Weight Percent Cyclodextrin (CD)			Removal Capacity[b] (mg/g) After Functionalization			
Materials	**Surface Area[a] (m^2/g)**	**α-CD**	**β-CD**	**γ-CD**	**α-CD**	**β-CD**	**γ-CD**	**–**
G-Fe_2O_3-CD	400(30)	17	18	16	80	140	160	120
MMS-CD	300(50)	19		17	4	–	8	5

[a]*Surface area is determined before functionalization with cyclodextrin. The value within parentheses indicates the surface area after being functionalized with γ-cyclodextrin.*
[b]*Removal capacity is defined by milligram (mg) of microcystin-LR removed per gram (g) of materials. It is determined from the initial concentration (C_O) and the final concentration (C_f) of microcystin-LR after treatment with nanocomposite, solution volume(V), and mass of nanocomposite (m) according to the following equation: removal capacity = $(C_O - C_f)V/m$.*

attributed to the most stable inclusion complex formation between γ-cyclodextrin and the Adda moiety of microcystin-LR (Chen et al., 2011a,b). Interestingly, dextran functionalization significantly decreases the removal efficiency of microcystin-LR, which implies that dextran significantly reduces the nonspecific binding of microcystin-LR with nanocomposites. Microcystin-LR separation efficiencies by G-Fe_2O_3-CD and MMS-CD have been compared, and it is observed that G-Fe_2O_3-CD offers better performance as compared to MMS-CD (Table 2.4 and Fig. 2.10) (Sinha et al., 2015a,b). For example, separation of similar concentration of microcystin-LR with the separation efficiency of >60% requires ~20 times of MMS-CD as compared to G-Fe_2O_3-γ-CD and it is shown in Fig. 2.10(E,F) (Sinha et al., 2015a,b). However, enhancement of separation performance after functionalization with γ-cyclodextrin is also observed in the case of MMS-CD. Removal capacity has been summarized for different nanocomposites. The values are in the range of 4–8 mg/g for MMS and MMS-CD and in the range 80–160 mg/g for G-Fe_2O_3 and G-Fe_2O_3-CD. The dextran functionalization of G-Fe_2O_3 significantly decreases the removal capacity to 10 mg/g as it decreases the nonspecific binding. It is interesting to note that although similar percent of cyclodextrin is present in both nanocomposites, the G-Fe_2O_3-CD performs better than the MMS-CD. This may be due to the accessibility of large molecular weight microcystin-LR to each of the MMS-bound cyclodextrin. In particular, the cyclodextrins that are bound inside the pores are not easily accessible to microcystin-LR. In contrast, graphene has a flat surface and thus all the cyclodextrins are accessible to microcystin-LR. So the high performance of microcystin-LR separation by G-Fe_2O_3-γ-CD is mainly due to high loading of cyclodextrin and the flat surface of graphene with dispersible property that offers easier accessibility to microcystin-LR. One of the most important aspects for practical application is if the materials can be regenerated and reused. This has been tested for removal of microcystin-LR using G-Fe_2O_3-γ-CD as representative material (Fig. 2.11) (Sinha et al., 2015a,b). Detailed procedures of G-Fe_2O_3-γ-CD for repeated use are shown in Fig. 2.11(A) (Sinha et al., 2015a,b). Regeneration of G-Fe_2O_3-γ-CD involves magnetic separation followed by repeated washing with ethanol to extract the adsorbed microcystin-LR. Results show that G-Fe_2O_3-γ-CD can be used several times with little loss of removal efficiency. The loss of removal efficiency can be due to aggregation of G-Fe_2O_3-γ-CD and detachment of some cyclodextrins during regeneration steps or poor microcystin-LR removal via regeneration steps. It is also important that the materials should be environmentally friendly or nontoxic so that they do not introduce any secondary pollutant

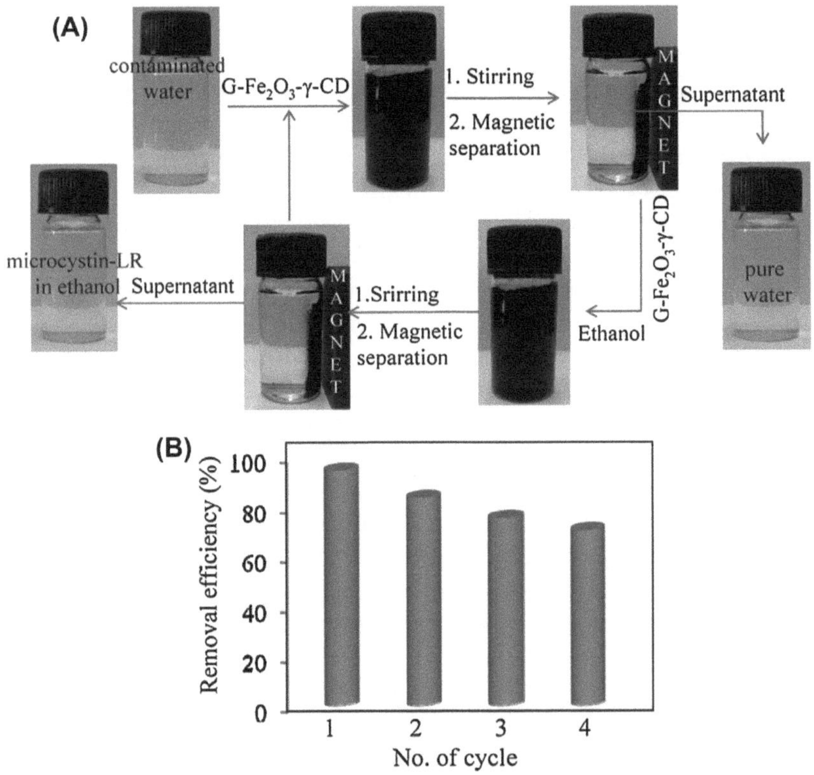

FIGURE 2.11

(A) Schematic representation of G-Fe_2O_3-γ-CD-based microcystin-LR separation and regeneration of G-Fe_2O_3-γ-CD for repeated use. (B) Microcystin-LR removal efficiency using successively reused G-Fe_2O_3-γ-CD. Typically, 0.25 mg of G-Fe_2O_3-γ-CD and 5 mL of fresh microcystin-LR solutions (5 μg/mL) have been used in this recycling experiment (Sinha et al., 2015a,b). *Copyright by the American Chemical Society.*

during the removal of microcystin-LR. To prove the nontoxic and environmentally friendly nature of G-Fe_2O_3-γ-CD, the cell viability study has been performed using Chinese hamster ovary cell as representative. Results show that about 80% of the cells are viable after 24 h of incubation with G-Fe_2O_3-γ-CD, suggesting that the material is nontoxic and environmentally friendly.

Selective separation performance of microcystin-LR from contaminated water has been investigated via separating microcystin-LR in the presence of NOM (Matilainen et al., 2010) and metal ions (Fig. 2.12). NOM and metal ions are present in most of the surface, ground, and soil waters, and they have negative effect on the removal efficiency of microcystin-LR. We have used a variety of NOM such as different amino acids, proteins (lysozyme), and polysaccharides (dextran) with the concentration of 20 μg/mL and different metal ions with the concentration of 20 μg/mL. We have observed that removal efficiency of microcystin-LR by G-Fe_2O_3-γ-CD is reasonably unaffected in the presence of NOM and metal ions. This result reveals that G-Fe_2O_3-γ-CD can selectively separate microcystin-LR from the contaminated water without significant loss of removal efficiency.

2.2.3.3.2 Functional Magnetic Graphene-Based Efficient Separation of Microcystin-LR

The high surface area of graphene has inspired researchers for adsorption-based separation application. These applications include separation of toxic metal

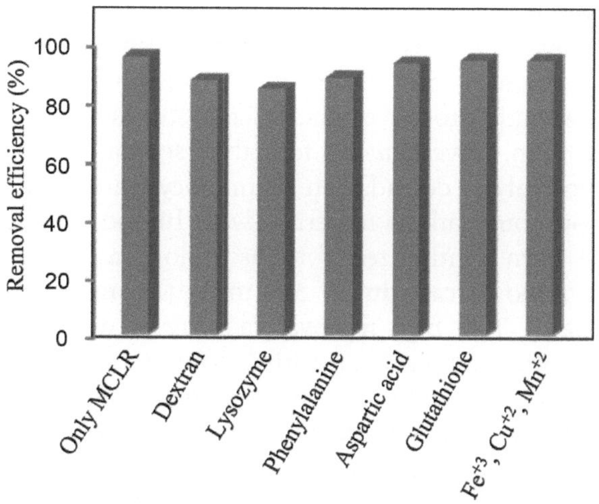

FIGURE 2.12

Removal efficiency of microcystin-LR by G-Fe_2O_3-γ-CD in the presence of natural organic matter and metal ions. Experimental conditions: 0.1 mg of G-Fe_2O_3-γ-CD is added to 2 mL of microcystin-LR (concentration 5 μg/mL) solution in the presence of different natural organic matter or metal ions (concentration 20 μg/mL) and is stirred for 2 h at room temperature, before the magnetic separation (Sinha et al., 2015a,b). *Copyright by the American Chemical Society.*

ions and organic pollutants. For example, ethylenediaminetetraacetic acid–functionalized GO has been used for separation of Pb(II) from water with removal capacity of 480 mg/g, thiol-functionalized GO (Gao et al., 2011) has been used for the separation of Hg(II) from water with removal capacity of 200 mg/g, polypyrrol-reduced GO (Chandra et al., 2011) has been used for separation of Hg(II) with removal capacity of 980 mg/g, ethylenediamine-functionalized rGO (Ma et al., 2012) has been used for Cr(VI) separation, and polydopamine- and polyacrylamide-functionalized graphene (Gao et al., 2013; Yang et al., 2013) have been used for the separation of toxic metal ions. Sulfonated graphene has been reported for separation of organic pollutants with removal capacity up to 347 mg/g, polydopamine- and polyacrylamide-functionalized graphenes (Gao et al., 2013; Yang et al., 2013) are reported for the separation of organic dye, and hydrophobic polymer–grafted GO (Zhao et al., 2014) has been used for separation of tetrabromobisphenol A with removal capacity of 22 mg/g. Sinha et al. (2015a,b) have used cyclodextrin-functionalized magnetic graphene for separation of microscystin-LR with the removal capacity up to 160 mg/g.

Various nanomaterials are under development for the separation of microcystin-LR with the focus of enhanced removal efficiency and improved selectivity. For example, mesoporous silica-based materials have been reported for the adsorption-based separation of microcystin-LR with the removal capacity of 6–13 mg/g (Teng et al., 2013a,b; Chen et al., 2009; Lu et al., 2011; Deng et al., 2008a,b). Bare GO has been used for separation of microcystin-LR with the removal capacity of 17 mg/g (Pavagadhi et al., 2013). This lower removal capacity by GO is due to low hydrophobic interaction between GO and microcystin-LR and electrostatic repulsion between negatively charged GO and anionic groups of the microcystin-LR. TiO_2-coated magnetic graphene has been investigated for the separation and UV light–dependent photocatalytic degradation of microcystin-LR (Liang et al., 2014). Recently, mesoporous carbon materials with bimodal mesopores of 2.8 and 5.8 nm have been synthesized for adsorption-based separation of microcystin-LR with removal capacity of 526 mg/g (Zhang et al., 2011a,b,c,d; Teng et al., 2013a,b). This high removal capacity of mesoporous carbon arises because of the high surface area, varied pore architectures, and hydrophobic surface of mesoporous carbon. In contrast, the presented G-Fe_2O_3-γ-CD offers selective binding of microcystin-LR with the hydrophobic cavity of cyclodextrin. In addition, Fe_2O_3 NPs present on the composite materials inhibit the graphene–graphene aggregation and further increase the adsorption site for microcystin-LR (Sinha et al., 2013). Compared to the reported nanomaterials, the presented G-Fe_2O_3-γ-CD has four distinct advantages. First, G-Fe_2O_3-γ-CD is dispersible in water, and cyclodextrins are attached on the flat surface of graphene. As a result, cyclodextrins are easily accessible for binding with microcystin-LR. Second, γ-Fe_2O_3 component in G-Fe_2O_3-γ-CD offers easier magnetic separation option of adsorbed microcystin-LR. Third, high surface area of graphene offers high loading of cyclodextrin on the graphene surface. Fourth, G-Fe_2O_3-γ-CD can selectively remove microcystin-LR from the water in the presence of NOM and metal ions.

2.3 CONCLUSION

The use of carbon-based nanomaterials and their nanocomposite materials can help to remove a wide range of organic and inorganic contaminants in the real field of applications. The better understanding of the formation of nanocomposites will certainly be a step toward improving the performance of multifunctional nanocomposites. The pattern of NPs within the host matrices and changes in the structures and properties of both nanomaterials and host matrices could be among the priority concerns in the real-field applications of the nanomaterials for removal of toxic materials. The literature in this chapter of the literature has shown that the nanomaterial is a promising field for the current and future research for removal of toxic materials.

References

Agnihotri, et al., 2005. Structural characterization of single walled carbon nanotube bundles by experiment and molecular simulation. Langmuir 21, 896–904.

Ahmed, F., et al., 2012. Antimicrobial applications of electroactive PVK-SWNT nanocomposites. Environ. Sci. Technol. 46 (3), 1804–1810.

Akasaka, T., Watari, F., 2009. Capture of bacteria by flexible carbon nanotubes. Acta Biomater. 5 (2), 607–612.

Al Momani, F., et al., 2008. Degradation of cyanobacteria toxin by advanced oxidation processes. J. Hazard. Mat. 150 (2), 238–249.

de Albuquerque Jr., E.C., et al., 2008. Removal of cyanobacteria toxins from drinking water by adsorption on activated carbon fibers. Mater. Res. 11 (3), 371–380.

Allen-King, R.M., et al., 2002. New modeling paradigms for the sorption of hydrophobic organic chemicals to heterogeneous carbonaceous matter in soils, sediments, and rocks. Adv. Water Resour. 25, 985–1016.

Amin, M.T., et al., 2014. A review of removal of pollutants from water/wastewater using different types of nanomaterials. Adv. Mater. Sci. Eng. 2014, 1–24, 825910.

Arias, L.R., Yang, L., 2009. Inactivation of bacterial pathogens by carbon nanotubes in suspensions. Langmuir 25 (5), 3003–3012.

Arjmandi, N., et al., 2009. CVD synthesis of small-diameter single-walled carbon nanotubes on silicon. Sci. Iran. 16 (1D), 61–64.

Asahi, R., et al., 2001. Visible-light photo-catalysis in nitrogen doped titanium oxides. Science 293, 269–271.

Bai, J., et al., 2010. Graphene nanomesh. Nat. Nanotech 5 (3), 190–194.

Balabanič, D., et al., 2011. Negative impact of endocrine-disrupting compounds on human reproductive health. Reprod. Fertil. Dev. 23, 403–416.

Ballesteros, E., et al., 2000. Analytical potential of fullerene as adsorbent for organic and organometallic compounds from aqueous solutions. J. Chromatogr. A 869, 101–110.

Basile, T., et al., 2011. Review of endocrine-disrupting-compound removal technologies in water and wastewater treatment plants: an EU perspective. Ind. Eng. Chem. Res. 50, 8389–8401.

Benny, et al., 2008. Effect of ozonolysis on the pore structure, surface chemistry, and bundling of single walled carbon nanotubes. J. Colloid Interface Sci. 317, 375–382.

Benotti, M.J., et al., 2009. Pharmaceuticals and endocrine disrupting compounds in U.S. drinking water. Environ. Sci. Technol. 43, 597–603.

Berry, D., et al., 2006. Microbial ecology of drinking water distribution systems. Curr. Opin. Biotechnol. 17 (3), 297–302.

Bhattarai, B., et al., 2014. Development of high efficiency silica coated β-cyclodextrin polymeric adsorbent for the removal of emerging contaminants of concern from water. J. Hazard. Mater. 273, 146–154.

Bhunia, et al., 2014. Reduced graphene oxide-silver nanoparticle composite as visible light photocatalyst for degradation of colorless endocrine disruptors. ACS Appl. Mater. Interfaces 6, 20085–20092.

Bibby, A., et al., 2003. Adsorption and separation of water-soluble aromatic molecules by cyclodextrin-functionalized mesoporous silica. Green Chem. 5, 15–19.

Boas, M., et al., 2012a. Thyroid effects of endocrine disrupting chemicals. Mol. Cell. Endocrionol. 355, 240–248.

Boas, M., et al., 2012b. Thyroid disruption in Zebrafish Larvae by short-term exposure to bisphenol AF. Mol. Cell. Endocrinol. 355, 240–248.

Bournea, D.G., et al., 2006. Biodegradation of the cyanobacterial toxin microcystin LR in natural water and biologically active slow sand filters. Water Res. 40, 1294–1302.

Brady-Estevez, A.S., et al., 2008. A single walled-carbon-nanotube filter for removal of viral and bacterial pathogens. Small 4 (4), 481–484.

Brady-Estevez, A.S., et al., 2010. SWNT-MWNT hybrid filter attains high viral removal and bacterial inactivation. Langmuir 26 (24), 19153–19158.

Briand, J.F., et al., 2003. Health hazards for terrestrial vertebrates from toxic cyanobacteria in surface water ecosystems. Vet. Res. 34, 361–377.

Cai, Y.-Q., et al., 2005. Multi-walled carbon nanotubes as a solid-phase extraction adsorbent for the determination of chlorophenols in environmental water samples. J. Chromatogr. A 1081 (2), 245–247.

Chandra, V., et al., 2010. Water-dispersible magnetite-reduced graphene oxide composites for arsenic removal. ACS Nano 4, 3979–3986.

Chandra, V., et al., 2011. Highly selective adsorption of Hg^{2+} by a polypyrrole-reduced graphene oxide composite. Chem. Commun. 47, 3942–3944.

Chen, et al., 2006. Interfacing carbon nanotubes with living cells. J. Am. Chem. Soc. 128 (19), 6292–6293.

Chen, H., et al., 2009. Facile synthesis of uniform microspheres composed of a magnetite core and copper silicate nanotube shell for removal of microcystins in water. J. Phys. Chem. C 113, 21068–21073.

Chen, L., et al., 2011a. Complexation of microcystins and nodularin by cyclodextrins in aqueous solution, a potential removal strategy. Environ. Sci. Technol. 45, 2293–2300.

Chen, G.C., et al., 2011b. Adsorption of diuron and dichlobenil on multiwalled carbon nanotubes as affected by lead. J. Hazard Mater. 188, 156–163.

Cheng, X., et al., 2005. Uptake and sequestration of naphthalene and 1,2-dichlorobenzene by C60. J. Nanopart. Res. 7, 555–567.

Choi, K.J., et al., 2006. Removal efficiencies of endocrine disrupting chemicals by coagulation/flocculation, ozonation, powdered/granular activated carbon adsorption, and chlorination. Korean J. Chem. Eng. 23, 399–408.

Christen, V., et al., 2013. Microcystin-LR induces endoplasmatic reticulum stress and leads to induction of NF κB, interferon-alpha and tumor necrosis factor-alpha. Environ. Sci. Technol. 47, 3378–3385.

Chung, C., et al., 2013. Biomedical applications of graphene and graphene oxide. Acc. Chem. Res. 46, 2211–2224.

Codd, G.A., et al., 2005. Cyanobacterial toxins: risk management for health protection. Toxicol. Appl. Pharmacol. 203, 264–272.

Cornelissen, G., et al., 2005. Extensive sorption of organic compounds to black carbon, coal, and kerogen in sediments and soils: mechanisms and consequences for distribution, bioaccumulation, and biodegradation. Environ. Sci. Technol. 39, 6881–6885.

Costi, R., et al., 2008. Visible light-induced charge retention and photocatalysis with hybrid CdSe–Au nanodumbbells. Nano Lett. 8, 637–641.

Dai, H., 2002. Carbon nanotubes: opportunities and challenges. Surf. Sci. 500, 218–241.

Dawson, R.M., 1998. The toxicology of microcystin. Toxicon 36, 953–962.

De Martino, A., et al., 2012. Removal of 4-chloro-2-methylphenoxyacetic acid from water by sorption on carbon nanotubes and metal oxide nanoparticles. RSC Adv. 2, 5693–5700.

De-Figueiredo, D.R., et al., 2004. Microcystin producing blooms a serious global public health issue. Ecotoxicol. Environ. Saf. 59, 151–163.

Del Valle, E.M.M., 2004. Cyclodextrins and their uses: a review. Process. Biochem. 39, 1033–1046.

Deng, Y., et al., 2008a. Superparamagnetic high-magnetization microspheres with an Fe_3O_4@SiO_2 core and perpendicularly aligned mesoporous SiO_2 shell for removal of microcystins. J. Am. Chem. Soc. 130, 28–29.

Deng, S., et al., 2008b. Adsorption equilibrium and kinetics of microorganisms on single-wall carbon nanotubes. IEEE Sensors J. 8 (6), 954–962.

Deng, J., et al., 2012. Multiwalled carbon nanotubes as adsorbents for removal of herbicide diuron from aqueous solution. Chem. Eng. J. 193, 339–347.

Di, Z.-C., et al., 2006. Chromium adsorption by aligned carbon nanotubes supported ceria nanoparticles. Chemosphere 62 (5), 861–865.

Donaldson, K., et al., 2006. Carbon nanotubes: a review of their properties in relation to pulmonary toxicology and workplace safety. Toxicol. Sci. 92 (1), 5–22.

Dong, P., et al., 2013. Ag_3PO_4/reduced graphite oxide sheets nanocomposites with highly enhanced visible light photocatalytic activity and stability. Appl. Catal. B Environ. 132, 133, 45–53.

Dreyer, D.R., et al., 2010. The chemistry of graphene oxide. Chem. Soc. Rev. 39, 228–240.

Dugan, N.R., Williams, D.J., 2006. Cyanobacteria passage through drinking water filters during perturbation episodes as a function of cell morphology, coagulant and initial filter loading rate. Harmful Algae 5 (1), 26–35.

Dumortier, H., et al., 2006. Functionalized carbon nanotubes are non-cytotoxic and preserve the functionality of primary immune cells. Nano Lett. 6 (7), 1522–1528.

El-Naas, M.H., et al., 2009. Biodegradation of phenol by *Pseudomonas putida* immobilized in polyvinyl alcohol (PVA) gel. J. Hazard. Mater. 164, 720–725.

Endo, M., et al., 2006. Large-scale production of carbon nanotubes and their applications. Pure Appl. Chem. 78 (9), 1703–1713.

Englert, J.M., et al., 2011. Covalent bulk functionalization of graphene. Nat. Chem. 3, 279–286.

Escobar, I.C., et al., 2001. Bacterial growth in distribution systems: effect of assimilable organic carbon and biodegradable dissolved organic carbon. Environ. Sci. Technol. 35 (17), 3442–3447.

Falconer, I.R., et al., 2006. Endocrine-disrupting compounds: a review of their challenge to sustainable and safe water supply and water reuse. Environ. Toxicol. 21, 181–191.

Fornasiero, F., et al., 2008. Ion exclusion by sub-2-nm carbon nanotube pores. Proc. Natl. Acad. Sci. U.S.A. 105 (45), 17250–17255.

Fritzmann, C., et al., 2007. State-of-the-art of reverse osmosis desalination. Desalination 216 (1–3), 1–76.

Gao, W., et al., 2011. Engineered graphite oxide materials for application in water purification. ACS Appl. Mater. Interfaces 3, 1821–1826.

Gao, H., et al., 2013. Mussel-inspired synthesis of polydopamine-functionalized graphene hydrogel as reusable adsorbents for water purification. ACS Appl. Mater. Interfaces 5, 425–432.

Geng, Z., et al., 2012. Highly efficient dye adsorption and removal: a functional hybrid of reduced graphene Oxide–Fe_3O_4 nanoparticles as an easily regenerative adsorbent. J. Mater. Chem. 22, 3527–3535.

Gotovac, S., et al., 2007. Adsorption of poly-aromatic hydrocarbons on single wall carbon nanotubes of different functionalities and diameters. J. Colloid Interface Sci. 314 (1), 18–24.

Griffete, N., et al., 2012. Magnetic nanocrystals coated by molecularly imprinted polymers for the recognition of bisphenol A. J. Mater. Chem. 22, 1807–1811.

Guo, Y., et al., 2010. Cyclodextrin functionalized graphene nano-sheets with high supramolecular recognition capability: synthesis and host guest inclusion for enhanced electrochemical performance. ACS Nano 4, 4001–4010.

Hedderman, T.G., et al., 2006. In-depth study into the interaction of single walled carbon nanotubes with anthracene and p-terphenyl. J. Phys. Chem. B 110 (9), 3895–3901.

Heo, J., et al., 2011. Removal of micro-pollutants and NOM in carbon nanotube–UF membrane system from seawater. Water Sci. Technol. 63, 2737–2744.

Heo, et al., 2012. Removal of bisphenol A and 17β-estradiol in single walled carbon nanotubes–ultrafiltration (SWNTs–UF) membrane systems. Sep. Purif. Technol. 90, 39–52.

Heymann, D., et al., 2003. Terrestrial and Extra-terrestrial Fullerenes. Fuller. Nanotub. Carbon Nanostruct. 11, 333–370.

Hinds, B., 2012. Dramatic transport properties of carbon nanotube membranes for a robust protein channel mimetic platform. Curr. Opin. Solid State Mater. Sci. 16 (1), 1–9.

Hoegera, S.J., et al., 2005. Occurrence and elimination of cyanobacterial toxins in drinking water treatment plants. Toxicol. Appl. Pharmacol. 203, 231–242.

Hoek, E.M.V., Ghosh, A.K., 2009. In: Nanotechnology-based Membranes for Water Purification. Nanotechnology Applications for Clean Water. William Andrew, Boston, Mass, USA, pp. 47–58 ch. 4.

Holt, J.K., et al., 2006. Fast mass transport through sub-2-nanometer carbon nanotubes. Science 312 (5776), 1034–1037.

Hou, Y., et al., 2013. $ZnFe_2O_4$ multi-porous microbricks/graphene hybrid photocatalyst: facile synthesis, improved activity and photocatalytic mechanism. Appl. Catal. B Environ. 142, 143, 80–88.

Hu, C., et al., 2006. Ag/AgBr/TiO_2 visible light photocatalyst for destruction of azodyes and bacteria. J. Phys. Chem. B 110, 4066–4072.

Hu, J., et al., 2010. Plasma-induced grafting of cyclodextrin onto multiwall carbon nanotube/iron oxides for adsorbent application. J. Phys. Chem. B 114, 6779–6785.

Hu, G., et al., 2013. Photocatalytic mechanism of graphene/titanate nanotubes photo-catalyst under visible-light irradiation. Mater. Chem. Phys. 138, 608–614.

Huang, W.-J., et al., 2007. Adsorption of microcystin-LR by three types of activated carbon. J. Hazard. Mater. 141 (1), 115–122.

Hummers, W.S., et al., 1958. Preparation of graphitic oxide. J. Am. Chem. Soc. 80, 1339.

Humplik, T., et al., 2011. Nanostructured materials for water desalination. Nanotechnology 22 (29), 292001, 1–19.

Jain, P., Pradeep, T., 2005. Potential of silver nanoparticle-coated polyurethane foam as an antibacterial water filter. Biotechnol. Bioeng. 90 (1), 59–63.

Jana, N.R., et al., 2007. Synthesis of water-soluble and functionalized nanoparticles by silica coating. Chem. Mater. 19, 5074–5082.

Jasionek, G., et al., 2010. Mitochondrial toxicity of microcystin-LR on cultured cells: application to the analysis of contaminated water samples. Environ. Sci. Technol. 44, 2535–2541.

Jia, G., et al., 2005. Cytotoxicity of carbon nanomaterials: single-wall nanotube, multi-wall nanotube, and fullerene. Environ. Sci. Technol. 39 (5), 1378–1383.

Jiang, G., et al., 2005. Synthesis and application of new fullerene derivative. J. Appl. Polym. Sci. 97, 2182–2195.

Jiang, D.-E., et al., 2009. Porous graphene as the ultimate membrane for gas separation. Nano Lett. 9 (12), 4019–4024.

Joseph, L., et al., 2011. Adsorption of bisphenol A and 17 alpha-ethinylestradiol on single walled carbon nanotubes from seawater and brackish water. Desalination 281, 68–74.

Jugan, M.L., et al., 2009. In vitro assessment of thyroid and estrogenic endocrine disruptors in wastewater treatment plants, rivers and drinking water supplies in the greater Paris area (France). Sci. Total Environ. 407 (11), 3579–3587.

Kale, M.J., et al., 2014. Direct photo-catalysis by plasmonic nanostructures. ACS Catal. 4, 116–128.

Kamat, P.V., 2010. Graphene-based nano-architectures. Anchoring semiconductor and metal nanoparticles on a two-dimensional carbon support. J. Phys. Chem. Lett. 1, 520–527.

Kang, et al., 2007. Single-walled carbon nanotubes exhibit strong antimicrobial activity. Langmuir 23 (17), 8670–8673.

Kang, et al., 2008a. Efficient control of population structure in model organism association mapping. Genetics 178 (3), 1709–1723.

Kang, S., et al., 2008b. Physicochemical determinants of multi-walled-carbon nanotube bacterial cytotoxicity. Environ. Sci. Technol. 42 (19), 7528–7534.

Kang, S., et al., 2008c. Antibacterial effects of carbon nanotubes: size does matter! Langmuir 24 (13), 6409–6413.

Kaniyankandy, S., et al., 2012. Ultrafast charge transfer dynamics in photoexcited CdTe quantum dot decoratedon graphene. J. Phys. Chem. C 116, 16271–16275.

Kar, S., et al., 2008. Potential of carbon nanotubes in water purification: an approach towards the development of an integrated membrane system. Int. J. Nucl. Desalin. 3 (2), 143–150.

Keane, D.A., et al., 2014. Solar photo-catalysis for water disinfection: materials and reactor design. Catal. Sci. Technol. 4, 1211–1226.

Kochuveedu, S.T., et al., 2012. Surface-plasmon-induced visible light photocatalytic activity of TiO_2 nanospheres decorated by Au nanoparticles with controlled configuration. J. Phys. Chem. C 116, 2500–2506.

Koelmans, A.A., et al., 2006. Biochar impacts on soil physical properties and greenhouse gas emissions. Chemosphere 63, 365–377.

Kronberg, L., et al., 1998. Ames mutagenicity and concentration of the strong mutagen chloro-4-(dichloromethyl)-5-hydroxy-2(5H)-furanone and of its geometric isomer e-2-chloro-3-(dichloromethyl)-4-oxobutenoic acid in chlorine-treated tap waters. Mutat. Res. Genet. Toxicol. 206, 177–1821.

Kuo, C.Y., 2009. Comparison with as-grown and microwave modified carbon nanotubes to removal aqueous bisphenol A. Desalination 249, 976–982.

Lawton, L.A., et al., 1999. Physico-chemical treatment methods for the removal of microcystins (cyanobacterial hepatotoxins) from potable waters. Chem. Soc. Rev. 28, 217–224.

Lee, D.-K., et al., 2004. Photocatalytic oxidation of microcystin-LR in a fluidized bed reactor having TiO_2-coated activated carbon. Sep. Purif. Technol. 34 (1–3), 59–66.

Lee, et al., 2005. Pore characterization of multi walled carbon nanotubes modified by KOH. Chem. Phys. Lett. 416, 251–255.

Li, Y.H., et al., 2002. Lead adsorption on carbon nanotubes. Chem. Phy. Lett. 357, 263–266.

Li, et al., 2003a. Adsorption of fluoride from water by aligned carbon nanotubes. Mater. Res. Bull. 38, 469–476.

Li, Y.H., et al., 2003b. Adsorption of cadmium (II) from aqueous solution by surface oxidized carbon nanotubes. Carbon 41, 1057–1062.

Li, Y.-H., et al., 2003c. Competitive adsorption of Pb^{2+}, Cu^{2+} and Cd^{2+} ions from aqueous solutions by multiwalled carbon nanotubes. Carbon 41 (14), 2787–2792.

Li, Y.-H., et al., 2005. Adsorption thermodynamic, kinetic and desorption studies of Pb^{2+} on carbon nanotubes. Water Res. 39 (4), 605–609.

Li, Q., et al., 2008a. Antimicrobial nanomaterials for water disinfection and microbial control: potential applications and implications. Water Res. 42 (18), 4591–4602.

Li, D., et al., 2008b. Effect of soil sorption and aquatic natural organic matter on the antibacterial activity of a fullerene water suspension. Environ. Toxicol. Chem. 27 (9), 1888–1894.

Li, X., et al., 2009. Surface modified graphite nanosheets used as adsorbent to remove 1,2-dichlorobenzene from water. Mat. Let. 63, 930–932.

Li, Z., et al., 2012. Mesoporous silica nanoparticles in biomedical applications. Chem. Soc. Rev. 41, 2590–2605.

Li, X., et al., 2015. The effects of chronic exposure to microcystin-LR on hepatocyte mitochondrial DNA replication in mice. Environ. Sci. Technol. 49, 4665–4672.

Liang, Y.L., et al., 2014. Preparation and characterization of TiO_2–Graphene@Fe_3O_4 magnetic composite and its application in the removal of trace amounts of microcystin-LR. RSC Adv. 4, 56883–56891.

Liao, et al., 2008. Adsorption of chlorophenols by multi walled carbon nanotubes treated with HNO_3 and NH_3. Carbon 46, 544–561.

Linsebigler, A.L., et al., 1995. Photocatalysis on TiO_2 surfaces-principles, mechanisms, and selected results. Chem. Rev. 95, 735–758.

Liu, Y., et al., 2006. Investigation of preparation and structures of activated carbon nanotubes. Mater. Res. Bull. 41, 1503–1512.

Liu, H., et al., 2009. Removal of persistent organic pollutants from micro-polluted drinking water by triolein embedded absorbent. Bioresour. Tech. 100 (12), 2995–3002.

Liu, Y.P., et al., 2012a. One-pot pyridine-assisted synthesis of visible-light-driven photocatalyst Ag/Ag_3PO_4. Appl. Catal. B 115, 116, 245–252.

Liu, Y.P., et al., 2012b. Highly efficient and stable Ag/Ag_3PO_4 plasmonic photocatalyst in visible light. Catal. Commun. 17, 200–204.

Liu, L., et al., 2012c. Graphene oxide enwrapped Ag_3PO_4 composite: towards a highly efficient and stable visible-light-induced photo-catalyst for water purification. Catal. Sci. Technol. 2, 2525–2532.

Lofgreen, J.E., et al., 2011. Molecularly imprinted mesoporous organosilica. ACS Nano 5, 2277–2287.

Long, R.Q., Yang, R.T., 2001. Carbon nanotubes as superior sorbent for dioxin removal. JACS 123 (9), 2058–2059.

Long, T.C., et al., 2006. Titanium dioxide (P25) produces reactive oxygen species in immortalized brain microglia (BV2): implications for nanoparticle neurotoxicity. Environ. Sci. Technol. 40, 4346–4352.

Lu, C., et al., 2005. Adsorption of trihalomethanes from water with carbon nanotubes. Water Res. 39 (6), 1183–1189.

Lu, C., et al., 2006a. Adsorption of zinc (II) from water with purified carbon nanotubes. Chem. Eng. Sci. 61, 1138–1145.

Lu, C., et al., 2006b. Adsorption thermodynamic and kinetic studies of trihalomethanes on multi-walled carbon nanotubes. J. Hazard. Mater. 138 (2), 304–310.

Lu, C., et al., 2006c. Removal of zinc (II) from aqueous solution by purified carbon nanotubes: kinetics and equilibrium studies. Industrial Eng. Chem. Res. 45 (8), 2850–2855.

Lu, X., et al., 2011. Facile synthesis and application of mesoporous silica coated magnetic carbon nanotubes. Chem. Commun. 47, 1210–1212.

Ma, H.L., et al., 2012. Chemical reduction and removal of Cr(VI) from acidic aqueous solution by ethylenediamine-reduced graphene oxide. J. Mater. Chem. 22, 5914–5916.

Machado, B.F., et al., 2012. Graphene-based materials for catalysis. Catal. Sci. Technol. 2, 54–75.

MacKintoshl, C., et al., 1990. Cyanobacterial microcystin-LR is a potent and specific inhibitor of protein phosphatases 1 and 2A from both mammals and higher plants. FEBS Lett. 264, 187–192.

Madadrang, C.J., et al., 2012. Adsorption behavior of EDTA-graphene oxide for Pb (II) removal. ACS Appl. Mater. Interfaces 4, 1186–1193.

Majsterek, I., et al., 2004. Toxicity of microcystin from cyanobacteria growing in a source of drinking water. Comp. Biochem. Physiol C Toxicol. Pharmacol. 139 (1–3), 175–179.

Mangun, C.L., et al., 2001. Adsorption of organic contaminants from water using tailored ACFs. Chem. Mater. 13 (7), 2356–2360.

Matilainen, A., et al., 2010. Removal of natural organic matter from drinking water by advanced oxidation processes. Chemosphere 80, 351–365.

Matsushima, R.N., et al., 1992. Liver tumor promotion by the cyanobacterial cyclic peptide toxin microcystin-LR. J. Cancer Res. Clin. Oncol. 118, 420–424.

Mauter, M.S., Elimelech, M., 2008. Environmental applications of carbon-based nanomaterials. Environ. Sci. Technol. 42 (16), 5843–5859.

Mohsen, M.S., et al., 2003. Desalination of brackish water by nanofiltration and reverse osmosis. Desalination 157 (1–3), 167.

Mondal, A., et al., 2012. Fluorescent detection of cholesterol using β-cyclodextrin functionalized graphene. Chem. Commun. 48, 7316–7318.

Mostafavi, S.T., et al., 2009. Preparation of nano-filter from carbon nanotubes for application in virus removal from water. Desalination 238 (1–3), 271–280.

Nakanishi, A., et al., 2002. Adsorption characteristics of bisphenol a onto carbonaceous materials produced from wood chips as organic waste. J. Colloid Interface Sci. 252, 393–396.

Nednoor, P., et al., 2005. Reversible biochemical switching of ionic transport through aligned carbon nanotube membranes. Chem. Mater. 17 (14), 3595–3599.

Nepal, D., et al., 2008. Strong antimicrobial coatings: single-walled carbon nanotubes armored with biopolymers. Nano Lett. 8 (7), 1896–1901.

Niu, J.J., et al., 2007. An approach to carbon nanotubes with high surface area and large pore volume. Microporous Mesoporous Mater. 100 (1–3), 1–5.

Nuzzo, J.B., 2006. The biological threat to U.S. water supplies: toward a national water security policy. Biosecur. Bioterrorism 4 (2), 147–159.

Ohko, Y., et al., 2001. Degradation of bisphenol A in water by TiO_2 photocatalyst. Environ. Sci. Technol. 35, 2365–2368.

Pavagadhi, S., et al., 2013. Removal of microcystin-LR and microcystin-RR by graphene oxide: adsorption and kinetic experiments. Water Res. 47, 4621–4629.

Pelaez, M., et al., 2013. Microcystin-LR removal from aqueous solutions using a magnetically separable N-doped TiO_2 nanocomposite under visible light irradiation. Chem. Commun. 49, 10118–10120.

Pendleton, P., et al., 2001. Microcystin-LR adsorption by activated carbon. J. Colloid Interface Sci. 240 (1), 1–8.

Peng, X.J., et al., 2003. Adsorption of 1,2-dichlorobenzene from water to carbon nanotubes. Chem. Phys. Lett. 376, 154–158.

Peng, X., et al., 2005. Ceria nanoparticles supported on carbon nanotubes for the removal of arsenate from water. Mater. Lett. 59 (4), 399–403.

Peng, H., et al., 2012. Adsorption of ofloxacin and norfloxacin on carbon nanotubes: hydrophobicity- and structure-controlled process. J. Hazard. Mater. 233–234, 89–96.

Pint, C.L., et al., 2008. Synthesis of high aspect-ratio carbon nanotube "flying carpets" from nanostructured flake substrates. Nano Lett. 8 (7), 1879–1883.

Punyamurtula, V.K., et al., 2007. Hysteresis of sorption isotherm of multiwall carbon nanotube in paraxylene. Mater. Res. Innov. 11, 37–39.

Rahaman, M.S., et al., 2012. Electrochemical carbon-nanotube filter performance toward virus removal and inactivation in the presence of natural organic matter. Environ. Sci. Tech. 46 (3), 1556–1564.

Rambabu, et al., 2012. Adsorption characteristics of atrazine on granulated activated carbon and carbon nanotubes. Chem. Eng. Technol. 35 (2), 272–280.

Rao, G.P., et al., 2007. Sorption of divalent metal ions from aqueous solution by carbon nanotubes: a review. Sep. Purif. Tech. 58 (1), 224–231.

Raty, J.Y., et al., 2005. Optical properties and structure of nanodiamonds. J. Electroanal. Chem. 584, 9–12.

Ray, S., et al., 2015. Copper incorporated nanorod like mesoporous silica for one pot aerobic oxidative synthesis of pyridines. Catal. Commun. 58, 97–102.

Rodriguez, E., et al., 2007. Oxidative elimination of cyanotoxins: comparison of ozone, chlorine, chlorine dioxide and permanganate. Water Res. 41, 3381–3393.

Roy, D., et al., 2003. Characterisation of carbon nano-onions using Raman spectroscopy. Chem. Phys. Lett. 373, 52–56.

Runnegar, M.T., et al., 1993. Protein phosphatase inhibition and in vivo hepatotoxicity of microcystins. Am. J. Physiol. 265, G224–G230.

Salam, N., et al., 2014. Synthesis of silver–graphene nanocomposite and its catalytic application for the one-pot three-component coupling reaction and one-pot synthesis of 1,4-disubstituted 1,2,3-triazoles in water. RSC Adv. 4, 10001–10012.

Saleh, N.B., et al., 2008. Aggregation kinetics of multi-walled carbon nanotubes in aquatic systems: measurements and environmental implications. Environ. Sci. Technol. 42 (21), 7963–7969.

Salipira, K.L., et al., 2007. Carbon nanotubes and cyclodextrin polymers for removing organic pollutants from water. Env. Chem. Lett. 5 (1), 13–17.

Savage, N., Diallo, M.S., 2005. Nanomaterials and water purification: opportunities and challenges. J. Nanoparticle Res. 7 (4–5), 331–342.

Schug, T.T., et al., 2011. Endocrine disrupting chemicals and disease susceptibility. J. Steroid. Biochem. Mol. Biol. 127, 204–215.

Schwarzenbach, R.P., et al., 2006. The challenge of micro-pollutants in aquatic systems. Science 313, 1072–1077.

Shi, et al., 2013. Facile low temperature hydrothermal synthesis of magnetic mesoporous carbon nanocomposite for adsorption removal of ciprofloxacin antibiotics. Ind. Eng. Chem. Res. 52 (7), 2604–2612.

Sinha, A., et al., 2012. Functional, mesoporous, superparamagnetic colloidal sorbents for efficient removal of toxic metals. Chem. Commun. 48, 9272–9274.

Sinha, A., et al., 2013. Graphene-based composite with γ-Fe_2O_3 nanoparticle for the high-performance removal of endocrine-disrupting compounds from water. Chem. Asian J. 8, 786–791.

Sinha, A., et al., 2015a. β-cyclodextrin functionalized magnetic mesoporous silica colloid for Cholesterol separation. ACS Appl. Mater. Interfaces 7, 1340–1347.

Sinha, A., et al., 2015b. Separation of microcystin-LR by cyclodextrin-functionalized magnetic composite of colloid graphene and porous silica. ACS Appl. Mater. Interfaces 7, 9911–9919.

Sint, K., et al., 2008. Selective ion passage through functionalized graphene nanopores. J. Am. Chem. Soc. 130 (49), 16448–16449.

Song, C., Corry, B., 2009. Intrinsic ion selectivity of narrow hydrophobic pores. J. Phy. Chem. B 113 (21), 7642–7649.

Song, W., et al., 2006. Ultrasonically induced degradation of microcystin-LR and -RR: identification of products, effect of pH, formation and destruction of peroxides. Environ. Sci. Technol. 40, 3941–3946.

Sreeprasad, T.S., et al., 2012. Graphene for environmental and biological applications. Int. J. Mod. Phys. B 26, 1242001–1242026.

Srinivasan, S., et al., 2008. Factors affecting bulk to total bacteria ratio in drinking water distribution systems. Water Res. 42 (13), 3393–3404.

Srivastava, A., et al., 2004. Carbon nanotube filters. Nat. Mater. 3 (9), 610–614.

Sui, Q., et al., 2011. Rapid removal of bisphenol A on highly ordered mesoporous carbon. J. Environ. Sci. 23, 177–182.

Takai, A., et al., 2011. Capture, store, and discharge. Shuttling photo-generated electrons across TiO_2-silver interface. ACS Nano 5, 7369–7376.

Teixeira, M.R., et al., 2005. Microcystins removal by nanofiltration membranes. Sep. Purif. Technol. 46, 192–201.

Teng, W., et al., 2013a. Ordered mesoporous carbons and their corresponding column for highly efficient removal of microcystin-LR. Energy Environ. Sci. 6, 2765–2776.

Teng, W., et al., 2013b. Rapid and efficient removal of microcystins by ordered mesoporous silica. Environ. Sci. Technol. 47, 8633–8641.

Tian, et al., 2013. Interactions between carbon nanotubes and sulfonamide antibiotics in aqueous solutions under various physicochemical conditions. J. Environ. Sci. Health A Tox. Hazard Subst. Environ. Eng. 48 (9), 1136–1144.

Tiraferri, A., et al., 2011. Covalent binding of single-walled carbon nanotubes to polyamide membranes for antimicrobial surface properties. ACS Appl. Mater. Interfaces 3 (8), 2869–2877.

Upadhyayula, et al., 2008a. Application of carbon nanotube technology for removal of contaminants in drinking water: a review. Sci. Total Environ. 408, 1–13.

Upadhyayula, V.K.K., et al., 2008b. Adsorption kinetics of *Escherichia coli* and *Staphylococcus aureus* on single-walled carbon nanotube aggregates. Water Sci. Tech. 58 (1), 179–184.

Upadhyayula, V.K.K., et al., 2008c. Single-walled carbon nanotubes as fluorescence biosensors for pathogen recognition in water systems. Res. Lett. Nanotechnol. 1–5, 156358.

Upadhyayula, V.K.K., et al., 2009. Application of carbon nanotube technology for removal of contaminants in drinking water: a review. Sci. Total Environ. 408 (1), 1–13.

Vecitis, C.D., et al., 2010. Electronic-structure-dependent bacterial cytotoxicity of single-walled carbon nanotubes. ACS Nano 4 (9), 5471–5479.

Vecitis, C.D., et al., 2011. Electrochemical multiwalled carbon nanotube filter for viral and bacterial removal and inactivation. Environ. Sci. Tech. 45 (8), 3672–3679.

Wang, X.K., et al., 2005. Sorption of ^{243}Am(III) to multiwall carbon nanotubes. Environ. Sci. Technol. 39, 2856–2860.

Wang, H., et al., 2007. Discriminating and assessing adsorption and biodegradation removal mechanisms during granular activated carbon filtration of microcystin toxins. Water Res. 41 (18), 4262–4270.

Wang, P., et al., 2009. Highly efficient visible-light plasmonic photocatalyst Ag@AgBr. Chem. Eur. J. 15, 1821–1824.

Wang, P.H., et al., 2013. Ag–AgBr/TiO_2/RGO nanocomposite for visible-light photocatalytic degradation of penicillin G. J. Mater. Chem. A 1, 4718–4727.

Wanga, C., et al., 2014. Photocatalytic degradation of bisphenol A and dye by graphene-oxide/Ag_3PO_4 composite under visible light irradiation. Ceram. Int. 40, 8061–8070.

Wei, et al., 2013. Regenerable granular carbon nanotubes/alumina hybrid adsorbents for diclofenac sodium and carbamazepine removal from aqueous solution. Water Res. 47, 4139–4147.

Wick, P., et al., 2007. The degree and kind of agglomeration affect carbon nanotube cytotoxicity. Toxicol. Lett. 168 (2), 121–131.

Wu, Z., et al., 2011a. Ordered mesoporous materials as adsorbents. Chem. Commun. 47, 3332–3338.

Wu, X., et al., 2011b. Mechanisms and factors affecting sorption of microcystins onto natural sediments. Environ. Sci. Technol. 45 (7), 2641–2647.

Wu, et al., 2012a. Oxidative removal of selected endocrine-disruptors and pharmaceuticals in drinking water treatment systems, and identification of degradation products of triclosan. Sci. Total Environ. 439, 18–25.

Wu, Q., et al., 2012b. Cadmium sulphide nanoparticles as a novel coating for solid-phase microextraction. J. Sep. Sci. 35, 193–199.

Wu, Q., et al., 2012c. Extraction of phthalate esters from water and beverages using a graphene-based magnetic nanocomposite prior to their determination by HPLC. Microchim. Acta 177, 23–30.

Xu, J., et al., 2012. Decontamination of bisphenol a from aqueous solution by graphene adsorption. Langmuir 28, 8418–8425.

Yan, H., et al., 2004. Effective removal of microcystins using carbon nanotubes embedded with bacteria. Chin. Sci. Bull. 49 (16), 1694–1698.

Yan, H., et al., 2006. Adsorption of microcystins by carbon nanotubes. Chemosphere 62 (1), 142–148.

Yan, X.M., et al., 2008. Adsorption and desorption of atrazine on carbon nanotubes. J. Colloid Inter. Sci. 321 (1), 30–38.

Yang, K., et al., 2006a. Adsorption of polycyclic aromatic hydrocarbons by carbon nanomaterials. Environ. Sci. Technol. 40 (6), 1855–1861.

Yang, F.H., et al., 2006b. Adsorption of spillover hydrogen atoms on single-wall carbon nanotubes. J. Phys. Chem. B 110, 6236–6244.

Yang, K., et al., 2010. Sorption of phenanthrene by nanosized alumina coated with sequentially extracted humic acids. Environ. Sci. Pollut. Res. 17 (2), 410–419.

Yang, X., et al., 2012. Graphene oxide-iron oxide and reduced graphene oxide-iron oxide hybrid materials for the removal of organic and inorganic pollutants. RSC Adv. 2, 8821–8826.

Yang, Y., et al., 2013. Preparation of reduced graphene oxide/poly(acrylamide) nanocomposite and its adsorption of Pb(II) and methylene blue. Langmuir 29, 10727–10736.

Yang, K., Xing, B., 2007. Desorption of polycyclic aromatic hydrocarbons from carbon nanomaterials in water. Environ. Pollut. 145, 529–537.

Ye, C., et al., 2007. Adsorption of middle molecular weight toxins on carbon nanotubes. Acta Phys. Chim. Sin. 23 (9), 1321–1324.

Yu, Z., et al., 2009. Adsorption of selected pharmaceuticals and an endocrine disrupting compound by granular activated carbon. 2. Model prediction. Environ. Sci. Technol. 43, 1474–1479.

Yu, et al., 2012. Aqueous adsorption and removal of organic contaminants by carbon nanotubes. Sci. Total Environ. 482, 483, 241–251.

Yuana, B.L., et al., 2002. Removal of cyanobacterial microcystin-LR by ferrate oxidation–coagulation. Toxicon 40, 1129–1134.

Zaib, I.A., et al., 2012. Removal of bisphenol A and 17 β-estradiol by single-walled carbon nanotubes in aqueous solution: adsorption and molecular modelling. Water Air Soil Pollut. 223, 3281–3293.

Zhang, Y., et al., 2006. Removal of bisphenol a by a nano filtration membrane in view of drinking water production. Water Res. 40, 3793–3799.

Zhang, Y.H., et al., 2010. TiO_2-graphene nanocomposites for gas-phase photocatalytic degradation of volatile aromatic pollutant: is TiO_2–graphene truly different from other TiO_2–carbon composite materials? ACS Nano 4, 7303–7314.

Zhang, et al., 2011a. Removal of phenolic endocrine disrupting compounds from waste activated sludge using UV, H_2O_2, and UV/H_2O_2 oxidation processes: effects of reaction conditions and sludge matrix. Sci. Total Environ. 493, 307–323.

Zhang, Y., et al., 2011b. Nanocomposite of Ag–AgBr–TiO_2 as a photoactive and durable catalyst for degradation of volatile organic compounds in the gas phase. Appl. Catal. B Environ. 106, 445–452.

Zhang, H., et al., 2011c. Graphene sheets grafted hybrid with enhanced plasmonic photocatalytic activity under visible light. Environ. Sci. Technol. 45, 5731–5736. Ag@AgCl.

Zhang, X., et al., 2011d. Fabrication of novel rattle-type magnetic mesoporous carbon microspheres for removal of microcystins. J. Mater. Chem. 21, 10653–10657.

Zhang, H.C., et al., 2012. Carbon quantum dots/Ag_3PO_4 complex photocatalysts with enhanced photocatalytic activity and stability under visible light. J. Mater. Chem. 22, 10501–10506.

Zhao, G., et al., 2011a. Colorimetric sensing of atrazine in rice samples using cysteamine functionalized gold nanoparticles after solid phase extraction. Anal. Chim. Acta 708, 155–159.

Zhao, G.X., et al., 2011b. Sulfonated graphene for persistent aromatic pollutant management. Adv. Mater. 23, 3959–3963.

Zhao, G., et al., 2012. Synthesis of graphene-based nanomaterials and their application in energy-related and environmental-related areas. RSC Adv. 2, 9286–9303.

Zhao, X., et al., 2014. Hydrophobic-polymer-grafted graphene oxide nano-sheets as an easily separable adsorbent for the removal of tetra-bromo-bisphenol A. Langmuir 30, 13699–13706.

Zhou, Q., et al., 2006a. Using multi-walled carbon nanotubes as solid phase extraction adsorbents to determine dichlorodiphenyltrichloroethane and its metabolites at trace level in water samples by high performance liquid chromatography with UV detection. J. Chromatogr. A 1125 (2), 152–158.

Zhou, Q., et al., 2006b. Determination of atrazine and simazine in environmental water samples using multiwalled carbon nanotubes as the adsorbents for pre-concentration prior to high performance liquid chromatography with diode array detector. Talanta 68 (4), 1309–1315.

Zhou, M., et al., 2007. New type of membrane material for water desalination based on a cross-linked bicontinuous cubic lyotropic liquid crystal assembly. J. Am. Chem. Soc. 129 (31), 9574–9575.

Zhu, Y., et al., 2006. Dependence of the cytotoxicity of multi-walled carbon nanotubes on the culture medium. Nanotechnology 17 (18), 4668–4674.

Zhu, Y., et al., 2011. Carbon-based supercapacitors produced by activation of graphene. Science 332, 1537–1541.

FURTHER READING

Lu, J., et al., 2009. Sorption of atrazine onto humic acids (HAs) coated nanoparticles. Colloids Surf. A Physicochem. Eng. Aspects 347 (1–3), 90–96.

Tan, X., et al., 2007. Effect of soil humic and fulvic acids, pH and ionic strength on Th (IV) sorption to TiO_2 nanoparticles. Appl. Radiat. Isot. 65 (4), 375–381.

Yu, J.G., et al., 2014. Aqueous adsorption and removal of organic contaminants by carbon nanotubes. Sci. Total Environ. 482, 483, 241–251.

CHAPTER 3

Application of Carbon-Based Nanomaterials as Biosensor

CONTENTS

3.1 INTRODUCTION

Biosensors are becoming an essential part of modern health care. A biosensor is an analytical device that incorporates a biological recognition element in direct spatial contact with a transduction element. This integration ensures the rapid and convenient conversion of the biological events to detectable signals (Thevenot et al., 2001). Biosensor development becomes more crucial due to the demand for personalized medicine, point-of-care devices, and cheaper diagnostic tools. Substantial advances in sensor technology are often fueled by the advent of new materials. With the discovery of rich nanomaterials and the development of exquisite nanofabrication tools, such as electron beam lithography, focused ion beam, and nanoimprint lithography, new avenues have been opened up in the field of biosensors in the last few decades (Liu et al., 2009a,b; Rosi et al., 2005). In particular, researchers around the world have been tailor-making a multitude of nanomaterial-based electrical biosensors and developing new strategies to apply them in ultrasensitive biosensing. The key issues in the development of all biosensors include design of the biosensing interface so that the analyte selectively interacts with the biosensing surface (Gooding et al., 2003a,b; Gooding, 2008) for achievements of efficient transduction of the biorecognition event (Heller, 1990; Bernhardt, 2006), increases the sensitivity and selectivity of the biosensor (Wang, 2007; Patolsky et al., 2006a,b,c,d), and improves the response times in very sensitive systems (Wang, 2008). More specific challenges include making biosensors compatible with biological matrices, so that they can be used in complex biological samples or even in vivo (Willner and Zayats, 2007; Amatore et al., 2008) fabrication of viable biosensors that can operate within confined environments such as inside cells (Amatore et al., 2008), and multiplexing biosensors so the multiple analytes can be detected on one device (Yu et al., 2007; Sadik et al., 2009; Stromberg et al., 2009). Various kinds of zero-, one-, two-, and three-dimensional nanomaterials are helping to meet these challenges. Examples of such materials include semiconductor quantum dots (Sapsford et al., 2006), metallic nanoparticles (NPs) (Pingarron et al., 2008), metallic or semiconductor nanowires (Wang, 2008; He et al., 2008), carbon nanotubes (CNTs) (Kauffman et al., 2008; Maehashi

Carbon Nanomaterials for Biological and Medical Applications. http://dx.doi.org/10.1016/B978-0-323-47906-6.00003-5

et al., 2009), nanostructured conductive polymers or nanocomposites thereof (Rajesh et al., 2009), mesoporous materials (Kilian et al., 2009), and various other nanomaterials (Qi et al., 2009; Sarma et al., 2009). Therefore, nanomaterials have motivated a large body of research, and such materials have been implemented into biosensor devices. Nanomaterials, particularly carbon nanomaterials that include CNTs (Huang et al., 2010; Yang et al., 2010; Wang, 2005a,b; Allen et al., 2007; Lahiff et al., 2010; Liu et al., 2009a,b, 2010; Roy and Gao, 2009; Gruner, 2006; Kauffman et al., 2008), nanowires (Roy and Gao, 2009; Arlett et al., 2011; Song et al., 2010; He et al., 2010a,b,c; Chen et al., 2011; Zheng et al., 2005; Stern et al., 2007, 2010; Tian et al., 2010), NPs (Allen et al., 2007; Medintz et al., 2005; Wilson, 2008; Nam et al., 2003; Alivisatos, 2004), nanopores (Dekker, 2007; Howorka and Siwy, 2009), nanoclusters (Soleymani et al., 2009), and graphene (Yang et al., 2010; Ratinac et al., 2011; Ohno et al., 2010a,b; Huang et al., 2011; He et al., 2012) and its derivatives have a significant role to play in new developments in each of the biosensor-sized domains. Compared with conventional optical, biochemical, and biophysical methods, nanomaterial-based electronic biosensing offers significant advantages, such as high sensitivity and new sensing mechanisms, high spatial resolution for localized detection, facile integration with standard wafer-scale semiconductor processing, and label-free and real-time detection in a nondestructive manner.

A chemical sensor is a device that quantitatively or semiquantitatively converts information about the presence of a chemical species to an analytically useful signal (Hulanicki et al., 1991). Sensors consist of two elements: a *receptor* and a *transducer* (Fig. 3.1; Pumera, 2011). A receptor can be any organic or inorganic material with (preferably) a specific interaction with one analyte or group of analytes. In the case of biosensors, the recognition element is a biomolecule. The second key element of the sensing platform is the transducer that converts chemical information into a measurable signal. Bioanalytical protocols usually include more than one processing step. In this chapter, we will describe biosensors and bioanalytical systems that utilize carbon nanomaterials as a key component.

However, we will focus in this chapter only on the use of biosensors on numerous classes of carbon nanomaterials such as CNTs, graphene and its derivatives, carbon dots (CDs), graphene quantum dots (GQDs), fullerene, carbon nanohorns (CNHs), and carbon nanoonions (CNOs). They have been explored for potential applications in the field of biology, owing to their unique electronic, optical, thermal, and mechanical properties.

3.2 APPLICATIONS OF BIOSENSOR

Owing to the sensitivity of the biological and chemical properties of carbon nanomaterials to the surrounding environment, they provide an exceptional

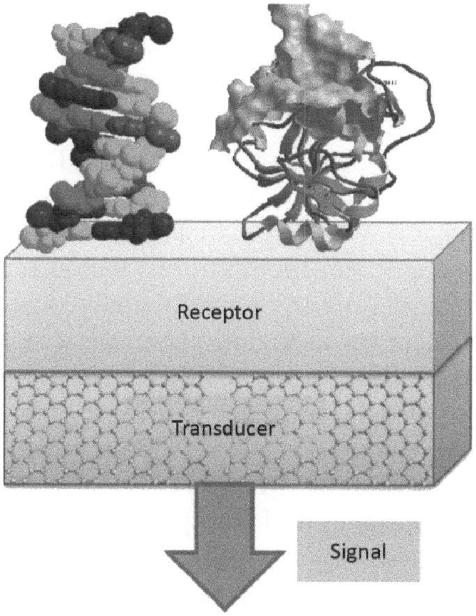

FIGURE 3.1
The biosensor consists of a receptor layer that consists of a biomolecule (e.g., DNA or protein) and a transducer, which is a carbon-based nanomaterial. *Copyright permission from Pumera, M., 2011. Graphene in bio-sensing. Mater. Today 14, 308–315.*

advantage for biosensors. In last 10 years, carbon nanomaterials have been used for sensing a variety of analytes including biomolecules, gases, and solvents. A majority of these are detected by means of fluorescence. Furthermore, most biological applications of carbon nanomaterials rely on modifications. To improve the properties of carbon nanomaterials, their modification is emerging. Many applications of carbon nanomaterials depend on their successful modifications, which are mainly classified into two categories: covalent modifications and noncovalent modifications. A very brief description of these two modifications of carbon nanomaterials is given in this section.

For CNTs, covalent modifications are mostly carried out via chemical reactions such as oxidation, halogenation, cycloaddition, or electrochemical reactions (Wu et al., 2010). These reactions can change the shape and even the structure or length of CNTs or create some chemical groups on the edges of CNTs (Liu et al., 1998). Covalent modifications mostly enhance the biocompatibility and hydrophilicity of CNTs and are hence widely used in biology and medical research. For graphene and its derivatives, some chemical groups, commonly carboxylic (−COOH) and hydroxyl (−OH) groups, can be covalently added onto their surface using strong acids and/or oxidants. Chemical groups created on the surface of graphene and its derivatives are used as chemical handles to graft functional molecules such as proteins, carbohydrates, and polymers via covalent bonding, hence increasing the biocompatibility, sensitivity, and specificity of graphene and its derivatives. For CDs, owing to their easy functionalization, there are many ways to modify CDs via surface chemistry or interactions. Modifications of CDs not only tune or enhance their luminescence but also make them superior for biological applications. GQDs can also be modified via many methods. Modification of GQDs can not only improve their properties further but also provide a way to modulate their properties (Qu et al., 2012; Li et al., 2012a,b,c). These covalent modifications will affect or even destroy the microstructure and properties of carbon nanomaterials to some extent. To avoid this drawback, noncovalent modification is emerging as an important way to modify carbon nanomaterials. Because of their benzene ring structures, CNTs can noncovalently interact with aromatic polymers or biomolecules via $\pi-\pi$ stacking, electrostatic interactions, van der Waals forces, hydrogen bonding, etc. These advantages provide approaches for controlling the biological behavior of CNTs, such as their toxicity and biocompatibility. Graphene and its derivatives, which are highly negatively charged, are able to electrostatically adsorb oppositely charged molecules. In addition, $\pi-\pi$ stacking, hydrophobic or van der Waals interactions may assist physical adsorption. Using this process, some biomolecules such as single-stranded DNA (ssDNA) can be anchored on graphene or its derivatives (Huang et al., 2011; Dong et al., 2011a,b; Mao et al., 2010; Park et al., 2010; Robinson et al., 2011; Hu et al., 2011; Liu et al., 2011a,b,c; Dubuisson et al., 2011; Yin et al., 2012). However, in this section, we will discuss fluorescence biosensing applications using carbon-based nanomaterials, mainly CNTs, graphene and its derivatives, CDs, and GQDs.

3.2.1 Biosensor Using Carbon Nanoparticles

3.2.1.1 Carbon Dots for Fluorescence Biosensing

CDs have excellent fluorescence properties and can be used as fluorescent labels for DNA, aptamers, proteins, glucose, phosphate, metal ions, etc. (Qian et al., 2014a,b; Lin et al., 2014; Liu et al., 2014a,b,c; Du et al., 2013; Wang et al., 2013a,b,c; Cayuela et al., 2013; Dong et al., 2013; Shi et al., 2013; Li et al., 2013a,b,c; Niu and Gao, 2014; Huang et al., 2013; Zheng et al., 2013; Mao et al., 2012; Zhou et al., 2012; Qu et al., 2013). Li et al. (2011) demonstrated an effective fluorescence sensing platform for the detection of nucleic acids using CDs. A dye-labeled ssDNA probe was adsorbed onto the surface of CDs via $\pi-\pi$ interaction and quenched the dye fluorescence. A double-stranded DNA (dsDNA) hybrid is formed in the presence of complementary oligonucleotide, which restored the fluorescence of the dye. Noh et al. (2013) successfully prepared a CD-based sensor for imaging miR124a with no evidence of cellular toxicity and a high level of self-promoted uptake into cells. The CD-based miR124a molecular beacon (CMB) was easily internalized into P19 cells and successfully visualized a gradual increase in miR124a expression during neuronal differentiation by providing signal-on imaging activity. A dsDNA oligonucleotide containing an miR124a binding site and Black Hole Quencher 1 (miR124a sensing oligo) was further conjugated with CDs to form a-CMB. P19 cells were incubated with miR124a-CMB to sense miR124a expression during neurogenesis. Xu et al. (2012) developed an aptamer–CD-based sandwich system for the sensitive and selective detection of thrombin with a limit of detection (LOD) of 1 nM (Fig. 3.2A). The presence of thrombin can induce aptamer-modified fluorescent CDs to form a sandwich structure with aptamer-functionalized silica NPs via specific protein–aptamer interaction. Maiti et al. (2012) developed a fluorometric technique for histone sensing with an LOD of 0.2 ng mL^{-1} using a quaternized carbon dot (QCD)–DNA nanobiohybrid for the first time. The QCD–dsDNA hybrid was prepared via electrostatic attraction. The emission of the QCD was quenched in the presence

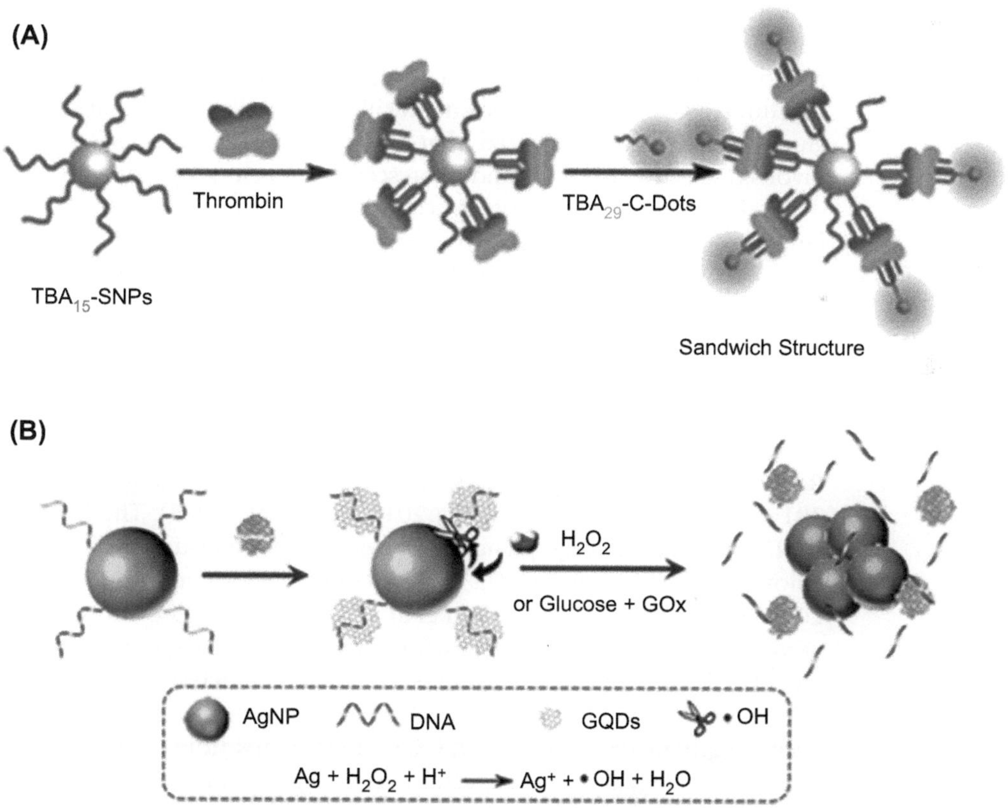

FIGURE 3.2

(A) Schematic illustration of sandwich-based thrombin detection principle using CDs (Xu et al., 2012). (B) Schematic description of H_2O_2 and glucose detection based on AgNP-DNA@GQDs (Wang et al., 2014a,b,c). *CDs*, carbon dots; *GQDs*, graphene quantum dots. *(A and B) Copyright by the Royal Society of Chemistry.*

of dsDNA but recovered with the addition of histone to the QCD–dsDNA hybrid due to strong binding affinity between histone and dsDNA.

In addition to biomolecules, CDs have also shown promising fluorescent probes in the detection of small bioanalytes such as antibacterial drugs, dopamine (DA), ascorbic acid (AA), glucose, etc. For example, Mao et al. (2012) synthesized a new type of eco-friendly molecularly imprinted polymer (MIP) by efficient one-pot room-temperature sol–gel polymerization and applied it as a molecular recognition element to construct a DA fluorescence opto-sensor. The new MIP-based DA sensing protocol was successfully applied to detect DA concentrations in aqueous solution with an LOD of 1.7 nM, as well as in human urine samples without the interference of other molecules and ions. Zheng et al. (2013) demonstrated an on–off fluorescent CD probe with the advantages of simplicity, convenience, rapid response, high selectivity, and sensitivity for detecting Cr(VI) based on the inner filter effect because the absorption bands of Cr(IV) fully cover the emission and excitation bands of CDs. They successfully employed AA as an example molecule to demonstrate this off–on fluorescent probe. Kiran et al. (2015) demonstrated a new class of "inert" nonenzymatic and boronic acid–functionalized CD-based sensors facilitating the intracellular detection of glucose. The study suggested that the mechanism of detection of glucose involved selective assembly and fluorescence quenching of CDs with an excellent dynamic response to varying concentrations of glucose within the biological range (1–100 mM). The strong dynamic response was related to the high selectivity for biomolecules and inertness of CDs.

3.2.1.2 Graphene Quantum Dots for Fluorescence Biosensing

GQDs not only have the good properties of CDs but also possess some of the excellent properties of graphene, such as high electron mobility and chemical stability. They are also widely used for biosensing (Razmi and Mohammad-Rezaei, 2013; Lu et al., 2013; Al-Ogaidi et al., 2014; Wang et al., 2013a,b,c, 2014a,b,c; Benítez-Martínez et al., 2014; Ran et al., 2013). The fluorescence of GQDs can be effectively quenched by selectively interacting with specifications, anions, or chemical groups (Li et al., 2012a,b,c; Zhou et al., 2014; Wang et al., 2014a,b,c; Ju et al., 2014a,b; Sun et al., 2013). This feature allows GQDs to be used as sensors to detect nucleic acids. Qian et al. (2014a,b) established a novel and effective fluorescence sensing platform for the detection of DNA based on fluorescence resonance energy transfer (FRET) by regulating the interaction between graphene oxide (GO) and GQDs for the first time. This can be used as a universal strategy for DNA detection, as well as distinguishing complementary and mismatched nucleic acid sequences with high sensitivity and good reproducibility. The GQDs can also be used as sensors to detect various biomolecules such as proteins. Ju et al. (2014a,b) synthesized a type of highly blue luminescent nitrogen-doped graphene quantum dots (N-GQDs) with high quantum yield via a facile one-step hydrothermal treatment of citric acid and 2-cyanoguanidine. The nitrogen functionalized-GQDs can be used as efficient fluorescent probes for the detection of glutathione (GSH) with an LOD of 87 nM. Wu et al. (2014) developed a facile method for the highly sensitive and selective sensing of biothiols based on GQDs with strong blue fluorescence in an aqueous buffer solution. It was observed that mercury (II) ions could efficiently bind to and quench the fluorescence of GQDs. When a biothiol compound (GSH, cysteine, or homocysteine) was added to an assay mixture of GQDs and mercury (II), it binds to mercury (II) ions. The Hg^{2+}—GQD complex is dissociated, and the fluorescence is restored. The changes in emission intensity of GQDs could be directly related to the amount of biothiol added to the assay solution. The LODs for GSH, Cys, and Hcy were 5, 2.5, and 5 nM, respectively. However, in addition to biomacromolecules, GQDs have also shown promise as fluorescent probes in the detection of small bioanalytes such as glucose. Wang et al. (2014a,b,c) proposed a DNA-mediated silver NP—GQD hybrid nanocomposite (Ag NP—DNA@GQDs) for the sensitive fluorescence detection of H_2O_2 and glucose (Fig. 3.2B; Huang et al., 2012).

The sensing mechanism was based on the etching effect of H_2O_2 on Ag NPs and the cleavage of DNA by as-generated hydroxyl radicals (•OH). The formation of an Ag NP—DNA@GQDs nanocomposite can result in fluorescence quenching of GQDs by Ag NPs via resonance energy transfer. Upon the addition of H_2O_2, the energy transfer between Ag NPs and GQDs mediated by DNA were reduced and obvious recovery of the fluorescence of GQDs could be observed. For the oxidation of glucose and formation of H_2O_2, this nanocomposite can be further extended to glucose sensing in human urine in combination with glucose oxidase (GO_x). Glucose concentrations in human urine were detected with satisfactory recoveries of 94.6—98.8%, which suggested potential for the ultrasensitive quantitative analysis of glucose.

3.2.2 Biosensor Using Carbon Nanotubes

Among carbon nanomaterials the CNTs are especially promising building blocks for biosensors due to their high aspect ratios, high mechanical strength, high surface areas, excellent chemical and thermal stability, and rich electronic and optical properties (Ajayan, 1999). Biosensing based on CNTs has attracted significant attention from scientists because of the CNTs' advantages such as a broad absorption spectrum, low background, high signal-to-noise ratio, label-free detection, real-time monitoring, high sensitivity, and simplicity of apparatus. CNTs also have an ultralarge surface area for loading multiple molecules to achieve multiplexed sensing. The excellent optical properties make CNTs important transducer materials in biosensors: high conductivity along their length means they are excellent nanoscale electrode materials (Heller et al., 2005; Krapf et al., 2006; Gooding et al., 2007); their semiconducting behavior makes them ideal for nanoscale field effect transistors (FETs) (Heller et al., 2008), and their optical properties are suitable for entirely nanoscale devices (Heller et al., 2006a,b, 2009a,b,c). This combination of properties has resulted in CNTs being used to address all of the biosensing issues. The combination of excellent conductivity, good electrochemical properties, and nanometer dimensions has seen CNTs being plugged directly into individual redox enzymes for better transduction in electrochemical enzyme biosensors (Guiseppi-Elie et al., 2002; Gooding et al. 2003a,b; Yu et al., 2003; Liu et al., 2005; Patolsky et al., 2004). Moreover, alignment of CNTs has created the potential for electrodes that resist nonspecific adsorption of proteins, but that can interface to individual biomolecules (Li et al., 2003; Koehne et al., 2003; Chen et al., 2003). FET biosensors based on CNTs (Chen et al., 2003; Besteman et al., 2003) hold the promise of detecting single-molecule events (Goldsmith et al., 2007). The sensitivity of the optical properties of CNTs to binding events has also been exploited to make entirely nanoscale, but highly sensitive, multiplexed optical biosensors that could be used inside cells or dispersed through a system to capture the small amount of analyte in a sample (Heller et al., 2009a,b,c). The success of CNTs in advancing biosensors is part of the reason for the incredible interest in graphene as a material that could potentially push the boundaries of this field even farther. CNTs are commonly referred to as rolled-up graphene sheets, and both allotropes have a meshwork of sp^2-hybridized carbon atoms, so the question arises as to whether graphene offers any real benefits in properties relative to CNTs. Given the identical composition of nanotubes and graphene, one could be forgiven for suspecting that their properties would also be similar; however, this is not always the case, as we shall see shortly, and the differences in structure and properties open new vistas for further developments in biosensors.

Moreover, CNTs attached to nucleic acids or proteins (Zhu et al., 2010) can protect these biomolecules from enzymatic digestion or degradation in a biological environment. Given these properties in relation to the design of fluorescence biosensing systems, CNTs have become promising candidates for fluorescence biosensing. Many research groups have been devoted to exploring CNT-based biosensing systems.

Huang et al. (2012) have developed an amplified chemiluminescence turn-on sensing platform for ultrasensitive detection of DNA, which depended on SWCNTs. The sensing platform was based on modulation of the efficiency of chemiluminescence resonance energy transfer (CRET) between a SWCNT acceptor and a chemiluminescent donor. The chemiluminescence of the sensor was switched on by exonuclease-recycled DNA cleavage and turned off by CRET on the SWCNT surface, which therefore resulted in amplification of the readout signal, attaining detection sensitivity with three orders of magnitude higher than that of traditional biosensors and higher specificity for the target molecules (Fig. 3.3). Meng et al. (2012) have used SWCNTs to quench the

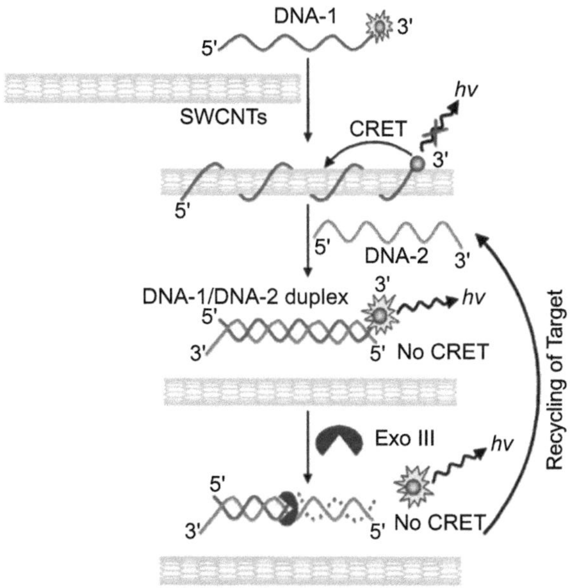

FIGURE 3.3
Single-walled carbon nanotube (SWCNT)-mediated chemiluminescence resonance energy transfer (CRET) platforms for the detection of DNA (Huang et al., 2012). *Copyright by the Royal Society of Chemistry.*

fluorescence of acridine orange (AO), due to the formation of a hybrid complex between AO and SWCNTs. Approximately, 18-fold enhancement in fluorescence can be observed after the addition of a certain amount of DNA to the complex mentioned earlier. The increase in fluorescence was linearly proportional to the amount of DNA added in the concentration range of 0—50.75 μM and the LOD of DNA was as low as 8.56×10^{-8} M. Wang et al. (2013a,b,c) have successfully constructed a novel and efficient method for the label-free turn-on fluorescence detection of the respiratory syncytial virus gene sequence with an LOD of 24 nM, based on FRET between MWCNTs and DNA-Ag NCs. The notable enhancement in the fluorescence of the DNA-Ag NCs resulted from specific binding of the DNA-Ag NCs to the target DNA and the quenching of the fluorescence of the DNA-Ag NCs with an extraordinarily high quenching efficiency (85.8%) resulted from MWCNTs. Nam et al. (2012) have fabricated horizontally aligned carbon nanotubes (ACNTs), which were functionalized with specific aptamers with the ability to specifically bind to biomolecules such as thrombin. The detection system was based on scanning probe microscopy imaging for ACNTs that specifically reacted with target biomolecules at an ultralow concentration with high detection sensitivity down to 1 pM.

Many biosensing systems are based on changes in the near-infrared (NIR) emission spectra of CNTs. Their fluorescence in the NIR region (between 820 and

1600 nm, where absorption of biological tissues is usually negligible), inherent photostability, and tissue transparency are exceptional characteristics for the design of in vitro and in vivo sensors. Iizumi et al. (2013) demonstrated an immunoassay using NIR CNT labels conjugated to immunoglobulin G (IgG) antibodies. The NIR emission of the conjugated CNTs at 1000–1200 nm confirmed that most of the CNT-conjugated IgGs had been successfully immune precipitated with magnetic beds attached to protein G and eluted from them. As a result, the photoluminescence intensity of the CNT labels was strong enough to detect antigens at 600 pM by the procedures mentioned earlier.

Moreover, some applications are based on the development of glucose sensors. Bhattacharyya et al. (2013) have exploited a lipid-functionalized SWCNT-based self-assembly supermicellar structure to trap glucose oxidase in a molecular cage for glucose sensing. The remarkable feature of such a molecular trap is that all components of this unique structure are reusable and rechargeable. Furthermore, glucose sensing was achieved without any hybrid fabrication. Similarly, CNTs are widely used to detect nitric oxide (NO) due to their large surface area. The ability to detect NO quantitatively may assist in the study of carcinogenesis and chemical signaling due to NO, as well as in medical diagnostics for inflammation. Zhang et al. (2011) reported the selective detection of single NO molecules based on a specific DNA sequence of d(AT)15 oligonucleotides, adsorbed onto an array of NIR-fluorescent semiconducting SWCNTs (AT15-SWCNTs). When the sensor was exposed to NO, a stepwise decrease in fluorescence was observed. This quenching process was described using a birth and death Markov model, of which the maximum likelihood estimator provided the adsorption and desorption rates of NO. The adsorption rate exhibited a linear dependence upon the NO concentration. In the following section, recent advances in biosensors made with CNTs and graphene are discussed.

3.2.2.1 Electrical Biosensor Using Carbon Nanotube

An ideal biosensor can directly translate the interactions between target biological molecules and the FET surface into readable electrical signals (Cui et al., 2001; Patolsky et al., 2006a,b,c,d; Wang et al., 2005; Chen et al., 2003; Besteman et al., 2003). In a standard FET, current flows along a semiconductor path (the channel) that is connected to two electrodes (the source and the drain). The channel conductance between the source and the drain is switched on and off by a third (gate) electrode that is capacitive coupled through a thin dielectric layer. In conventional complementary metal oxide semiconductor-fabricated transistors, the conducting channel is buried inside the substrate; in FET-based biosensors, the channel is in direct contact with the environment and this gives better control over the surface charge. This implies that surface FET-based biosensors might be more sensitive: biological events occurring at the channel surface could result in the surface potential variation of the semiconductor channel and then modulate the channel conductance. In conjunction with the ease of on-chip integration of device arrays and the cost-effective device fabrication, the surface ultrasensitivity places FET-based biosensors as attractive alternatives to existing biosensor technologies. Here the recent progress in ultrasensitive biosensors formed from CNTs and graphene-based FETs are summarized. Some important aspects are highlighted including strategies to increase sensitivity, dynamic detection in cells and liquid environment, DNA hybridization and single-molecule detection, as these have been neglected in most previous reviews. Fortunately, there are a number of excellent previous review papers in the literature covering various aspects of carbon nanomaterial-based biosensors, which can amend these deficiencies (Huang and Chen, 2010; Yang et al., 2010; Wang, 2005a,b; Allen et al., 2007; Lahiff et al., 2010; Liu et al., 2009a,b, 2010; Roy and Gao, 2009; Gruner, 2006; Kauffman and Star, 2008; Cui et al., 2001; Patolsky et al., 2006a,b,c,d).

3.2.2.1.1 Dynamic Detection in Living Cells

The dynamic detection of the release of biomolecules from living cells in real time is important both in fundamental studies and in the evaluation of drugs for the treatment of secretion-related diseases. Huang et al. (2009) utilized an SWCNT network to directly interface with living neuroglial astrocytes and detect the triggered release of ATP from these cells without labels. This detection scheme showed high temporal resolution. Highly charged ATP molecules secreted from the astrocyte diffused into the conductive channel of the FET and electrostatically modulated the SWCNT conductance, leading to measurable current responses. Heller et al. (2009a,b,c) used SWCNTs in a contact-passivated, suspended layout to allow close contact between the cell and the SWCNT. They followed the process of phagocytosis in real time by simultaneously monitoring both changes in transistor conductance (FET signal) and changes in the electrochemical current (signal), which suggests successful detection of cellular activity. They also demonstrated that the sensitivity for certain electrochemical processes could be enhanced when the SWCNT was coated with catalytic platinum NPs. Sudibya et al. (2009) improved the biocompatible interactions between SWCNTs and living cells. They demonstrated that noncovalent functionalization of SWCNTs with bioactive sugar moieties conferred biocompatibility without compromising the sensing capabilities of their devices. The SWCNT network was first surface-functionalized via $\pi-\pi$ interactions with bioactive sugars (N-acetyl-D-glucosamine) to allow $PC1_2$ cells to adhere and grow on the SWCNT net substrate. When the solution with high K^+ content was administrated to the cells to evoke Ca^{2+} influx through voltage-gated Ca^{2+} channels and to consequently trigger exocytosis of secretory vesicles, single-cell secretion of catecholamine molecules occurred, and this resulted in current responses (spikes) of glycosylated

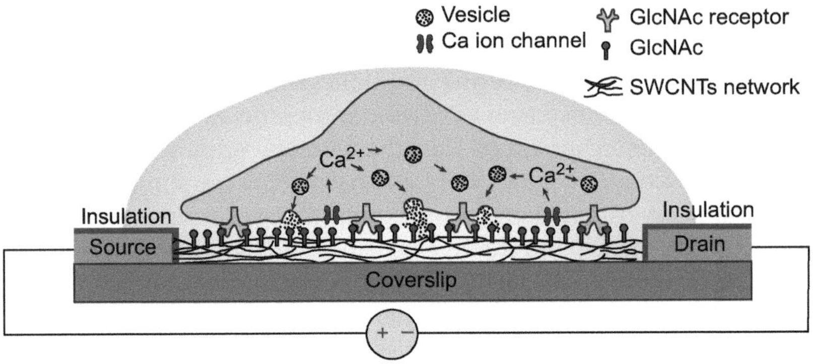

FIGURE 3.4
Triggered exocytosis and SWCNT net detection (Sudibya et al., 2009). *Copyright Wiley-VCH Verlag GmbH & Co. KGaA. Reproduced with permission.*

SWCNT net FETs (Fig. 3.4; Sudibya et al., 2009). This is because the aromatic rings of the catecholamine at the cell–nanotube junction attach noncovalently to the nanotube sidewall and thereby impose a p-doping effect that increases the nanotube conductance. A similar approach of surface functionalization was also reported previously by Wang et al. (2007a,b,c), who used single SWCNT–based FETs to detect the acute release of chromogranin A at low concentration (1 nM) from living cortical neurons and monitored dose-dependent chromogranin A release from a single bovine chromaffin cell positioned above the sensing region by a micropipette and stimulated by histamine (Tsai et al., 2008). These studies provide a real-time and noninvasive measurement platform to examine subtle cellular activities from living cells with high temporal resolution and ease of detection.

3.2.2.1.2 Single-Molecule Detection

Goldsmith et al. (2007) have developed an electrochemical method to create single-point defects in SWCNTs in a controllable manner and then covalently bind biomolecules at this scattering site. Owing to the real-time monitoring of conductance during the defect generation, these point-functionalized SWCNT–FETs can be prepared in high yield. This sensitivity is due to the Coulomb interaction between the molecule and the defect that modulates scattering in the 1D channel (Goldsmith et al., 2008). This approach provides a new electronic platform for studying biomolecular interactions and kinetics that are hidden in ensemble measurements, as demonstrated by Sorgenfrei et al. (2011). In this study, they covalently attached a single-stranded probe DNA sequence, which was terminated with an amine group, to a carboxylic acid–functionalized point defect in a CNT using a standard amide formation coupling reaction. After probe DNA was attached, these devices were used to study the kinetics and thermodynamics of DNA hybridization (Sorgenfrei et al., 2011). In the absence of target DNA, the devices did not show any particular features in a conductance dominated by flicker (1/f) noise (Sorgenfrei et al., 2011). When the device was immersed in buffer containing complementary target DNA, however, reproducible large amplitude two-level fluctuations appeared at different temperatures (Sorgenfrei et al., 2011). Conductance differences reached $\approx$60–100 ns and the signal-to-noise ratio reached 43 (over the 1/f noise background) over a time interval of 30 s. This observation can be explained by the proposed model: the device conductance is controlled by probe-target hybridization that decreases the device conductance because of increased scattering and charge transfer at the position of target DNA binding. This effect would be partially offset by the Debye screening from the dissolved solution counter ions for longer DNA strands. Further kinetic investigations of the system as a function of temperature demonstrated non-Arrhenius behavior; this agrees with DNA hybridization experiments using fluorescence correlation spectroscopy. This technique is label-free and could be used to probe single-molecule dynamics at microsecond timescales.

Another system was developed by Guo et al. (2006) for measuring the conductance of a single molecule covalently immobilized within a nanotube gap. In this system, gaps are formed in carboxylic acid–functionalized SWCNTs that can be reconnected by one or a few molecules attached to both sides of the gap through amide bond formation. Consequently, the devices are sufficiently robust so that a wide range of chemistries and conditions can be applied. By using this method, Feldman et al. (2008) have made molecular devices that detect the binding between proteins and substrates at the single-event level and probe the dependence of charge transport of a single DNA duplex on its P-stacking integrity. However, biomolecular interactions cannot be measured in real time. To do this, a useful strategy has been recently developed to create an integrated system that can combine rapid real-time measurements with single-molecule sensitivity (Liu et al., 2011a,b,c). In this study, individual DNA aptamers were coupled with SWCNTs as point contacts to form single-molecule devices that allow us to selectively and reversibly detect a target protein, thrombin (Liu et al., 2011a,b,c). After further thrombin treatment, these fresh aptamer-functionalized devices showed consistent conductance increases originating from the enhanced DNA charge transfer that is due to the rigidification of DNA conformation by DNA–thrombin interactions. To achieve real-time measurements, a repeating pattern that consists of 79 identical SWCNT transistors by a double photolithographic process was designed and fabricated (Liu et al., 2011a,b,c). Combining this design with microfluidics allowed us to detect proteins and monitor stochastic DNA–protein interactions in real time (Liu et al., 2011a,b,c). Reversible and equivalent conductance changes at different thrombin concentrations (from 2.6 fM to 2.6 pM and 2.6 nM) were observed by Liu et al. (2011a,b,c), thus demonstrating single-molecule sensitivity. Further delivery of elastase (3.4 nM) did not lead to any detectable conductance change in the same device. In a separate experiment, Liu et al. (2011a,b,c) observed negligible conductance changes upon thrombin injection using the devices reconnected with a different DNA (Con-A) that could not bind human thrombin. Both control experiments demonstrated that this protein detection scheme has excellent selectivity. These results distinguished this method as a valuable platform to achieve real-time, label-free, reversible detection of DNA–protein interactions with high selectivity and real single-molecule sensitivity.

3.2.2.2 Electrochemical Biosensors of Carbon Nanotubes

It is essential to understand the three main types of nanotube-derived electrodes before discussing the use of electrochemical biosensors. The first, and the most commonly used, electrode has nanotubes "randomly distributed" on its surface (which often means an unknown configuration rather than genuinely randomized configuration). To fabricate random networks of chemical vapor deposited SWCNTs have produced electrodes that are significantly faster than conventional metal disc–based ultramicroelectrodes (Dumitrescu et al., 2008). The second class of electrodes use aligned nanotubes to optimize electrode performance. This geometry can be achieved by self-assembly (Liu et al., 2000; Chou et al., 2009) or by growing aligned nanotubes directly from a surface (Dai and Mau, 2001); in the latter approach, growth of aligned SWCNT "forests" is an especially interesting development (Qu et al., 2008; Wei et al., 2002; Dai et al., 2003). Electrodes made with nanotubes aligned normal to the electrode surface exhibit faster heterogeneous electron transfer compared with randomly distributed arrays (Chou et al., 2009; Diao and Liu, 2005). This effect occurs because the nanotube tips typically facilitate more rapid electron transfer than sidewalls and the electrons are only required to travel down one tube, rather than having to jump from tube to tube, to be transferred to the bulk electrode (Gooding et al., 2007). The third type of electrode avoids the use of ensembles of many tubes with variable properties and instead uses a single CNT as a nanoelectrode. This is probably the most attractive design of CNT electrode, despite the challenges of fabricating and manipulating a single-CNT probe. These types of electrodes can be made with single MWCNTs (Campbell et al., 1999) or single SWCNTs (Heller et al., 2006a,b), which give different electrochemical performance.

When it comes to electrochemical biosensing, CNT-modified electrodes appear to offer substantially improved ampere-metric biosensors, with particularly enhanced sensitivity to H_2O_2 and NADH. Wang et al. (2003) used Nafion, a sulfonated tetrafluoroethylene-based polymer, to incorporate MWCNTs into composite electrodes for glucose oxidase–based detection of glucose, a process that involves the oxidation of glucose by the oxidase enzyme and then measurement of the resulting H_2O_2 concentration. The composite electrodes offered substantially greater sensitivity to glucose, in particular at low potentials (−0.05 V), with negligible interference from DA, uric acid, or AA, which are biological molecules that commonly interfere with electrochemical detection of glucose. It was also found that CNT-modified electrodes can accelerate electron transfer from NADH molecules; decreasing the over potential and minimizing surface fouling, which are properties that are particularly useful for addressing the limitations of NADH oxidation at ordinary electrodes (Musameh et al., 2002). Similar improvements in electrode performance were more recently observed for composite electrodes made with CNTs and ionic liquids, which offer high stability, high electrical conductivity, and extremely low vapor pressure (Wang et al., 2007a,b,c; Kachoosangi et al., 2009). However, caution is needed when interpreting these results. The mechanism of favorable electrochemistry for CNT-based electrodes remains controversial because, as we discussed earlier, most CNTs contain metal impurities derived from the catalysts used in their growth, which are at least partially responsible for the observed electrochemical activity. Although they complicated the fundamental electrochemistry, such remnant metal NPs had one benefit: they provided a clear indication that the electrochemical properties of sensors could be enhanced by deliberately integrating catalytic NPs within CNTs.

CNTs also offer more efficient ways of communicating between sensor electrodes and the redox-active sites of biological molecules, which are frequently embedded deep inside surrounding peptides. The high aspect ratio and small diameters of SWCNTs make them suitable for penetrating through the molecule to the internal electroactive sites, while the rapid electron-transfer kinetics at the tip of oxidized tubes can enhance electron transfer. A major step in this direction was accomplished when microperoxidase-11 (MP-11), an 11-amino acid sequence that contains a heme-center and is derived from the proteolytic digestion of heme-proteins, was attached to the ends of SWCNTs, which were self-assembled normal to the electrode surface to produce a nanoelectrode array (Gooding et al., 2003a,b). The high efficiency of the nanotubes as molecular wires were demonstrated by the calculated rate constant of heterogeneous electron transfer is 3.9 s^{-1}, between electrode and the MP-11 molecules. Similarly, by using enzymes covalently attached to the ends of aligned SWCNT "forest" arrays, Yu et al. (2003) reported quasireversible FeIII/FeII voltammetry for the heme-enzymes myoglobin and horseradish peroxidase (HRP).

Another elegant application of CNTs to immune assays involved forming a "forest" of SWCNTs oriented perpendicularly to the basal plane of abraded pyrolytic graphite and exploiting the high surface areas of MWCNTs for delivery of the label molecules (Yu et al., 2006). In this electrochemical-based sandwich immunoassay, the CNTs were used both as "nanoelectrodes," which coupled primary antibodies (Ab1) to the pyrolytic graphite electrode and as "vectors" in suspension that hosted multiple secondary antibodies (Ab2) and multiple copies of the electrochemical label HRP. Amplified sensing signals resulted from using Ab2-MWCNT-HRP bioconjugates that had high HRP/Ab2 ratios, instead of conventional single-HRP labeled Ab2. The sensing process occurred in three steps. First, the Ab1 recognized and bound the prostate-specific antigen (PSA), a prostate-cancer biomarker, present in serum. Second, the application of Ab2-MWCNT-HRP bioconjugates targeted the now surface-tethered PSA, binding through Ab2. Third, addition of H_2O_2 allowed the indirect detection of PSA by measuring the electrochemical voltage derived from the action between the added H_2O_2 and the HRP on the nanotube complexes. Two points are particularly noteworthy about this approach. The first point is that the MWCNTs can bind multiple HRP molecules in contrast to Ab2 molecules, which have only a limited labeling and binding capacity because of their size and chemistry. So this approach could increase the detection sensitivity for PSA about 10−100 times compared to the commercial clinical immune assays presently available. The second point is that the clinical potential of this biosensor was demonstrated by direct measurement of PSA concentration in samples of human serum from patients with cancer and from healthy subjects.

Another interesting development in electrochemical biosensors is the use of aptamers. These structures are oligonucleotide sequences that can be generated to have affinity for a variety of specific biomolecular targets such as drugs, proteins, and other relevant molecules. Aptamers even hold potential for use in novel therapies and are also considered as highly suitable receptors for selective detection of a wide range of molecular targets, including bacteria (Shamah et al., 2008). Furthermore, aptamers can self-assemble on CNTs through π stacking between the nucleic acid bases and the nanotube walls. Consequently, considerable efforts have been directed toward incorporating aptamers and CNTs into the design of biosensors (Willner and Zayats, 2007). Zelada et al. (2009) reported a novel potentiometric biosensor made with aptamer-modified SWCNTs that allowed specific real-time detection of one single colony-forming unit, effectively a single bacterium, of *Salmonella* Typhi. In this elegant study, it was demonstrated the potential of SWCNTs to detect the highly virulent *Salmonella* Typhi pathogen at the single bacterium level. In contrast, classical microbiological tests currently take between 24 and 48 h before a diagnosis for Salmonellosis can be made (because of the need to grow cultures), thus illustrating the strong potential of microbiological diagnostic sensors. Early diagnosis can be life saving because serious dehydration from diarrhea can lead to death, especially in tropical countries.

3.2.2.3 *Optical Biosensors Based on Carbon Nanotubes*

The optical nanoscale biosensors could operate in confined environments such as inside cells. Such systems typically rely on either the use of the nanotubes on which a classical sandwich-type optical assay is performed (Cui et al., 2008) or the ability of CNTs to quench fluorescence (Engel et al., 2008) or the NIR photoluminescence exhibited by semiconducting nanotubes (Connell et al., 2002; Avouris et al., 2008a,b). The NIR luminescence of semiconducting SWCNTs is particularly interesting for biosensing because NIR radiation is not absorbed by biological tissue and hence can be used for biosensing within biological samples or organisms.

The ability of CNTs to quench fluorescence has been explored by a number of research groups. Yang et al. (2008) used the preference for single-stranded oligonucleotides to wrap around SWCNTs compared with the related duplexes. SWCNTs and the sample, which may contain the complementary DNA, were added to oligonucleotides labeled with the fluorophore 6-carboxyfluorescein solution. If no complementary DNA is present, the fluorescently labeled DNA will wrap around the SWCNTs, and the fluorescence will be quenched. If the complementary strand of DNA is present in the sample, hybridization with the fluorescently labeled probe DNA will give a rigid duplex that does not wrap around the nanotubes, and hence a fluorescence signal will be observed. Satishkumar et al. (2007) employed a dye–ligand conjugate in which the dye was complexed with the SWCNTs, thus causing its fluorescence to be quenched. Interaction of the nanotube-bound receptor ligand and the analyte caused the displacement of the dye–ligand conjugate from the nanotubes and the recovery of fluorescence. Such a strategy resulted in nanomolar sensitivity.

Infrared luminescence was used by Heller et al. (2006a,b) for biosensors in which semiconducting SWCNTs are wrapped in dsDNA. The change in conformation of the DNA from its B to Z forms results in a change of the dielectric environment of the SWCNTs with a concomitant shift in the wavelength of the SWCNT fluorescence. In this initial study (Heller et al., 2006a,b), the shift in optical properties upon the change in dsDNA structure was used to detect metal ions that induced such changes in DNA structure. Divalent metal ions of mercury, cobalt, calcium, and magnesium are all known to cause transitions from B to Z in dsDNA, and the DNA-wrapped SWCNT biosensors were shown to be able to detect all these metals with the sensitivity decreasing in the order $Hg^{2+} > Co^{2+} > Ca^{2+} > Mg^{2+}$. Changes in the structure of dsDNA wrapped around nanotubes has also been exploited for the detection of Hg^{2+} ions by circular dichroism, as the Hg^{2+} ions are believed to cause a weakening of the DNA–SWCNTs interaction, with a resultant decrease in the circular dichroism signal induced by the association of the nanotubes with the DNA (Gao et al., 2008).

Wrapping the nanotubes with ssDNA has also been explored for monitoring DNA hybridization (Jeng et al., 2006, 2007) and small-molecule interactions with the DNA (Fig. 3.5; Heller et al., 2009a,b,c). The latter is a particularly exciting aspect of the earlier study by Heller et al. (2006a,b) because it is an

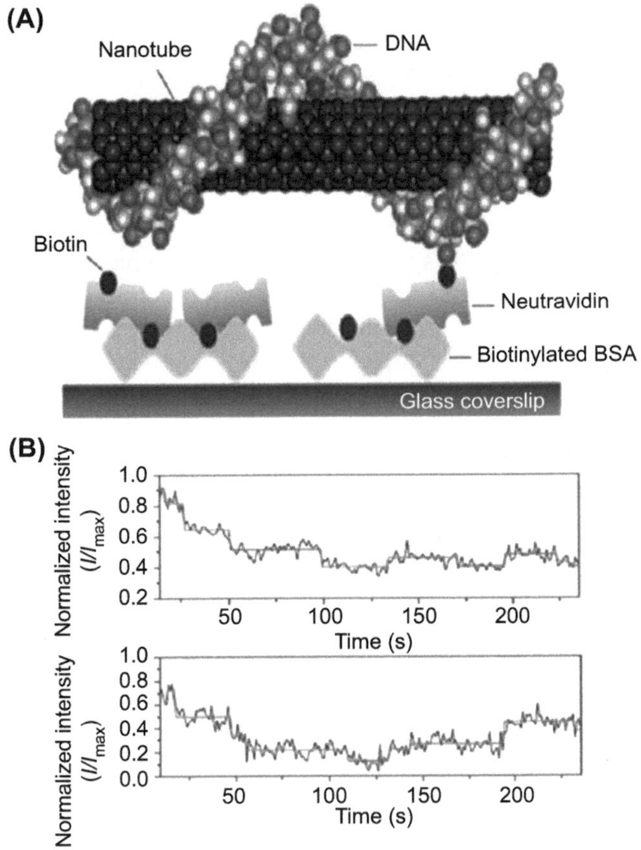

FIGURE 3.5
Immobilized DNA–SWCNT complexes for the detection of H_2O_2. (A) Schematic of DNA–SWCNT binding to a glass surface with bovine serum albumin (BSA)-biotin and NeutrAvidin. (B) Fitted traces from a near-infrared movie that shows single-step quenching of SWCNT emission upon perfusion of H_2O_2 (Heller et al., 2009a,b,c). *Copyright by the Nature Publishing Group.*

extension of the concept to multimodal optical sensing. In this way, Heller et al. (2009a,b,c) simultaneously detected up to six genotoxic analytes, including chemotherapeutic alkylating agents and reactive oxygen species such as H_2O_2, singlet oxygen, and hydroxyl radicals. The ability to detect multiple different analytes on the same sample of ssDNA-wrapped SWCNTs is due to the differing optical responses of (6, 5) and (7, 5) SWCNTs. For example, the chemotherapeutic DNA-alkylating agent melphalan causes a red shift in the photoluminescence of both the (6, 5) and (7, 5) nanotubes; H_2O_2 and Cu^{2+} cause a red shift in the (6, 5) band, but no change in the (7, 5) band; H_2O_2 and Fe^{2+} damage the DNA, causing an attenuation of both bands, but particularly the (7, 5) band. Hence, because of the differing effects of various analytes on the optical signature of an SWCNT mixture, chemometric analysis enables multiple analytes to be detected simultaneously. Some sequence specificity was also reported as sequences with more guanine bases are more susceptible to singlet oxygen, while metal ion responses are greater for DNA sequences with stronger metal binding.

The final aspect of this study illustrated the ability of the DNA–SWCNTs to detect drugs and reactive oxygen species inside living cells. The DNA–SWCNTs had been shown to be able to enter 3T3 fibroblasts by endocytosis without being genotoxic and retain their photoluminescence (Jin et al., 2008). Perfused drugs or reactive oxygen species were observed to induce spectral changes in the SWCNTs inside the living cells (Heller et al., 2009a,b,c).

An important feature of using the NIR luminescence of DNA–SWCNTs is that it has been reported to be able to detect single-molecule interactions when wrapped either in DNA (Heller et al., 2009a,b,c) or collagen (Jin et al., 2008), in common with nanotube FET-type devices (Besteman et al., 2003). In many ways, this system looks almost like the ideal biosensor, as it has nanoscale dimensions and can detect multiple analytes with exquisite sensitivity in biological media.

3.2.3 Biosensor Using Graphene/Graphene Oxide/Reduced Graphene Oxide

Graphene, emerging as a true 2-dimensional (2D) material, has received increasing attention due to its unique physicochemical properties (high surface area, excellent conductivity, high mechanical strength, and ease of functionalization and mass production). This section is discussed with graphene-based biosensors. In particular, graphene for direct electrochemistry of enzyme, its electrocatalytic activity toward small biomolecules (hydrogen peroxide, NADH, DA, etc.), and graphene-based enzyme biosensors have been summarized in more detail.

CNTs are rolled-up cylinders of carbon monolayers called graphene. They can be chemically modified in such a way that biologically relevant molecules can be detected with high sensitivity and selectivity. On the basis of their fluorescence quenching abilities, graphene and its derivatives can serve as either energy donors or acceptors in a FRET sensor. They have been extensively investigated for the sensing of DNA, proteins, or other biomolecules; detection of single-base mismatches; analysis of the melting of DNA duplexes; etc. (He et al., 2013; Guo et al., 2013; Zhang et al., 2014; Lin et al., 2011). For the detection of DNA, Lin et al. (2011) reported a GO-based fluorescence quenching recovery sensor for detecting ssDNA with an LOD of nM range. Because ssDNA retained on the GO surface was indigestible by DNAase, their sensors can perform even in the presence of DNAase. Two similar DNA sensors, with the ability to detect single-base mismatches, have also been exploited (Jeong et al., 2010; Lu et al., 2010). A hybrid graphene-ZnAl-LDH nanocomposite has been fabricated via a one-step process and used as a facile platform of a $Ru(phen)_3Cl_2$ (tris(1,10-phenanthroline)ruthenium(II)dichloride) sensor to selectively detect DNA. Moreover, both the platform and the sensor can be easily collected and used for the next sample if no DNA was present in the solution (Li et al., 2015). Some researchers have combined graphene and its derivatives with noble metal NPs to induce a double quenching effect that resulted in an increase in the achievable signal-to-noise ratio and hence provided amplification of the achievable sensitivity. Tao et al. (2012) reported a DNA–silver nanocluster–GO nanohybrid material for the detection of multiple nucleic acid targets with a low LOD and high sensitivity and selectivity, which was attributed to the high achievable signal-to noise ratio resulting from the high quenching efficiency of GO.

Graphene and its derivatives are also widely used for fluorescence biosensing of proteins. Li et al. (2013a,b,c) utilized chemically converted graphene (CCG) to effectively quench the fluorescence emission of Cy3 dye 1 (the intensity was reduced to 1/38 that of 1 alone) in aqueous solution (Fig. 3.6). After the addition of a certain amount of bovine serum albumin (BSA), ~60-fold enhancement in fluorescence was observed for the hybrid CCG-1. This was employed to detect

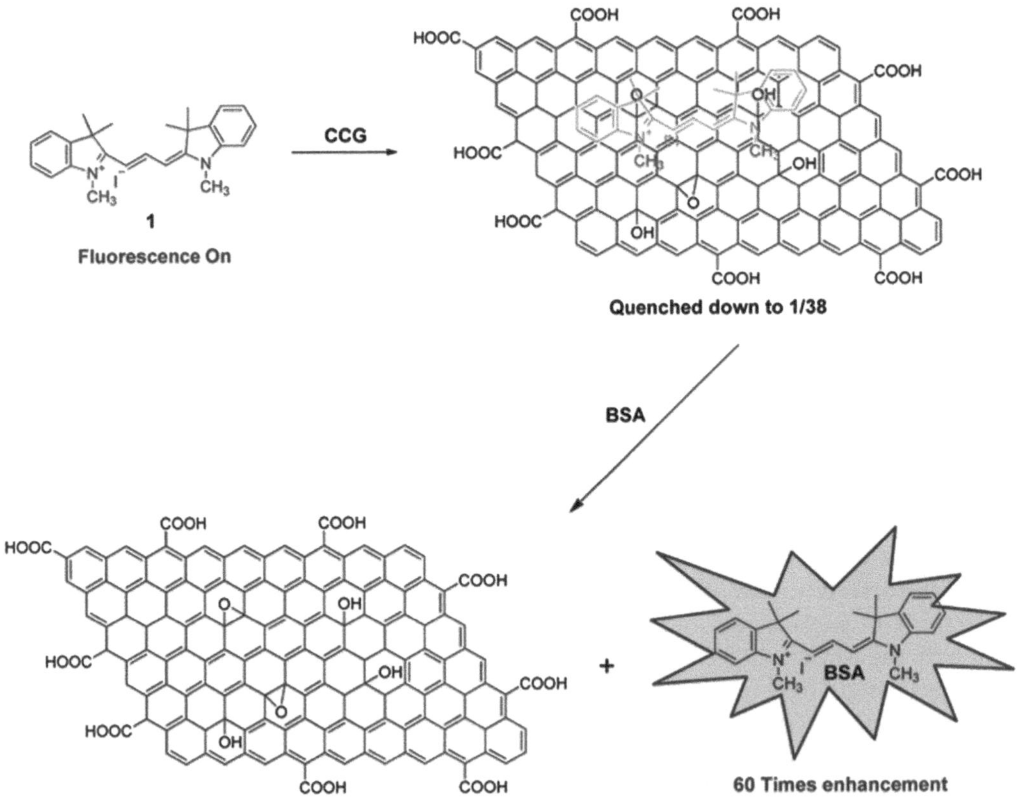

FIGURE 3.6
The fluorescence detection of bovine serum albumin (BSA) using the hybrid of 1 and chemically converted graphene (CCG) (Li et al., 2013a,b,c). *Copyright by the Royal Society of Chemistry.*

BSA: fluorescence intensity was found to be proportional to BSA added in the concentration range from 0 to 8×10^{-6} M and the LOD of BSA was as low as 5×10^{-8} M.

Zhuang et al. (2013) designed a simple, selective, and sensitive fluorescent GO-based molecular aptamer beacon (MAB) for detection of PrP^C using GO as a quenching reagent. As a result, the TAMRA-labeled MAB moved away from the surface of GO and the fluorescence of MAB was recovered. Owing to the high energy transfer efficiency between GO and the fluorophore, the background signal was significantly reduced. Also for the detection of PrP^C, the authors then developed a new FRET strategy using QDs as energy donor and GO as energy acceptor by means of specific recognition between the two binding aptamers and PrPC with high sensitivity and good selectivity (Zhen et al., 2013). The detection signals were greatly enhanced by the high FRET efficiency between QDs and GO.

Graphene and its derivatives have been extensively used in other biosensing applications such as enzymatic reaction monitoring and detection of biomacromolecules. Zhou et al. (2013) proposed a novel and versatile biosensing platform for the detection of protein kinase activity based on a GO-peptide nanocomplex and phosphorylation-induced suppression of cleavage of carboxypeptidase Y (CPY). Kinase-catalyzed phosphorylation protected the fluorophore-labeled peptide probe against CPY digestion and induced the formation of a GO-peptide nanocomplex, which resulted in fluorescence quenching, while the non–phosphorylated peptide was degraded by CPY to release free fluorophore, thus restoring fluorescence. This GO-based nanosensor has been successfully applied to sensitively detect two model kinases, casein kinase (CKII) and cAMP-dependent protein kinase (PKA), with low LODs of 0.0833 and 0.134 mU mL^{-1}, respectively. Li et al. (2013a,b,c) reported a versatile biosensing platform capable of achieving ultrasensitive detection of both small-molecule and macromolecular targets. The system consisted of three parts: a nanomaterial (graphene), a biomaterial (DNA aptamers), and an isothermal signal amplification technique (RCA). Graphene was chosen for its ability to adsorb ssDNA molecules nonspecifically. The key to the design was grafting a short primer onto an aptamer sequence, which resulted in a small DNA probe that allowed both effective adsorption of the probe onto the graphene surface to mask the primer domain in the absence of the target and efficient release of the probe in the presence of the target to make the primer available for template binding and RCA. The detection was highly sensitive and feasible for protein targets, DNA sequences, and small-molecule analytes.

Glucose detection is clinically significant for the diagnosis and management of diabetes and can be achieved using graphene and its derivatives as a mediator. Wang et al. (2014a,b,c) demonstrated an efficient biosensing system for glucose detection based on the enzyme-like activity of GO integrated with chitosan. The chitosan-functionalized graphene oxide (CS–GO) hybrid was demonstrated to be a good enzyme mimetic for the oxidation of a typical substrate (TMB) under visible-light (l Z 400 nm) stimulation and did not require destructive hydrogen peroxide. Mondal and Jana (2012) developed a fluorescence-based cholesterol detection method, which used competitive host–guest interaction between graphene-bound β-cyclodextrin with rhodamine 6G (R6G) and cholesterol. The fluorescence of R6G incorporated in β-cyclodextrin was quenched by graphene but was restored by cholesterol as it displaced R6G from the β-cyclodextrin host. Cholesterol is an important component of animal cell membranes (Ikonen, 2008) and the main precursor for synthesis of different biomolecules such as bile acids, steroid hormones, and vitamin D (Myant, 1981). A desirable amount of cholesterol in healthy human serum is 200 mg dL^{-1} (Motonaka and Faulkner, 1993). Excess cholesterol in blood serum forms plaques in the arteries of blood vessels, which prevent the blood circulation and cause cardiovascular diseases (Raines, 1995). Thus the levels of total cholesterol in serum and food are major parameters for diagnostic treatment. Various analytical methods have been developed, and detection selectivity in most of the methods relies on use of cholesterol selective enzymes (Amundson and Zhou, 1999; Devadoss and Burgess, 2002; Dey and Raj, 2010) and antibodies (Luthi et al., 2012), which are expensive and prone to denaturation. Mondal and Jana (2012) used β-CD-graphene (β-CD-G)-based hybrid system for optical detection of cholesterol where the β-CD component offers detection selectivity via selective host–guest interaction and graphene translates it into an optical signal. The unique property of graphene offers superior optical response compared to the earlier reported gold NP-β-CD-based hybrid (Zhang et al., 2008). The advantage of β-CD functionalization is that it offers high water solubility to graphene and guest molecules incorporated into β-CD are easily accessible to graphene. The basic principle of cholesterol detection is shown in Fig. 3.7, which uses competitive host–guest interaction between graphene-bound β-CD (β-CD–G) with R6G and cholesterol. When R6G enters into the β-CD host its fluorescence is quenched by graphene but this fluorescence "turns on" after cholesterol replaces the R6G.

The R6G present inside the β-CD-G host can be selectively replaced by cholesterol. This is because cholesterol has high binding affinity to the β-CD cavity due to its hydrophobic nature. This replacement of R6G by cholesterol releases R6G in bulk solution with the resultant fluorescence "turning on." Addition of cholesterol into the solution of β-CD-G–R6G increases the fluorescence of R6G, and the increased fluorescence is directly related to the amount of cholesterol added. The observed fluorescence response can be used for naked eye–based semiquantitative detection and spectrometer-based quantitative detection of cholesterol, and the detection sensitivity can reach up to the nanomolar

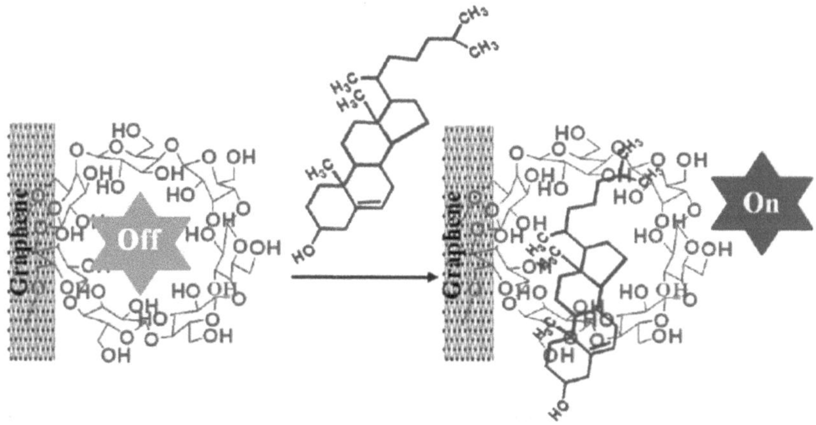

FIGURE 3.7

Strategy for fluorescence-based cholesterol detection using β-CD-G via competitive host–guest interaction. Fluorescence of R6G inside β-CD is quenched by graphene but it "turns on" as cholesterol replaces R6G (Mondal and Jana, 2012). *Copyright by the Royal Society of Chemistry.*

concentration range. The observed sensitivity is comparable to most of the existing cholesterol detection methods (Jauhiainen and Dolphin, 1986; Linsel-Nitschke and Tall, 2005; Kishi et al., 2002; Zhou et al., 1997; Mohanty et al., 1997; Amundson and Zhou, 1999).

3.2.3.1 Graphene-Based Electrochemical Biosensors

The excellent electrochemical behaviors of graphene indicate graphene is a promising electrode material in electroanalysis (McCreery, 2008; Wang et al., 2005). Several electrochemical sensors based on graphene and graphene composites for bioanalysis and environmental analysis have been developed (Shan et al., 2009a,b; Zhou et al., 2009; Kang et al., 2009).

3.2.3.1.1 Graphene-Based Enzyme Biosensors

On the basis of the high electrocatalytic activity of graphene toward H_2O_2 and the excellent performance for direct electrochemistry of GOD, graphene could be an excellent electrode material for oxidase biosensors. Several graphene-based glucose biosensors have been reported (Shan et al., 2009a,b; Zhou et al., 2009; Kang et al., 2009; Wang et al., 2009; Lu et al., 2007).

Shan et al. (2009a,b) reported the first graphene-based glucose biosensor with graphene-/polyethylenimine-functionalized ionic liquid nanocomposite–modified electrode which exhibits wide linear glucose response (2–14 mM, $R = 0.994$), good reproducibility (relative standard deviation of the current response to 6 mM glucose at −0.5 V was 3.2% for 10 successive measurements), and high stability (response current +4.9% after 1 week) (Shan et al., 2009a,b).

Zhou et al. (2009) reported a glucose biosensor based on chemically reduced graphene oxide (CR-GO). Graphene (CR-GO)-based biosensor exhibits substantially enhanced ampere-metric signals for sensing glucose: wide linear range (0.01–10 mM), high sensitivity (20.21 μA mM cm^{-2}), and low detection limit of 2.00 μM (S/N = 3). The linear range for glucose detection is wider than that on other carbon material–based electrodes, such as CNTs (Liu et al., 2006) and carbon nanofibers (Wu et al., 2007). The detection limit for glucose at the GOD/CR-GO/GC electrode (2.00 μM at −0.20 V) is lower than that of some reported carbon material–based biosensors, such as CNT paste (Rubianes and Rivas, 2003), CNT nanoelectrode (Lin et al., 2004), carbon nanofiber (Wu et al., 2007), exfoliated graphite nanoplatelets (Lu et al., 2007), and highly ordered mesoporous carbon (Zhou et al., 2008). The response at the GOD/CR-GO/GC electrode to glucose is very fast (9 ± 1 s to steady-state response) and highly stable (91% signal retention for 5 h), which makes GOD/CR-GO/GC electrode a potential fast and highly stable biosensor to continuously measure the plasma glucose level for the diagnosis of diabetes.

Kang et al. (2009) employed biocompatible chitosan to disperse graphene and construct glucose biosensors. It was found that chitosan helped to form a well-dispersed graphene suspension and immobilize the enzyme molecules, and the graphene-based enzyme sensor exhibited excellent sensitivity (37.93 μA mM^{-1} cm^{-2}) and long-term stability for measuring glucose.

Graphene-/metal NP-based biosensors have also been developed. Shan et al. (2009a,b) reports a graphene/AuNPs/chitosan composite film–based biosensor which exhibited well electrocatalytic activity toward H_2O_2 and O_2. Wu et al. (2009) reports GOD-/graphene-/PtNPs-/chitosan-based glucose biosensor with detection limit of 0.6 μM glucose. These enhanced performances were attributed to the large surface area and good electrical conductivity of graphene, and the synergistic effect of graphene and metal NPs (Wu et al., 2009; Shan et al., 2009a,b).

The excellent catalytic activity of functionalized graphene toward NADH oxidation indicates that graphene is a promising material for dehydrogenase biosensors. Zhou et al. (2009) reports an ethanol biosensor based on graphene-ADH. The ADH-graphene-GC electrode exhibits faster response, wider linear range, and lower detection limit for ethanol detection compared with ADH-graphite/GC and ADH/GC electrodes. This enhanced performance can be explained by the effective transfer of substrate and products through graphene matrixes containing enzymes as well as the inherent biocompatibility of graphene (Chen et al., 2008; Zhou et al., 2009).

Bhunia and Jana (2011) used the peptide-functionalized graphene for the detection of enzyme and found that peptide-functionalized graphene becomes responsive to enzyme and quenched fluorescence returns back in presence of chymotrypsin. It was also found that, when peptide-functionalized graphene mixed with different amount of chymotrypsin, the emission intensity of solution mixture increases with the increasing concentration of chymotrypsin with their detection sensitivity in the nanomolar range. In the higher enzyme concentration range the color change is visible to naked eye. Control experiments show that this responsive action is specific and does not occur for other enzyme or protein. It was also used in in vitro application when mixed this peptide-functionalized graphene probe with cultured cell lines that produce such enzymes and observed the enzyme response. The graphene probe responds to the enzyme present in culture media/cell and fluorescence turns on. In this case no significant fluorescence was observed from cells possibly because fluorescein gets detached from graphene by enzyme action. This study suggests that functional graphene has low cytotoxicity and can be used for cellular assay of enzymes. Functionalized graphene used here is very robust because peptide is covalently linked with coated graphene and gives detection advantage in complex in vitro conditions.

3.2.3.1.2 Graphene-Based Electrochemical DNA Biosensors

Electrochemical DNA sensors offer high sensitivity, high selectivity, and low cost for the detection of selected DNA sequences or mutated genes associated with human disease and promise to provide a simple, accurate, and inexpensive platform of patient diagnosis (Sassolas et al., 2008; Drummond et al., 2003). Electrochemical DNA sensors also allow device miniaturization for samples with a very small volume (Zhou et al., 2009). Among all kinds of electrochemical DNA sensors, the one based on the direct oxidation of DNA is the simplest (Niwa et al., 2006; Zhou et al., 2009; Drummond et al., 2003).

Zhou et al. (2009) reported an electrochemical DNA sensor based on graphene [chemically reduced graphene oxide (rGO)]. The current signals of the four free bases of DNA [i.e., guanine (G), adenine (A), thymine (T), and cytosine (C)] on the CR-GO/GC electrode are all separated efficiently, indicating that CR-GO/GC can simultaneously detect four free bases, but neither graphite nor glassy carbon can. This is attributed to the antifouling properties and the high electron-transfer kinetics for base oxidation on CR-GO/GC electrode (Zhou et al., 2009), which results from high density of edge-plane-like defective sites and oxygen-containing functional groups on CR-GO that provide many active sites and are beneficial for accelerating electron transfer between the electrode and species in solution (Banks et al., 2004, 2005). The CR-GO/GC electrode is also able to efficiently separate all four DNA bases in both ssDNA and dsDNA, which are more difficult to oxidize than free bases, at physiological pH without the need of a prehydrolysis step, which allows to detect a single-nucleotide-polymorphism (SNP) site for short oligomers with a particular sequence at the CR-GO/GC electrode without any hybridization or labeling processes (Zhou et al., 2009). This is attributed to the unique physicochemical properties of CR-GO (the single sheet nature, high conductivity, large surface area, antifouling properties, high electron-transfer kinetics, etc.) (Zhou et al., 2009).

3.2.3.1.3 Graphene-Based Electrochemical Sensors for Heavy Metal Ions

Graphene-based electrochemical sensors have been developed for environmental analysis for the detection of heavy metal ions (Pb^{2+} and Cd^{2+}) (Li et al., 2009a,b). Li et al. (2009a,b) report that Nafion–graphene composite film–based electrochemical sensors not only exhibit improved sensitivity for the metal ion (Pb^{2+} and Cd^{2+}) detections, but also alleviate the interferences due to the synergistic effect of graphene nanosheets and Nafion. The stripping current signal is greatly enhanced on graphene electrodes. The linear range for the detection of Pb^{2+} and Cd^{2+} is wide (0.5–50 and 1.5–30 $\mu g\ L^{-1}$ for Pb^{2+} and Cd^{2+}, respectively). The detection limits ($S/N = 3$) are 0.02 $\mu g\ L^{-1}$ for both Pb^{2+} and Cd^{2+}, which are more sensitive than those of Nafion film–modified bismuth electrode (Kefala et al., 2004) and ordered mesoporous carbon coated GCE (Zhu et al., 2008) and comparable to Nafion-/CNT-coated bismuth film electrode (Xu et al., 2008). The enhanced performance is attributed to the unique properties of graphene (nanosized graphene sheet, nanoscale thickness of these sheets, and high conductivity), which endowed the capability to strongly adsorb target ions, enhanced the surface concentration, improved the sensitivity, and alleviates the fouling effect of surfactants (Li et al., 2009a,b).

3.2.3.2 Graphene-Based Biosensors for Environmental Sensing

Based on the electrocatalytic activity of graphene and the performance for direct electrochemistry of glucose oxidase, graphene proved to be, till this moment, a good electrode material for oxidase biosensors (Shan et al., 2009a,b). Several glucose biosensors were reported in the last 2 years. Lu et al. (2007) reported the first example of glucose biosensor based on graphitic nanoplatelets (xGnP) with good properties, and these properties were lately improved by introducing metal NPs on the graphitic nanoplatelets and keeping in this way the NPs were made extremely small and well distributed (Lu et al., 2008; Wu et al., 2009). Chitosan was used by Kang et al. (2009) for a better graphene dispersion and a better immobilization of the enzyme molecules. A composite film deposited on gold electrode shows enhanced performances due to the large surface area and good electrical conductivity of graphene (Shan et al., 2009a,b).

A new acetylcholinesterase (AChE) biosensor based on the immobilization of exfoliated graphitic xGnPs in chitosan and glutaraldehyde for organophosphate pesticides was proposed by Ion et al. (2010). Glutaraldehyde is used as a cross-linker to bond AChE to a composite of cross-linked chitosan and xGnPs leading to a new acetylthiocholine iodide (ATCI) sensor. The presence of xGnPs on the electrode surface leads to enhanced electron-transfer rate with reduced surface fouling (Kachoosangi et al., 2009). xGnPs are highly conductive nanomaterials with interesting possible future application in biochemical sensing. The proposed sensor combines for the first time the highly conductive and electroanalytic behavior of xGnPs with the biocompatibility of chitosan, leading to good stability and increased sensitivity for detection of ATCI. It will be further applied to the analysis of organophosphate pesticides for environmental monitoring. The detection limit of this sensor was 1.58×10^{-10} M, with a simple fabrication, a fast response, and an acceptable stability.

Networked sensing systems can monitor environmental parameters and provide data maintaining water and soil quality. CNT-based sensors present advantages in sensor platforms in simultaneous determinations of several kinds of on-field contaminants (Valentini et al., 2007; Chopra et al., 2007; Allen et al., 2007). The improved characteristics of these sensors lie in covalent and supramolecular functionalization with enzymes, metals, and chemical groups. The environmental applications of CNT-based biosensors were presented in several reviews (Lin et al., 2005; Rogers, 2006). Based on the models offered by CNTs (considered as enrolled graphene), graphene opens the way of ultrasensitive and ultrafast electronic sensors due to their low electrical noise materials. Even though CNTs have almost ideal properties for electronic applications, they have one dimensional structure which is not suitable in electronic devices; but this problem was solved after the discovery of grapheme (Novoselov et al., 2004) that is 2D structure of one atomic thick carbon. Together with the interesting properties of CNTs, graphene can be considered as very challenging materials for environmental sensors.

3.2.3.3 Graphene-Based FET biosensors

Graphene-based detection in liquid environment: Graphene FETs have operated either under vacuum or atmospheric conditions, not in solution. In some cases, high-quality graphene can be grown epitaxial on SiC substrates. Because of the thickness of the intrinsic substrate on SiC, to achieve field-effect responses, top gating of the epitaxial graphene is required in most cases except that based on nitrogen implantation into a SiC wafer before graphene growth (Ang et al., 2011). Ang et al. (2011) first demonstrated the use of solution-gated epitaxial graphene as a pH sensor. Ohno et al. (2009) reported on electrolyte-gated graphene field-effect transistors for detecting pH and protein adsorption. In another report by Ohno et al. (2010a,b), they demonstrated label-free immune sensing based on aptamer-modified graphene FETs. Immunoglobulin E aptamers with an approximate height of 3 nm were successfully immobilized on a graphene surface, as confirmed by atomic force microscopy. The aptamer-modified graphene FETs showed selective electrical detection of immunoglobulin E protein. From the dependence of the drain current variation on the immunoglobulin E concentration, the dissociation constant was estimated to be 47 nM, indicating good affinity. In a similar way, Agarwal et al. (2010) demonstrated the biocompatibility of rGOs with proteins and further used them after protein functionalization to create biosensors for detecting various metals in real time with high sensitivity. These reports already confirm the potential of graphene for sensing in aqueous electrolytes; however, detailed understandings of the graphene–electrolyte interface and the effect of the electrolyte on the electronic transport in graphene are still lacking. Dankerl et al. (2010) developed a facile method for the scalable fabrication of graphene FET arrays and provided a comprehensive characterization of operation of these devices in aqueous electrolytes. By using in-solution Hall-effect measurements and taking into account the microscopic structure of water at the interface, they demonstrated that charge carrier mobilities and concentrations as a function of electrolyte gate potential can be directly determined. They also showed that graphene FETs exhibited a high transconductance and correspondingly high sensitivity together with an effective gate noise as low as tens of millivolts. These studies demonstrated that graphene FETs, with their ease of fabrication, high transconductance, and low noise, hold great promise for biosensor and bioelectronic applications.

Graphene-based detection in cells: To investigate the biocompatibility of graphene with live cells or tissue, Cohen-Karni et al. (2010) demonstrated, for the first time, the recording from eletrogenic cells using single-layer graphene formed by mechanical exfoliation from graphite and carried out simultaneous recording using graphene and silicon nanowire FETs. They found that graphene FET conductance signals recorded from spontaneously beating embryonic chicken cardiomyocytes yielded well-defined extracellular signals with signal-to-noise ratio routinely >4. Water gate (V_{wg})–dependent experiments demonstrated that the conductance signal amplitude could be tuned over nearly an order of magnitude, thus showing a robust graphene–cell interface. Furthermore, by varying V_{wg} across the Dirac point, they achieved the expected signal polarity flip, thus allowing both n- and p-type recording to be achieved with the same device simply by offsetting V_{wg}. Finally, they compared peak-to-peak recorded signal widths (made as a function of graphene FET device size) with those made using silicon nanowire FETs and showed that the widths increased with the area of graphene FET devices. This indicates that they were measuring a signal that was averaged from different points across the outer membrane of the beating cells. In another work, Agarwal et al. (2010) demonstrated the biocompatibility of rGOs with live cells, such as neuroendocrine PC12 cells and further used them to create biosensors for detecting the dynamic secretion of the hormonal catecholamine molecules from living cells (Cohen-Karni et al., 2010). Pursuing the development of a graphene-based technology that can detect action potentials from electrically active cells, Hess et al. (2011) reported using arrays of CVD-grown graphene FETs for the extracellular detection of action potentials from electrogenic cells. The action potentials of cardiomyocyte-like HL-1 cells could be effectively resolved and tracked across the transistor array. The low-noise characteristic of graphene FETs together with the large transconductive sensitivity of these devices clearly indicates an advantage of graphene FETs in terms of signal-to-noise ratio. These studies suggest that the outstanding performance of graphene FETs together with the feasibility of easily integrating graphene electronics with flexible substrates can pave the way for a true breakthrough in bioelectronics, in particular for electrically functional neural prostheses.

3.3 CONCLUSIONS AND PERSPECTIVES

Based on the unique properties of carbon, nanomaterials with different size, shape, and compositions have been introduced into biosensing. The nanomaterials can be functionalized with biomolecules via noncovalent interaction and covalent route for specific recognition. The biofunctional nanomaterials can produce a synergic effect among catalytic activity, conductivity, and biocompatibility. Therefore, the biofunctional nanomaterials have been used as carriers or tracers for design of a new generation of electronic, optical, and photoelectrochemical biosensing devices. Many considerations, such as the good biocompatibility, the sufficient binding sites for functionalization, capacity in the multiple analyses, and so on, should be emphasized in the development of ultrasensitive bioassay based on the biofunctional nanomaterial systems. In addition, the photoelectrochemical assays, which hold the advantages of both optical and electrochemical detections, should be a promising direction for constructing an ultrasensitive tool. Signal amplification strategies based on nanomaterials not only provide an ultrasensitive assay in detection of trace analytes but also a concept for basic research in nanobiosensing. We feel these biosensors have had substantial impact and have genuine potential for future applications. By taking advantage of the outstanding electrical properties and the environmental ultrasensitivity of carbon nanomaterials many strategies have been developed to create SWCNT or graphene FET-based biosensors for directly detecting DNA–DNA hybridizations, DNA–protein interactions, protein functions and cellular activities in real time with high sensitivity and excellent selectivity. In particular, because of the size comparability and the surface compatibility with biological molecules, an amazing feature of using SWCNT or graphene FETs in biosensing is their ability to detect biomolecules at the single-molecule level, as well as at the single-cell level. This chapter has mainly focused on the modifications, as well as fluorescence biosensing of carbon nanomaterials such as CNTs, graphene, CDs, GQDs, fullerene, CNHs, and CNOs. Furthermore, the modifications of these carbon nanomaterials are widely used for better applications in fluorescence biosensors. Biological applications of carbon nanomaterials are significantly impacting current biotechnology. Especially, carbon nanomaterials enable the development of biosensors with enhanced sensitivity, better selectivity, and a wide range of detection. Multiple detections have also been achieved with a low LOD and high sensitivity. The biocompatible properties of carbon nanomaterials make them applicable for in situ detection of living cells.

References

Agarwal, S., et al., 2010. Interfacing live cells with nanocarbon substrates. Langmuir 26, 2244–2247.

Ajayan, P.M., 1999. Nanotubes from carbon. Chem. Rev. 99, 1787–1800.

Al-Ogaidi, I., et al., 2014. Detection of the ovarian cancer biomarker CA-125 using chemiluminescence resonance energy transfer to graphene quantum dots. Chem. Commun. 50, 1344–1346.

Alivisatos, P., 2004. The use of nanocrystals in biological detection. Nat. Biotechnol. 22, 47–52.

Allen, B.L., et al., 2007. Carbon nanotube field-effect-transistor-based biosensors. Adv. Mater. 19, 1439–1451.

Amatore, C., et al., 2008. Electrochemical monitoring of single cell secretion: vesicular exocytosis and oxidative stress. Chem. Rev. 108, 2585–2621.

Amundson, D.M., Zhou, M., 1999. Fluorometric method for the enzymatic determination of cholesterol. J. Biochem. Biophys. Methods 38 (1), 43–52.

Ang, et al., 2011. Probing the input-output behavior of biochemical and genetic systems system identification methods from control theory. Methods Enzymol. 487, 279–317.

Arlett, J.L., et al., 2011. Comparative advantages of mechanical biosensors. Nat. Nanotechnol. 6, 203–215.

Avouris, P., Freitag, M., Perebeinos, V., 2008a. Carbon-nanotube photonics and optoelectronics. Nat. Photonics 2, 341–350.

Avouris, P., Freitag, M., Perebeinos, V., 2008b. Carbon-nanotube optoelectronics. Carbon Nanotub. 111, 423–454.

Banks, C.E., et al., 2004. Investigation of modified basal plane pyrolytic graphite electrodes: definitive evidence for the electrocatalytic properties of the ends of carbon nanotubes. Chem. Commun. 1804–1805.

Banks, C.E., et al., 2005. Electrocatalysis at graphite and carbon nanotube modified electrodes: edge-plane sites and tube ends are the reactive sites. Chem. Commun. 829–841.

Benítez-Martínez, S., et al., 2014. Graphene quantum dots sensor for the determination of graphene oxide in environmental water samples. Anal. Chem. (86), 12279–12284.

Bernhardt, P.V., 2006. Enzyme electrochemistry–biocatalysis on an electrode. Aust. J. Chem. 59, 233–256.

Besteman, K., et al., 2003. Enzyme-coated carbon nanotubes as single-molecule biosensors. Nano Lett. 3, 727–730.

Bhattacharyya, T., et al., 2013. Reusable glucose sensing using carbon nanotube-based self-assembly. Nanoscale 5, 9231–9237.

Bhunia, S.K., Jana, N.R., 2011. Peptide-functionalized colloidal graphene via interdigited bilayer coating and fluorescence turn-on detection of enzyme. ACS Appl. Mater. Interfaces 3, 3335–3341.

Campbell, J.K., et al., 1999. Electrochemistry using single carbon nanotubes. J. Am. Chem. Soc. 121, 3779–3780.

Cayuela, A., et al., 2013. Strong luminescence of carbon dots induced by acetone passivation: efficient sensor for a rapid analysis of two different pollutants. Anal. Chim. Acta 804, 246–251.

Chen, R.J., et al., 2003. Noncovalent functionalization of carbon nanotubes for highly specific electronic biosensors. Proc. Nat. Acad. Sci. U.S.A. 100, 4984–4989.

Chen, H., et al., 2008. Mechanically strong, electrically conductive, and biocompatible graphene paper. Adv. Mater. 20, 3557–3561.

Chen, K.-I., et al., 2011. Silicon nanowire field-effect transistor-based biosensors for biomedical diagnosis and cellular recording investigation. Nano Today 6, 131–154.

Chopra, N., et al., 2007. Functional one-dimensional nanomaterials: applications in nanoscale biosensors. Anal. Lett. 40, 2067–2096.

Chou, A., et al., 2009. Self-assembled carbon nanotube electrode arrays: effect of length of the linker between nanotubes and electrode. J. Phys. Chem. C 113, 3203–3211.

Cohen-Karni, T., et al., 2010. Graphene and nanowire transistors for cellular interfaces and electrical recording. Nano Lett. 10, 1098–1102.

O'Connell, M.J., et al., 2002. Band gap fluorescence from individual single-walled carbon nanotubes. Science 297, 593–596.

Cui, Y., et al., 2001. Nanowire nanosensors for highly sensitive and selective detection of biological and chemical species. Science 293, 1289–1292.

Cui, D., et al., 2008. Self-assembly of quantum dots and carbon nanotubes for ultrasensitive DNA and antigen detection. Anal. Chem. 80, 7996–8001.

Dai, L.M., Mau, A.W.H., 2001. Controlled synthesis and modification of carbon nanotubes and C_{60}: carbon nanostructures for advanced polymeric composite materials. Adv. Mater. 13, 899–913.

Dai, L.M., et al., 2003. Aligned nanotubes. Chem. Phys. Chem. 4, 1150–1169.

Dankerl, M., et al., 2010. Graphene solution-gated field effect transistor array for sensing applications. Adv. Funct. Mater. 20, 3117–3124.

Dekker, C., 2007. Solid-state nanopores. Nat. Nanotechnol. 2, 209–215.

Devadoss, A., Burgess, J.D., 2002. Detection of cholesterol through electron transfer to cholesterol oxidase in electrode-supported lipid bilayer membranes. Langmuir 18 (25), 9617–9621.

Dey, R.S., Raj, C.R., 2010. Development of an amperometric cholesterol biosensor based on graphene–Pt nanoparticle hybrid material. J. Phys. Chem. C 114, 21427–21433.

Diao, P., Liu, Z.F., 2005. Electrochemistry at chemically assembled single-wall carbon nanotube arrays. J. Phys. Chem. B 109, 20906–20913.

Dong, X.C., et al., 2011a. In situ synthesis of reduced graphene oxide and gold nanocomposites for nanoelectronics and biosensing. Nanoscale Res. Lett. 6 (60), 1–6.

Dong, H.F., et al., 2011b. The use of polyethylenimine-grafted graphene nanoribbon for cellular delivery of locked nucleic acid modified molecular beacon for recognition of micro RNA. Biomaterials 32, 3875–3882.

Dong, Y.Q., et al., 2013. Electrochemiluminescence emission from carbon quantum dot-sulfite coreactant system. Carbon 56, 12–17.

Drummond, T.G., Hill, M.G., Barton, J.K., 2003. Electrochemical DNA sensors. Nat. Biotechnol. 21, 1192–1199.

Du, F., et al., 2013. Carbon dots-based fluorescent probes for sensitive and selective detection of iodide. Microchim. Acta 180, 453–460.

Dubuisson, E., et al., 2011. Optimizing label-free DNA electrical detection on graphene platform. Anal. Chem. 83, 2452–2460.

Dumitrescu, I., et al., 2008. Single-walled carbon nanotube network ultra-microelectrodes. Anal. Chem. 80, 3598–3605.

Engel, M., et al., 2008. Thin film nanotube transistors based on self-assembled, aligned, semiconducting carbon nanotube arrays. ACS Nano 2, 2445–2452.

Feldman, A.K., et al., 2008. Molecular electronic devices based on single-walled carbon nanotube electrodes. Acc. Chem. Res. 41, 1731–1741.

Gao, X., et al., 2008. Detection of trace Hg^{2+} via induced circular dichroism of DNA wrapped around single-walled carbon nanotubes. J. Am. Chem. Soc. 130, 9190–9191.

Goldsmith, B.R., et al., 2007. Conductance-controlled point functionalization of single-walled carbon nanotubes. Science 315, 77–81.

Goldsmith, B.R., et al., 2008. Monitoring single-molecule reactivity on a carbon nanotube. Nano Lett. 8, 189–194.

Gooding, J.J., et al., 2003a. Self-assembled monolayers into the 21st century: recent advances and applications. Electroanalysis 15, 81–96.

Gooding, J.J., et al., 2003b. Protein electrochemistry using aligned carbon nanotube arrays. J. Am. Chem. Soc. 125, 9006–9007.

Gooding, J.J., et al., 2007. The effects of the lengths and orientations of single-walled carbon nanotubes on the electrochemistry of nanotube-modified electrodes. Electrochem. Commun. 9, 1677–1683.

Gooding, J.J., 2008. Advances in interfacial design for electrochemical biosensors and sensors: aryl diazonium salts for modifying carbon and metal electrodes. Electroanalysis 20, 573–582.

Gruner, G., 2006. Carbon nanotube transistors for biosensing applications. Anal. Bioanal. Chem. 384, 322–335.

Guiseppi-Elie, A., et al., 2002. Direct electron transfer of glucose oxidase on carbon nanotubes. Nanotechnology 13, 559–564.

Guo, X.F., et al., 2006. Covalently bridging gaps in single-walled carbon nanotubes with conducting molecules. Science 311, 356–359.

Guo, S., et al., 2013. PNA-assembled graphene oxide for sensitive and selective detection of DNA. Analyst 138, 3216–3220.

He, B., et al., 2008. Nanowire sensors for multiplexed detection of biomolecules. Curr. Opin. Chem. Biol. 12, 522–528.

He, et al., 2010a. A graphene nanoprobe for rapid, sensitive, and multicolor fluorescent DNA analysis. Adv. Funct. Mater. 20, 453–459.

He, Y., et al., 2010b. Silicon nanostructures for bio-applications. Nano Today 5, 282–295.

He, Q.Y., et al., 2010c. Centimeter-long and large-scale micro-patterns of reduced graphene oxide films: fabrication and sensing applications. ACS Nano 4, 3201–3208.

He, Q.Y., et al., 2012. Graphene-based electronic sensors. Chem. Sci. 3, 1764–1772.

He, Y., et al., 2013. Quenching the chemiluminescence of acridinium ester by graphene oxide for label-free and homogeneous DNA detection. ACS Appl. Mater. Interfaces 5, 11336–11340.

Heller, A., 1990. Electrical wiring of redox enzymes. Acc. Chem. Res. 23, 128–134.

Heller, I., et al., 2005. Individual single-walled carbon nanotubes as nanoelectrodes for electrochemistry. Nano Lett. 5, 137–142.

Heller, D.A., et al., 2006a. Optical detection of DNA conformational polymorphism on single-walled carbon nanotubes. Science 311, 508–511.

Heller, I., et al., 2006b. Electrochemistry at single-walled carbon nanotubes: the role of band structure and quantum capacitance. J. Am. Chem. Soc. 128, 7353–7359.

Heller, I., et al., 2008. Identifying the mechanism of biosensing with carbon nanotube transistors. Nano Lett. 8, 591–595.

Heller, I., et al., 2009a. Optimizing the signal-to-noise ratio for biosensing with carbon nanotube transistors. Nano Lett. 9, 377–382.

Heller, I., et al., 2009b. Probing macrophage activity with carbon-nanotube sensors. Small 5, 2528–2532.

Heller, D.A., et al., 2009c. Multimodal optical sensing and analyte specificity using single-walled carbon nanotubes. Nat. Nanotechnol. 4, 114–120.

Hess, L.H., et al., 2011. Graphene transistor arrays for recording action potentials from electrogenic cells. Adv. Mater. 23, 5045–5049.

Howorka, S., Siwy, Z., 2009. Nanopore analytics: sensing of single molecules. Chem. Soc. Rev. 38, 2360–2384.

Hu, W.B., et al., 2011. Protein corona-mediated mitigation of cytotoxicity of graphene oxide. ACS Nano 5, 3693–3700.

Huang, Y.X., Chen, P., 2010. Nanoelectronic biosensing of dynamic cellular activities based on nanostructured materials. Adv. Mater. 22, 2818–2823.

Huang, Y.X., et al., 2009. Label-free detection of ATP release from living astrocytes with high temporal resolution using carbon nanotube network. Biosens. Bioelectron. 24, 2716–2720.

Huang, Y.X., et al., 2010. Nanoelectronic biosensors based on CVD grown graphene. Nanoscale 2, 1485–1488.

Huang, X., et al., 2011. Graphene-based materials: synthesis, characterization, properties, and applications. Small 7, 1876–1902.

Huang, Y., et al., 2012. An amplified single-walled carbon nanotube-mediated chemiluminescence turn-on sensing platform for ultrasensitive DNA detection. Chem. Commun. 48, 9400–9402.

Huang, Q., et al., 2013. Carbon dots and chitosan composite film based biosensor for the sensitive and selective determination of dopamine. Analyst 138, 5417–5423.

Hulanicki, A., et al., 1991. Chemical sensors definitions and classification. Pure Appl. Chem. 63, 1247–1250.

Iizumi, Y., et al., 2013. Immunoassay with single-walled carbon nanotubes as near-infrared fluorescent labels. ACS Appl. Mater. Interfaces 5, 7665–7670.

Ikonen, E., 2008. Cellular cholesterol trafficking and compartmentalization. Nat. Rev. Mol. Cell Biol. 9, 125–138.

Ion, A.C., et al., 2010. Acetylcholinesterase voltammetric biosensors based on carbon nanostructure-chitosan composite material for organophosphate pesticides. Mat. Sci. Eng. C30, 817–821.

Jauhiainen, M., Dolphin, P.J., 1986. Human plasma lecithin-cholesterol acyltransferase. J. Biol. Chem. 261, 7032–7043.

Jeng, E.S., et al., 2006. Detection of DNA hybridization using the near-infrared band-gap fluorescence of single-walled carbon nanotubes. Nano Lett. 6, 371–375.

Jeng, E.S., et al., 2007. Hybridization kinetics and thermodynamics of DNA adsorbed to individually dispersed single-walled carbon nanotubes. Small 3, 1602–1609.

Jeong, H.S., et al., 2010. Structural diversity induced by pyrene intercalators in homogeneous oligodeoxyguanylates. Mol. BioSyst. 6, 951–953.

Jin, H., et al., 2008. Stochastic analysis of stepwise fluorescence quenching reactions on single-walled carbon nanotubes: single molecule sensors. Nano Lett. 8, 4299–4304.

Ju, J., et al., 2014a. Synthesis of highly fluorescent nitrogen-doped graphene quantum dots for sensitive, label-free detection of Fe (III) in aqueous media. Biosens. Bioelectron. 58, 219–225.

Ju, J., et al., 2014b. Nitrogen-doped graphene quantum dots-based fluorescent probe for the sensitive turn-on detection of glutathione and its cellular imaging. RSC Adv. 4, 52583–52589.

Kachoosangi, R.T., et al., 2009. Carbon nanotube-ionic liquid composite sensors and biosensors. Anal. Chem. 81, 435–442.

Kang, X.H., et al., 2009. Glucose oxidase-graphene-chitosan modified electrode for direct electrochemistry and glucose sensing. Biosens. Bioelectron. 25, 901–905.

Kauffman, Star, 2008. Carbon nanotube gas and vapor sensors. Angew. Chem. Int. Ed. 47 (35), 6550–6570.

Kauffman, D.R., et al., 2008. Electronically monitoring biological interactions with carbon nanotube field-effect transistors. Chem. Soc. Rev. 37, 1197–1206.

Kefala, G., et al., 2004. A study of nafion-coated bismuth-film electrodes for the determination of trace metals by anodic stripping voltammetry. Analyst 129, 1082–1090.

Kilian, K.A., et al., 2009. The importance of surface chemistry in mesoporous materials: lessons from porous silicon biosensors. Chem. Commun. 630, 630–640.

Kiran, S., et al., 2015. Mechanism of intracellular detection of glucose through nonenzymatic and boronic acid functionalized carbon dots. J. Biomed. Mat. Res. A 103A (9), 2888–2897.

Kishi, K., et al., 2002. Highly sensitive cholesterol assay with enzymatic cycling applied to measurement of remnant lipoprotein-cholesterol in serum. Clin. Chem. 48, 737–741.

Koehne, J., et al., 2003. Ultrasensitive label-free DNA analysis using an electronic chip based on carbon nanotube nanoelectrode arrays. Nanotechnology 14, 1239–1245.

Krapf, D., et al., 2006. Experimental observation of nonlinear ionic transport at the nanometer scale. Nano Lett. 6, 2531–2535.

Lahiff, E., et al., 2010. The increasing importance of carbon nanotubes and nanostructured conducting polymers in biosensors. Anal. Bioanal. Chem. 398, 1575–1589.

Li, J., et al., 2003. Carbon nanotube nanoelectrode array for ultrasensitive DNA detection. Nano Lett. 3, 597–602.

Li, J., et al., 2009a. High-sensitivity determination of lead and cadmium based on the nafion-graphene composite film. Anal. Chim. Acta 649, 196–201.

Li, J., et al., 2009b. Nafion–graphene nanocomposite film as enhanced sensing platform for ultrasensitive determination of cadmium. Electrochem. Commun. 11, 1085–1088.

Li, H.L., et al., 2011. Nucleic acid detection using carbon nanoparticles as a fluorescent sensing platform. Chem. Commun. 47, 961–963.

Li, L.L., et al., 2012a. Facile microwave avenue to electrochemiluminescent two-color graphene quantum. Adv. Funct. Mater. 22, 2971–2979.

Li, H.T., et al., 2012b. Carbon nanodots: synthesis, properties and applications. J. Mater. Chem. 22, 24230–24253.

Li, H.T., et al., 2012c. Carbon quantum dots/Cu_2O composites with protruding nanostructures and their highly efficient (near) infrared photocatalytic behaviour. J. Mater. Chem. 22, 17470–17475.

Li, et al., 2013a. Large scale synthesis of photoluminescent carbon nanodots and their application for bioimaging. Nanoscale 5, 1967–1971.

Li, J.Z., et al., 2013b. Electrogenerated chemiluminescence detection of trace level pentachlorophenol using carbon quantum dots. Analyst 138, 2038–2043.

Li, H., et al., 2013c. Interaction of Cy3 dye with CCG and its application for BSA detection. J. Mater. Chem. B 1, 693–697.

Li, H., et al., 2015. Facile synthesis of a nanocomposite based on graphene and ZnAl layered double hydroxides as a portable shelf of a luminescent sensor for DNA detection. RSC Adv. 5, 9341–9347.

Lin, Y.H., et al., 2004. Glucose biosensors based on carbon nanotube nanoelectrode ensembles. Nano Lett. 4, 191–195.

Lin, Y., et al., 2005. Carbon nanotubes (CNTs) for the development of electrochemical biosensors. Front. Biosci. 10, 492–505.

Lin, Y., et al., 2011. Combination of graphene oxide and thiol-activated DNA metallization for sensitive fluorescence turn-on detection of cysteine and their use for logic gate operations. Adv. Funct. Mater. 21, 4565–4572.

Lin, X.M., et al., 2014. Encapsulation of strongly fluorescent carbon quantum dots in metal–organic frameworks for enhancing chemical sensing. Anal. Chem. 86, 1223–1228.

Linsel-Nitschke, P., Tall, A.R., 2005. HDL as a target in the treatment of atherosclerotic cardiovascular disease. Nat. Rev. Drug Discov. 4, 193–205.

Liu, J., et al., 1998. Fullerene pipes. Science 280, 1253–1256.

Liu, Z.F., et al., 2000. Organizing single-walled carbon nanotubes on gold using a wet chemical self-assembling technique. Langmuir 16, 3569–3573.

Liu, J., et al., 2005. Achieving direct electrical connection to glucose oxidase using aligned single walled carbon nanotube arrays. Electroanalysis 17, 38–46.

Liu, et al., 2006. Facile preparation of amperometric laccase biosensor with multifunction based on the matrix of carbon nanotubes–chitosan composite. Biosens. Bioelectron. 21, 2195–2201.

Liu, J.W., et al., 2009a. Functional nucleic acid sensors. Chem. Rev. 109, 1948–1998.

Liu, Z., et al., 2009b. Carbon nanotubes in biology and medicine: in vitro and in vivo detection, imaging and drug delivery. Nano Res. 2, 85–120.

Liu, S., et al., 2010. Chemical functionalization of single-walled carbon nanotube field-effect transistors as switches and sensors. Coord. Chem. Rev. 254, 1101–1116.

Liu, et al., 2011a. Direct electrodeposition of graphene enabling the one-step synthesis of graphene-metal nanocomposite films. Small 7 (9), 1203–1206.

Liu, S., et al., 2011b. Single-molecule detection of proteins using aptamer-functionalized molecular electronic devices. Angew. Chem. Int. Ed. 50, 2496–2502.

Liu, K.P., et al., 2011c. Green and facile synthesis of highly biocompatible graphene nanosheets and its application for cellular imaging and drug delivery. J. Mater. Chem. 21, 12034–12040.

Liu, et al., 2014a. Intracellular detection of ATP using an aptamer beacon covalently linked to graphene oxide resisting nonspecific probe displacement. Anal. Chem. 86, 12229–12235.

Liu, X.J., et al., 2014b. Carbon dots based dual-emission silica nanoparticles as a ratiometric nanosensor for Cu^{2+}. Anal. Chem. 86, 2289–2296.

Liu, M., et al., 2014c. A graphene-based biosensing platform based on the release of DNA probes and rolling circle amplification. ACS Nano 8, 5564–5573.

Lu, J., et al., 2007. Simple fabrication of a highly sensitive glucose biosensor using enzymes immobilized in exfoliated graphite nanoplatelets nafion membrane. Chem. Mat. 19, 6240–6246.

Lu, J., et al., 2008. Nanometal-decorated exfoliated graphite nanoplatelet based glucose biosensors with high sensitivity and fast response. ACS Nano 2, 1825–1832.

Lu, C.H., et al., 2010. Increasing the sensitivity and single-base mismatch selectivity of the molecular beacon using graphene oxide as the "nanoquencher". Chem. Eur. J. 16, 4889–4894.

Lu, J., et al., 2013. Electrochemiluminescence of blue-luminescent graphene quantum dots and its application in ultrasensitive aptasensor for adenosine triphosphate detection. Biosens. Bioelectron. 47, 271–277.

Luthi, A.J., et al., 2012. Tailoring of biomimetic high-density lipoprotein (HDL) nanostructures changes cholesterol binding and efflux. ACS Nano 6, 276–285.

Maehashi, K., et al., 2009. Label-free electrical detection using carbon nanotube-based biosensors. Sensors 9, 5368–5378.

Maiti, et al., 2012. Multiplex targeted in vivo cancer detection using sensitive near-infrared SERS nanotags. Nano Today 7, 85–93.

Mao, S., et al., 2010. Specific protein detection using thermally reduced graphene oxide sheet decorated with gold nanoparticle-antibody conjugates. Adv. Mater. 22, 3521–3526.

Mao, Y., et al., 2012. Efficient one-pot synthesis of molecularly imprinted silica nanospheres embedded carbon dots for fluorescent dopamine optosensing. Biosens. Bioelectron. 38, 55–60.

McCreery, R.L., 2008. Advanced carbon electrode materials for molecular electrochemistry. Chem. Rev. 108, 2646–2687.

Medintz, I.L., et al., 2005. Quantum dot bioconjugates for imaging, labelling and sensing. Nat. Mater. 4, 435–446.

Meng, Y., et al., 2012. Interaction study on DNA, single-wall carbon nanotubes and acridine orange. Mater. Sci. Eng. B 177, 887–891.

Mohanty, et al., 1997. A highly sensitive fluorescent micro-assay of H_2O_2 release from activated human leukocytes using a dihydroxyphenoxazine derivative. J. Immunol. Methods 202 (2), 133–141.

Mondal, A., Jana, N.R., 2012. Fluorescent detection of cholesterol using β-cyclodextrin functionalized graphene. Chem. Commun. 48, 7316–7318.

Motonaka, J., Faulkner, L.R., 1993. Determination of cholesterol and cholesterol ester with novel enzyme microsensors. Anal. Chem. 65, 3258–3261.

Musameh, M., et al., 2002. Low-potential stable NADH detection at carbon-nanotube-modified glassy carbon electrodes. Electrochem. Commun. 4, 743–746.

Myant, N.B., 1981. The Biology of Cholesterol and Related Steroids. William Heinemann Medical Books, London.

Nam, J.-M., et al., 2003. Nanoparticle-based bio-bar codes for the ultrasensitive detection of proteins. Science 301, 1884–1886.

Nam, K., et al., 2012. Aptamer-functionalized nano-pattern based on carbon nanotube for sensitive, selective protein detection. J. Mater. Chem. 22, 23348–23356.

Niu, J.J., Gao, H., 2014. Synthesis and drug detection performance of nitrogen-doped carbon dots. J. Lumin. 149, 159–162.

Niwa, O., et al., 2006. Electrochemical performance of angstrom level flat sputtered carbon film consisting of sp^2 and sp^3 mixed bonds. J. Am. Chem. Soc. 128, 7144–7145.

Noh, E.H., et al., 2013. Carbon nanodot-based self-delivering microRNA sensor to visualize microRNA124a expression during neurogenesis. J. Mater. Chem. B 1, 4438–4445.

Novoselov, K.S., et al., 2004. Electric field effect in atomically thin carbon films. Science 306, 666–669.

Ohno, Y., et al., 2009. Electrolyte-gated graphene field-effect transistors for detecting pH and protein adsorption. Nano Lett. 9, 3318–3322.

Ohno, et al., 2010a. Label-free biosensors based on aptamer-modified graphene field-effect transistors. J. Am. Chem. Soc. 132 (51), 18012–18013.

Ohno, Y., et al., 2010b. Chemical and biological sensing applications based on graphene field-effect transistors. Biosens. Bioelectron. 26, 1727–1730.

Park, S., et al., 2010. Biocompatible, robust free-standing paper composed of a TWEEN/graphene composite. Adv. Mater. 22, 1736–1740.

Patolsky, F., et al., 2004. Long-range electrical contacting of redox enzymes by SWCNT connectors. Angew. Chem. Int. Ed. 43, 2113–2117.

Patolsky, et al., 2006a. Nanowire sensors for medicine and the life sciences. Nanomedicine 1, 51–65.

Patolsky, F., et al., 2006b. Detection, stimulation, and inhibition of neuronal signals with high-density nanowire transistor arrays. Science 313, 1100–1104.

Patolsky, F., et al., 2006c. Fabrication of silicon nanowire devices for ultrasensitive, label-free, real-time detection of biological and chemical species. Nat. Protoc. 1, 1711–1724.

Patolsky, F., et al., 2006d. Nanowire-based biosensors. Anal. Chem. 78, 4260–4269.

Pingarron, J.M., et al., 2008. Gold nanoparticle-based electrochemical biosensors. Electrochim. Acta 53, 5848–5866.

Pumera, M., 2011. Graphene in bio-sensing. Mater. Today 14, 308–315.

Qi, H.L., et al., 2009. Applications of nanomaterials in electrogenerated chemiluminescence biosensors. Sensors 9, 674–695.

Qian, Z.S., et al., 2014a. Highly luminescent N-doped carbon quantum dots as an effective multifunctional fluorescence sensing platform. Chem. Eur. J. 20, 2254–2263.

Qian, Z.S., et al., 2014b. A universal fluorescence sensing strategy based on biocompatible graphene quantum dots and graphene oxide for the detection of DNA. Nanoscale 6, 5671–5674.

Qu, L.T., et al., 2008. Preferential syntheses of semiconducting vertically aligned single-walled carbon nanotubes for direct use in FETs. Nano Lett. 8, 2682–2687.

Qu, S.N., et al., 2012. A biocompatible fluorescent ink based on water-soluble luminescent carbon nanodots. Angew. Chem. Int. Ed. 51, 12215–12218.

Qu, K.G., et al., 2013. Carbon dots prepared by hydrothermal treatment of dopamine as an effective fluorescent sensing platform for the label-free detection. Chem. Eur. J. 19, 7243–7249.

Raines, 1995. The essential function of protein-disulfide isomerase is to unscramble non-native disulfide bonds. J. Biol. Chem. 270 (47), 28006–28009.

Rajesh, T.A., et al., 2009. Recent progress in the development of nano-structured conducting polymers/nanocomposites for sensor applications. Sens. Actuators B 136, 275–286.

Ran, X., et al., 2013. Ag nanoparticle-decorated graphene quantum dots for label-free, rapid and sensitive detection of Ag^+ and biothiols. Chem. Commun. 49, 1079–1081.

Ratinac, K.R., et al., 2011. Graphene and related materials in electrochemical sensing. Electroanalysis 23, 803–826.

Razmi, H., Mohammad-Rezaei, R., 2013. Graphene quantum dots as a new substrate for immobilization and direct electrochemistry of glucose oxidase: application to sensitive glucose determination. Biosens. Bioelectron. 41, 498–504.

Robinson, J.T., et al., 2011. Ultrasmall reduced graphene oxide with high near-infrared absorbance for photothermal therapy. J. Am. Chem. Soc. 133, 6825–6831.

Rogers, K.R., 2006. Recent advances in biosensor techniques for environmental monitoring. Anal. Chim. Acta 568, 222–231.

Rosi, N.L., et al., 2005. Nanostructures in biodiagnostics. Chem. Rev. 105, 1547–1562.

Roy, S., Gao, Z.Q., 2009. Nanostructure-based electrical biosensors. Nano Today 4, 318–334.

Rubianes, M.D., Rivas, G.A., 2003. Carbon nanotubes paste electrode. Electrochem. Commun. 5, 689–694.

Sadik, O.A., et al., 2009. Status of biomolecular recognition using electrochemical techniques. Biosens. Bioelectron. 24, 2749–2765.

Sapsford, K.E., et al., 2006. Biosensing with luminescent semiconductor quantum dots. Sensors 6, 925–953.

Sarma, A.K., et al., 2009. Recent advances in material science for developing enzyme electrodes. Biosens. Bioelectron. 24, 2313.

Sassolas, A., et al., 2008. DNA biosensors and microarrays. Chem. Rev. 108, 109–139.

Satishkumar, B.C., et al., 2007. Reversible fluorescence quenching in carbon nanotubes for biomolecular sensing. Nat. Nanotechnol. (2), 560–564.

Shamah, S.M., Healy, J.M., Cload, S.T., 2008. Complex target SELEX. Acc. Chem. Res. 41, 130–138.

Shan, C., et al., 2009a. Direct electrochemistry of glucose oxidase and biosensing for glucose based on graphene. Anal. Chem. 81, 2378–2382.

Shan, C., et al., 2009b. Graphene/AuNPs/chitosan nanocomposites film for glucose biosensing. Biosens. Bioelectron. 25, 1070–1074.

Shi, J.X., et al., 2013. High selectivity sensing of cobalt in HepG2 cells based on necklace model microenvironment-modulated carbon dot-improved chemiluminescence in fenton-like system. Biosens. Bioelectron. 45, 58–64.

Soleymani, L., et al., 2009. Programming the detection limits of biosensors through controlled nanostructuring. Nat. Nanotechnol. 4, 844–848.

Song, S.P., et al., 2010. Functional nanoprobes for ultrasensitive detection of biomolecules. Chem. Soc. Rev. 39, 4234–4243.

Sorgenfrei, S., et al., 2011. Label-free single-molecule detection of DNA-hybridization kinetics with a carbon nanotube field-effect transistor. Nat. Nanotechnol. 6, 125–132.

Stern, E., et al., 2007. Label-free immune-detection with CMOS-compatible semiconducting nanowires. Nature 445, 519–522.

Stern, E., et al., 2010. Label-free biomarker detection from whole blood. Nat. Nanotechnol. 5, 138–142.

Stromberg, M., et al., 2009. Multiplex detection of DNA sequences using the volume-amplified magnetic nanobead detection assay. Anal. Chem. 81, 3398–3406.

Sudibya, H.G., et al., 2009. Interfacing glycosylated carbon-nanotube-network devices with living cells to detect dynamic secretion of biomolecules. Angew. Chem. Int. Ed. 48, 2723–2726.

Sun, H., et al., 2013. Recent advances in graphene quantum dots for sensing. Mater. Today 16, 433–442.

Tao, Y., et al., 2012. DNA-templated silver nanoclusters-graphene oxide nanohybrid materials: a platform for label-free and sensitive fluorescence turn-on detection of multiple nucleic acid targets. Analyst 137, 2588–2592.

Thevenot, D.R., et al., 2001. Electrochemical biosensors: recommended definitions and classification. Biosens. Bioelectron. 16, 121–131.

Tian, B.Z., et al., 2010. Three-dimensional, flexible nanoscale field-effect transistors as localized bioprobes. Science 329, 830–834.

Tsai, C.-C., et al., 2008. Exocytosis of a single bovine adrenal chromaffin cell: the electrical and morphological studies. J. Phys. Chem. B 112, 9165–9173.

Valentini, F., et al., 2007. The electrochemical detection of ammonia in drinking water based on multi-walled carbon nanotube/copper nanoparticle composite paste electrodes. Sens. Actuators B 128, 326–333.

Wang, J., 2005a. Nanomaterial-based electrochemical biosensors. Analyst 130, 421–426.

Wang, J., 2005b. Carbon-nanotube based electrochemical biosensors: A review. Electroanalysis 17, 7–14.

Wang, J., 2007. Nanoparticle-based electrochemical bioassays of proteins. Electroanalysis 19, 769–776.

Wang, J., 2008. Electrochemical glucose biosensors. Chem. Rev. 108, 814–825.

Wang, J., et al., 2003. Solubilization of carbon nanotubes by nafion toward the preparation of amperometric biosensors. J. Am. Chem. Soc. 125, 2408–2409.

Wang, W.U., et al., 2005. Label-free detection of small-molecule-protein interactions by using nanowire nanosensors. Proc. Natl. Acad. Sci. U.S.A. 102, 3208–3212.

Wang, et al., 2007a. Carbon nanotubes for biological and biomedical applications. Nanotechnology 18 (412001), 1–12.

Wang, Q., et al., 2007b. Room-temperature ionic liquids/multi-walled carbon nanotubes/chitosan composite electrode for electrochemical analysis of NADH. Electrochim. Acta 52, 6630–6637.

Wang, C.-W., et al., 2007c. In situ detection of chromogranin a released from living neurons with a single-walled carbon-nanotube field-effect transistor. Small 3, 1350–1355.

Wang, Z.J., et al., 2009. Direct electrochemical reduction of single-layer graphene oxide and subsequent functionalization with glucose oxidase. J. Phys. Chem. C (113), 14071–14075.

Wang, R.X., et al., 2013a. Carbon quantum dot-functionalized aerogels for NO_2 gas sensing. Anal. Chem. 85, 8065–8069.

Wang, W., et al., 2013b. Label-free DNA detection on the basis of fluorescence resonance energy transfer from oligonucleotide-templated silver nanoclusters to multi-walled carbon nanotubes. Anal. Methods 5, 5555–5559.

Wang, Y., et al., 2013c. Using graphene quantum dots as photoluminescent probes for protein kinase sensing. Anal. Chem. 85, 9148–9155.

Wang, G.L., et al., 2014a. Visible-light-stimulated enzymelike activity of graphene oxide and its application for facile glucose sensing. J. Phys. Chem. C 118, 28109–28117.

Wang, F., et al., 2014b. Graphene quantum dots as a fluorescent sensing platform for highly efficient detection of copper (II) ions. Sens. Actuators B 190, 516–522.

Wang, L., et al., 2014c. AgNP-DNA@GQDs hybrid: new approach for sensitive detection of H_2O_2 and glucose via simultaneous AgNP etching and DNA cleavage. Anal. Chem. 86, 12348–12354.

Wei, B.Q., et al., 2002. Microfabrication technology: organized assembly of carbon nanotubes. Nature 416, 495–496.

Willner, I., Zayats, M., 2007. Electronic aptamer-based sensors. Angew. Chem. Int. Ed. 46, 6408–6418.

Wilson, R., 2008. The use of gold nanoparticles in diagnostics and detection. Chem. Soc. Rev. 37, 2028–2045.

Wu, L., et al., 2007. Amperometric glucose sensor based on catalytic reduction of dissolved oxygen at soluble carbon nanofiber. Biosens. Bioelectron. 23, 479–484.

Wu, H., et al., 2009. Glucose biosensor based on immobilization of glucose oxidase in platinum nanoparticles/graphene/chitosan nanocomposite film. Talanta 80, 403–406.

Wu, H.C., et al., 2010. Chemistry of carbon nanotubes in biomedical applications. J. Mater. Chem. 20, 1036–1052.

Wu, Z., et al., 2014. A graphene quantum dot-based method for the highly sensitive and selective fluorescence turn on detection of biothiols. Talanta 119, 538–543.

Xu, H., et al., 2008. Ultrasensitive voltammetric detection of trace lead(II) and cadmium(II) using MWCNTs-nafion/bismuth composite electrodes. Electroanalysis 20, 2655–2662.

Xu, B., et al., 2012. Aptamer carbon nanodot sandwich used for fluorescent detection of protein. Analyst 137, 5483–5486.

Yang, R.H., et al., 2008. Noncovalent assembly of carbon nanotubes and single-stranded DNA: an effective sensing platform for probing biomolecular interactions. Anal. Chem. 80, 7408–7413.

Yang, W.R., et al., 2010. Carbon nanomaterials in biosensors: should you use nanotubes or graphene? Angew. Chem. Int. Ed. 49, 2114–2138.

Yin, Z.Y., et al., 2012. Real-time, DNA detection using Pt nanoparticle-decorated reduced graphene oxide field-effect transistors. Nanoscale 4, 293–297.

Yu, X., et al., 2003. Peroxidase activity of enzymes bound to the ends of single-wall carbon nanotube forest electrodes. Electrochem. Commun. 5, 408–411.

Yu, X., et al., 2006. Carbon nanotube amplification strategies for highly sensitive immunodetection of cancer biomarkers. J. Am. Chem. Soc. 128, 11199–11205.

Yu, C., et al., 2007. Multiplex biosensor using gold nanorods. Anal. Chem. 79, 572–579.

Zelada, G.A., et al., 2009. Immediate detection of living bacteria at ultralow concentrations using a carbon nanotube based potentiometric aptasensor. Angew. Chem. Int. Ed. 48, 7334–7337.

Zhang, N., et al., 2008. A novel assembly of Au NPs–β-CDs–FL for the fluorescent probing of cholesterol and its application in blood serum. Analyst 133, 1176–1181.

Zhang, J., et al., 2011. Single molecule detection of nitric oxide enabled by $d(AT)_{15}$ DNA adsorbed to near infrared fluorescent single-walled carbon nanotubes. J. Am. Chem. Soc. 133, 567–581.

Zhang, W., et al., 2014. Perylene derivative-bridged Au–graphene nanohybrid for label-free HpDNA biosensor. J. Mater. Chem. B 2, 3142–3148.

Zhen, S.J., et al., 2013. Dual-aptamer-based sensitive and selective detection of prion protein through the fluorescence resonance energy transfer between quantum dots and graphene oxide. Anal. Methods 5, 6904–6907.

Zheng, G.F., et al., 2005. Multiplexed electrical detection of cancer markers with nanowire sensor arrays. Nat. Biotechnol. 23, 1294–1301.

Zheng, M., et al., 2013. On–off–on fluorescent carbon dot nanosensor for recognition of chromium(VI) and ascorbic acid based on the inner filter effect. ACS Appl. Mater. Interfaces 5, 13242–13247.

Zhou, M., et al., 1997. A stable non fluorescent derivative of resorufin for the fluorometric determination of trace hydrogen peroxide: applications in detecting the activity of phagocyte NADPH oxidase and other oxidases. Anal. Biochem. 253, 162–168.

Zhou, M., et al., 2008. Highly ordered mesoporous carbons as electrode material for the construction of electrochemical dehydrogenase- and oxidase-based biosensors. Biosens. Bioelectron. 24, 442–447.

Zhou, M., et al., 2009. Electrochemical sensing and biosensing platform based on chemically reduced graphene oxide. Anal. Chem. 81, 5603–5613.

Zhou, L., et al., 2012. Carbon nanodots as fluorescence probes for rapid, sensitive, and label-free detection of Hg^{2+} and biothiols in complex matrices. Chem. Commun. 48, 1147–1149.

Zhou, J., et al., 2013. Graphene oxide–peptide nanocomplex as a versatile fluorescence probe of protein kinase activity based on phosphorylation protection against carboxypeptidase digestion. Anal. Chem. 85, 5746–5754.

Zhou, Y., et al., 2014. A novel composite of graphene quantum dots and molecularly imprinted polymer for fluorescent detection of paranitrophenol. Biosens. Bioelectron. 52, 317–323.

Zhu, L.D., et al., 2008. Anodic stripping voltammetric determination of lead in tap water at an ordered mesoporous carbon/nafion composite film electrode. Electroanalysis 20, 527–533.

Zhu, Z., et al., 2010. Single-walled carbon nanotube as an effective quencher. Anal. Bioanal. Chem. 396, 73–83.

Zhuang, H.L., et al., 2013. Sensitive detection of prion protein through long range resonance energy transfer between graphene oxide and molecular aptamer beacon. Anal. Methods 5, 208–212.

FURTHER READING

Ang, P.K., et al., 2008. Solution-gated epitaxial graphene as pH sensor. J. Am. Chem. Soc. 130, 14392–14393.

Baby, T.T., et al., 2010. Metal decorated graphene nanosheets as immobilization matrix for amperometric glucose biosensor. Sens. Actuator B145, 71–77.

Banks, C.E., Compton, R.G., 2005. Exploring the electrocatalytic sites of carbon nanotubes for NADH detection: an edge plane pyrolytic graphite electrode study. Analyst 130, 1232–1239.

Cao, X.H., et al., 2011. Graphene oxide as a carbon source for controlled growth of carbon nanowires. Small 7, 1199–1202.

Gooding, J.J., 2008. Diazonium salts for modifying carbon and metal electrodes. Electroanalysis 20, 573–582.

Liu, G.D., Lin, Y.H., 2006. Amperometric glucose biosensor based on self-assembling glucose oxidase on carbon nanotubes. Electrochem. Commun. 8, 251–256.

Maiti, S., et al., 2013. Label-free fluorimetric detection of histone using quaternized carbon dot–DNA nanobiohybrid. Chem. Commun. 49, 8851–8853.

Mohanty, J.G., et al., 1999. A highly sensitive fluorescent micro-assay of H_2O_2 release from activated human leukocytes using a dihydroxyphenoxazine derivative. J. Immunol. Methods 202, 133–141.

Raines, W.E., Ross, R., 1995. Biology of atherosclerotic plaque formation: possible role of growth factors in lesion development and the potential impact of soy. J. Nutr. 125, 624S–630S.

Sudibya, H.G., et al., 2011. Electrical detection of metal ions using filed-effect transistors based on micropatterned reduced graphene oxide films. ACS Nano 5, 1990–1994.

Tang, Z., 2010. Constraint of DNA on functionalized graphene improves its biostability and specificity. Small 6, 1205–1209.

Waldmann, D., et al., 2011. Bottom-gated epitaxial graphene. Nat. Mater. 10, 357–360.

CHAPTER 4

Application of Carbon-Based Nanomaterials as Bioimaging Probe

CONTENTS

4.1 INTRODUCTION

Bioimaging is a term that covers the complex chain of acquiring, processing, and visualizing structural or functional images of living objects or systems, including extraction and processing of image-related information. Medical imaging and microscope/fluorescence image processing are important parts of bioimaging referring to the techniques and processes used to create images of the human body, anatomical areas, tissues, and so on, down to the molecular level, for clinical purposes, seeking to reveal, diagnose, or examine diseases, or medical science, including the study of normal anatomy and physiology. Image processing methods, such as denoising, segmentation, deconvolution, and registration methods, feature recognition and classification represent an indispensable part of bioimaging, as well as related data analysis and statistical tools. The carbon-based nanomaterials as important agents for biological imaging applications have emerged in the past few years due to intrinsic optical properties. They exhibit a wide absorption spectrum in the visible and near-infrared (NIR) range, photoluminescence (PL) in the NIR region, and strong resonance Raman scattering. These characteristics have been utilized for in vitro and in vivo biological imaging. Carbon nanomaterials have unique optical, electronic, mechanical, and chemical properties. Many of these applications rely on successful surface modifications. Numerous classes of carbon nanomaterials such as carbon nanotubes (CNTs), graphene, graphene oxide and its derivatives, carbon dots (CDs), graphene quantum dots (GQDs), fullerene, carbon nanohorns (CNHs), and carbon nanoonions (CNOs) have been explored for bioimaging applications. Among these, CNTs, graphene and its derivatives, CDs, and GQDs as well as their nanocomposites are subjects of fundamental research in bioimaging process. A pristine graphene sheet is a zero band gap semiconductor with the conduction band and valence band touching at the K points and should thus be nonfluorescent (Geim, 2009; Dresselhaus and Dresselhaus, 2002). When the size of the graphene sheet shrinks and functional groups are introduced to the edges of graphene, band gap opening is predicted and experimentally confirmed due to quantum confinement and edge effects (Nakada et al., 1996; Son et al.,

Carbon Nanomaterials for Biological and Medical Applications. http://dx.doi.org/10.1016/B978-0-323-47906-6.00004-7

2006; Han et al., 2007), and as a result fluorescence can be observed across the band gap of narrow graphene nanoribbons and oxidized graphene sheets (i.e., graphene oxide or GO) (Sun et al., 2008; Chernov et al., 2013). Owing to the facile chemical synthesis of GO in solution phase, GO is the most common type of fluorescent graphene materials widely used for bioimaging. CDs generally refer to small carbon nanoparticles (CNPs) with sizes below 10 nm; moreover, oxygen, hydrogen, and other elements may exist in the small dots. The up- and downconversion PL properties of CDs are similar to those of semiconductor quantum dots (QDs). Superior to QDs, CDs are biocompatible and environmentally friendly, without heavy metal ions and toxic elements. There have been many functional groups and/or passivation agents used to cover the surface of the CDs outside the carbon core, endowing CDs with high quantum yield (QY), chemical stability, and good water solubility. CDs can be easily conjugated with target molecules to expand their functionality. In addition, starting materials for CDs are abundantly available, resulting in the possibility of low cost and mass production of the CDs. Another class of carbon-based nanomaterials, diamond nanoparticles (NPs), has gained wide acceptance in the biomedical field. Due to the lack of sp^2 graphitic carbon, nanodiamond emits fluorescence based on a completely different mechanism, arising from the presence of the defect centers. Defect centers are point defects commonly found in diamond, among which the well-studied one is the N—V center owing to its unique and useful PL in the visible window (Dresselhaus and Dresselhaus, 2002; Aharonovich et al., 2011). The fluorescence emission spectrum of the N—V centers in nanodiamond is usually composed of three peaks: a peak located at 576 nm assigned to the zero-phonon line (ZPL) of neutral N—V center ($N{-}V^0$), a peak located at 638 nm assigned to the ZPL of negatively charged N—V center ($N{-}V^-$), and another broad peak centered at ~700 nm assigned to the phonon-coupled side band (Fu et al., 2007a,b). Their beneficial properties such as the chemical inertness of the core, interesting electronic and mechanical properties, the stability of luminescent lattice defects and other characteristics make nanoscale diamond an ideal material for imaging applications.

Distinguished by "bottom-up" dehydration and "top-down" cutting routes, a large number of methods have been developed to synthesize carbon nanomaterials. Researchers are attempting to obtain high-quality CDs via simple methods. CDs mainly consist of three kinds of fluorescent dots: GQDs, carbon nanodots (CNDs), and polymer dots (PDs), which are mainly categorized by the inner structure of the carbon in the small dots. GQDs are defined as nanographene fragments usually with diameters less than 10 nm. Although some other elements (e.g., oxygen, hydrogen, and nitrogen) may exist on the edge, the main body of GQDs is comprised of conjugated sp^2 carbon. Carbon nanodots represent a wide range of fluorescent spherical carbon materials, which are mainly of two types. The first one is made up of an sp^2 carbon core and surrounding chemical groups, while the other one consists of amorphous aggregations. In a sense, GQDs are a special kind of CNPs with a large conjugated domain and regular structure. Synthesis of different carbon nanomaterials, *viz.*, CDs, GQDs, CNTs, graphene, graphene oxide, and their derivatives has already been discussed in Chapter 1.

4.2 APPLICATION FOR BIOIMAGING PROBE

Carbon nanomaterials have high aqueous solubility, good biocompatibility, and low cytotoxicity, as well as superior resistance to photobleaching that show great potential for fluorescence bioimaging and multimodal bioimaging of cells and tissues. Many kinds of carbon nanomaterials can emit blue/green fluorescence when excited by UV light; thus, they are suitable for cellular imaging under a fluorescence microscope. Some of the carbon nanomaterials can be excited by long wavelength light and/or emit upconversion PL light; therefore, they can be utilized for in vivo imaging. In this section, we will discuss the use of different carbon nanomaterials for bioimaging probes.

4.2.1 Bioimaging Probe Using Carbon Nanoparticles

4.2.1.1 Carbon Dots and Carbon Nanoparticles for Bioimaging

CDs in a broad definition generally refer to small CNPs in aqueous or other suspensions. The original hint on nanoscale carbon particles being PL could be traced back to the observation that the surface defects in both single-walled CNTs (SWCNTs) and multiwalled CNTs (MWCNTs) became brightly emissive under some specific conditions (Riggs et al., 2000), especially when the nanotubes were functionalized for the purpose of dispersion at the individual nanotube level without bundling or aggregation (Riggs et al., 2000; Sun et al., 2002; Lin et al., 2005; Guldi et al., 2003; Lacerda et al., 2006). In a conceptually and experimentally similar fashion, Sun et al. (2006) functionalized small CNPs with various oligomeric species to target an effective passivation of the particle surface, resulting in bright and colorful fluorescence emissions, thus the finding of CDs. It was subsequently found that CDs are also strongly multiphoton fluorescent, with extremely large two-photon cross sections on pulsed laser excitation (800–900 nm) (Cao et al., 2007; Yang et al., 2009a,b).

The effective carbon particle surface passivation has apparently been a necessary condition for CDs of high fluorescence performance, despite the rather broad definition on CDs. "Naked" CNPs without deliberate surface functionalization in relatively stable aqueous or other suspensions may also exhibit colorful fluorescence emissions, as reported by many research groups (Liu et al., 2007; Zhou et al., 2007; Ray et al., 2009; Cao et al., 2011; Li et al., 2011; Shen et al., 2012a,b,c), but with generally low to very low emission QYs. Since the presence of solvents around the naked particles has been necessary for the fluorescence emissions, it may be argued that the solvent molecules provide relatively weak but nevertheless meaningful surface passivation for the suspended CNPs (Cao et al., 2011). Therefore, there seems no need to differentiate conceptually CDs with and without deliberate surface functionalization (or other forms of more effective passivation effects), as the surface passivation simply alters the fluorescence performance, dramatically may be, of the various CDs. The CNPs in CDs may come from a variety of sources, some even from really unconventional ones. In the first study (Sun et al., 2006), carbon soot from laser ablation of a graphitic target, which typically has a higher population of smaller CNPs, was used as a precursor for surface passivation in the preparation of brightly fluorescent CDs. Since then, many different production strategies have been practiced (Li et al., 2011), including the use of commercially available carbon nanopowders as the starting material to harvest the desired smaller nanoparticles (Wang et al., 2011a,b,c). However, CDs have multiple advantages including excellent optical properties and good chemical and photochemical stability. CDs are attractive candidates for bioimaging because of their excellent PL properties and low toxicity. Traditional QDs, such as CdTe and related core–shell nanoparticles, have been widely used in in vitro and in vivo optical imaging experiments (Gao et al., 2005; Hardman, 2006; Jaiswal and Simon, 2004). However, QDs contain toxic heavy metals and their application may cause health and environmental problems (Hardman, 2006), so using CDs for bioimaging has set a trend. Moreover, because of their high photostability and biocompatibility, CDs are also well suited for cellular imaging. These CNPs were transformed into various functionalized nanoprobes for use as cell-imaging probes with hydrodynamic diameters of 5–15 nm (Bhunia et al., 2013). Chen et al. (2004) synthesized CDs with high yield (41.8%), high fluorescence QY (21.6%), and excellent stability in one simple step by carbonization of sucrose with oleic acid. The obtained CDs can be effectively used for cell imaging. Zhu et al. (2014) used a simple and effective route, which employed lithium-intercalated graphite from lithium-ion batteries as a carbon source to prepare CDs. These CDs are water soluble, nanosized, and biocompatible and can easily enter into HeLa cells to act as a cell-imaging agent without any further functionalization. Tong et al. (2015) developed a simple hydrothermal approach for preparing amorphous CDs with high two-photon fluorescence from hyperbranched poly(amino amine) and citric acid without further modification. The as-synthesized CDs exhibited excellent fluorescence properties and excitation-dependent fluorescence behavior with a corresponding QY of 17.1% in aqueous solution. Moreover, the CDs displayed low cytotoxicity against L929 normal cells. As fluorescent nanomaterials, CDs can deliver drugs with high loading efficiency, as well as observing the distribution of drugs and monitoring their effects (He et al., 2014; Pandey et al., 2013a,b; Hu et al., 2014; Cheng et al., 2014). Gogoi and Chowdhury (2014) successfully prepared novel carbon dot–coated alginate beads (CA–CD) with distinctive stability and high loading efficiency in comparison to calcium alginate beads. Kim et al. (2013) coupled CDs with gold NPs to produce an assembly, which was then conjugated to polyethylenimine-plasmid DNA (pDNA) for delivering DNA to cells. The fluorescence emissions from the assembly of CD-gold NPs could be quenched by pDNA while the release of pDNA could be probed via recovery of the fluorescence signals. Pandey et al. (2013a,b) used phenylalanine-derived nontoxic CDs as vehicles for the delivery of the antipsychotic drug haloperidol (HaLO). Cysteamine hydrochloride (CysHCl) as a linker can offer controlled release under physiological conditions for more than 40 h following the Hixson–Crowell model under standardized conditions. Moreover, the CDs-CysHCl-HaLO conjugate was found to exhibit much higher compatibility with MDCK cells at pH 7.2 in comparison to bare HaLO. As a type of fluorescent nanomaterial, CDs can also combine medical therapy and bioimaging diagnostics to visualize the distribution of drugs and monitor their effects (Pandey et al., 2013a,b; Wang et al., 2014a,b,c,d,e). Zheng et al. (2014a,b,c) prepared a multifunctional theranostic agent (CD-Oxa) by conjugation of an anticancer agent (oxidized oxaliplatin, Oxa(IV)-COOH) to the surface of CDs (Fig. 4.1). CD-Oxa successfully combined the optical properties of CDs and the therapeutic performance of Oxa. In vitro, CD-Oxa exhibited good biocompatibility, bioimaging functions, and anticancer effects. In vivo, the distribution of the drug can be monitored via the fluorescence signal of CD-Oxa.

Analogously, Wang et al. (2014a,b,c,d,e) successfully synthesized multifunctional hybrid NPs (B100 nm), which combined with magnetic Fe_3O_4

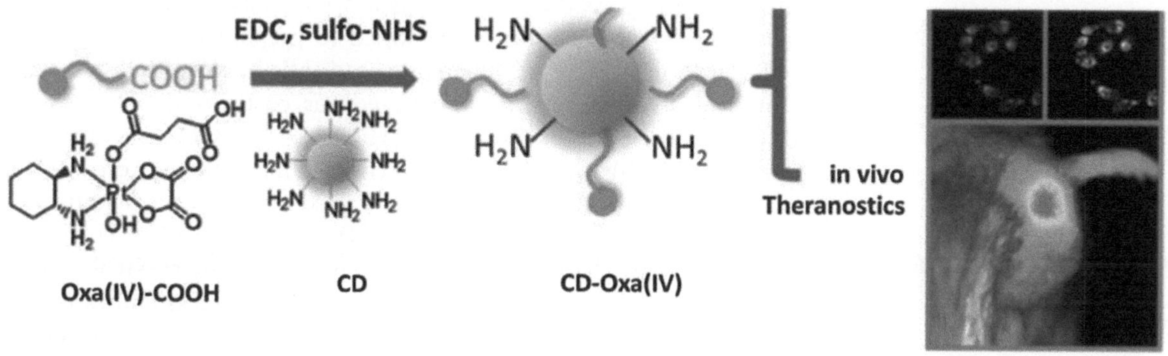

FIGURE 4.1
Synthetic scheme for CD-Oxa and its applications in bioimaging and theranostics (Zheng et al. 2014a,b,c). *Copyright Wiley-VCH Verlag GmbH & Co. KGaA. Reproduced with permission.*

nanocrystals and CDs in porous carbon (C) via a one-pot solvothermal method by simply increasing the H_2O_2 concentration. Their mesoporous carbon shell and hydrophilic surface functional groups endowed the hybrid NPs with high loading capacity. Moreover, Fe_3O_4@C-CDs hybrid NPs can absorb and convert NIR light into heat due to the presence of CDs and therefore can provide NIR-controlled drug release and combined photothermal treatment with high therapeutic efficacy.

Sun et al. (2006) obtained 5-nm nonfluorescent CDs via laser ablation of a carbon target and then modified them with polyethylene glycol (PEG) to get fluorescent CDs with fluorescence QY of 4–10%. The PL of these CDs was broad and strongly dependent on the excitation wavelength, which could be attributed to the different sizes in the sample and different emission sites on the passivated particle surfaces. After fractionation with gel column chromatography, most of the fluorescent fractions could achieve emission yields close to 60% (Wang et al., 2010a,b,c). Interestingly, their optical properties resemble band gap transitions, which are found in nanoscale semiconductors, suggesting that CDs have essentially semiconductor-like characteristics. Bhunia et al. (2013) prepared hydrophobic and hydrophilic CDs with tuneable size and visible emissions by dehydrating carbohydrate in octadecene in the presence of octadecylamine, or in concentrated sulfuric acid. Their method produced gram-scale fluorescent CDs with a QY of 6–30%. Zhang et al. (2013c) also reported a rapid and high-output hydrothermal approach to prepare polymer-like CDs with QYs as high as 80%.

In addition to solid fluorescent CDs, there are some reports on hollow fluorescent CDs (Fang et al., 2012; Wang et al., 2013a,b,c,d). Fang et al. (2012) simply mixed acetic acid, water, and diphosphorus pentoxide to obtain cross-linked hollow fluorescent CNPs. By reducing the release of heat, they also obtained solid fluorescent NPs. So far, many approaches, such as arc discharge, laser ablation, electrochemical oxidation, combustion/pyrolysis, and hydrothermal and microwave methods, have been developed to prepare solid and hollow fluorescent CDs (Ding et al., 2013). The preparation is inexpensive on a large scale without the need for stringent, intricate, tedious, costly, or inefficient steps (Baker and Baker, 2010). Others researchers (Luo et al., 2013; Ding et al., 2013; Baker and Baker, 2010) also synthesized and characterized the fluorescent CDs for bioimaging applications. Cao et al. (2007) used poly-(propionylethylenimine-*co*-ethylenimine) (with ethyleneimine (EI) fraction ~20%) to modify the CDs and then applied them to label human breast cancer MCF-7 cells. These labeled cells exhibited bright fluorescence in both cell membrane and cytoplasm regions under an excitation of 800-nm laser pulses. The results demonstrate that CDs exhibit strong luminescence with two-photon excitation in the NIR, and moreover, large two-photon absorption cross sections, comparable to those of available high-performance semiconductor QDs

(Cao et al., 2007). The authors further demonstrated the in vivo imaging of fluorescent CDs (Yang et al., 2009a,b). They compared the imaging capability of CDs and ZnS-doped CDs and found that the latter dots emitted more strongly than the former dots both in solution and in mice. The fluorescence from the bladder area was observed, and 3 h after injection, the fluorescence could be detected in the urine, but it completely faded 24 h after injection. They analyzed the biodistribution of CDs and found that the CDs accumulated in the kidney and, to a small extent, in the liver (Yang et al., 2009a,b). This is attributed to the surface PEG, which likely reduces the protein adsorption. Huang et al. (2013) investigated the effects of injection routes on the biodistribution, clearance, and tumor uptake of CDs (Fig. 4.2). They prepared fluorescent CDs through a laser ablation approach and then functionalized CDs with the NIR dye ZW-800 and the isotope ^{64}Cu. They injected the conjugates into mice in three different manners, i.e., intravenous, intramuscular, and subcutaneous injection. They found that the CDs were efficiently and rapidly

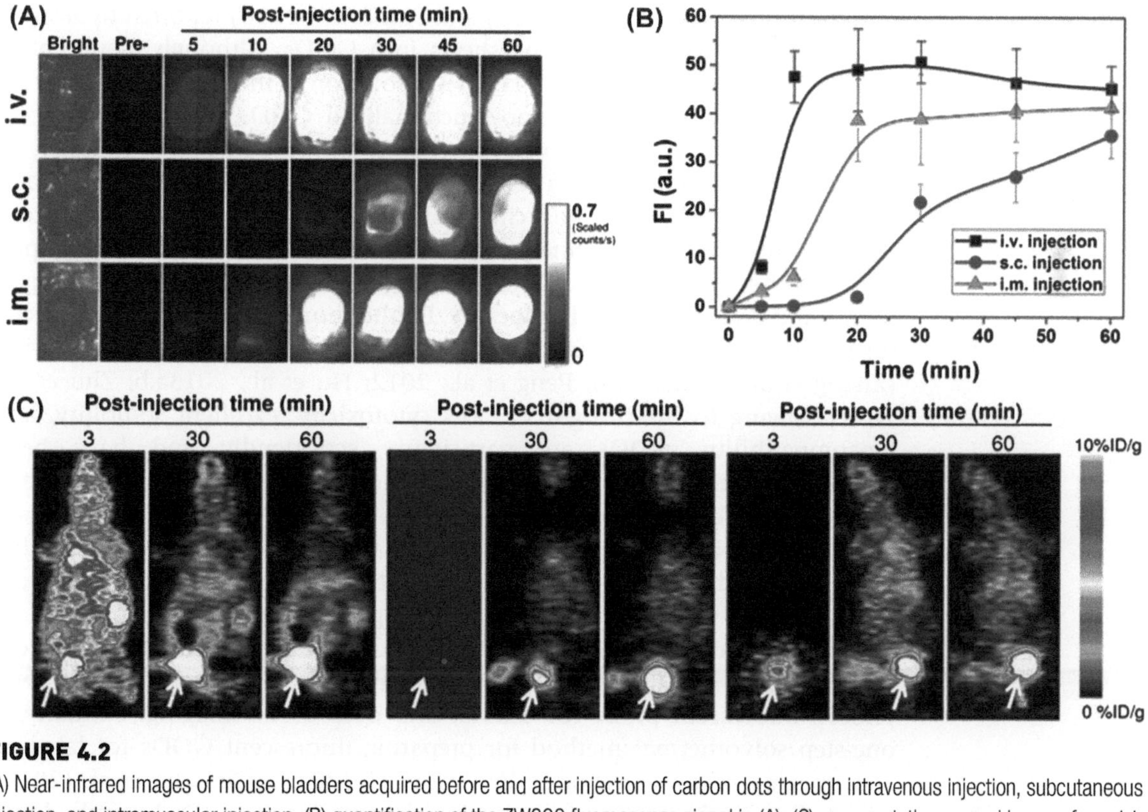

FIGURE 4.2

(A) Near-infrared images of mouse bladders acquired before and after injection of carbon dots through intravenous injection, subcutaneous injection, and intramuscular injection; (B) quantification of the ZW800 fluorescence signal in (A); (C) representative coronal images from 1 h dynamic positron emission tomography (PET) imaging (Huang et al., 2013). *Copyright by the American Chemical Society.*

excreted from body after injection, and the clearance rate of CDs decreased when the administration was varied from intravenous, to intramuscular, and then to subcutaneous injection (Fig. 4.2).

4.2.1.2 Graphene Quantum Dots for Bioimaging

The fluorescent GQDs, the analogs of CDs, have also attracted considerable attention (Shen et al., 2012a,b,c; Li et al., 2013a,b,c). Similar to CDs, GQDs can be prepared by top-down and bottom-up approaches, and their fluorescence can be enhanced via surface modification. GQDs are composed of a single or a few layers of small graphene fragments. The top-down methods usually refer to cutting larger size carbon materials such as CNTs, graphene, or graphene oxide sheets, and carbon fibers into small GQDs, through strong acid oxidation, hydrothermal or solvothermal treatment, or microwave and sonication treatment (Li et al., 2013a,b,c). GQDs can be dissolved in most polar solvents without further chemical modifications and have high stability compared with other fluorescent dyes, so GQDs can be used for bioimaging. The PL of GQDs is affected by the preparation method, size, and the surface modifications. Tetsuka et al. (2012) and Pan et al. (2010) reported hydrothermal routes for cutting graphene sheets into GQDs. Although they applied different vapors (H_2O and $NH_3 \cdot H_2O$ solution), the produced carboxy-functionalized GQDs and amino-functionalized GQDs were all fluorescent blue. Zhang et al. (2012a,b,c,d) prepared yellow-PL GQDs through an electrochemical method. Peng et al. (2012) derived GQDs from carbon fibers during the acid treatment and chemical exfoliation of a traditional process. They produced blue, green, and yellow GQDs by changing the reaction temperature. All the synthesized GQDs showed high solubility, excellent biocompatibility, and favorable optical properties. Furthermore, they can be used directly for intracellular imaging without any surface processing or functionalization (Zhang et al., 2012a,b,c,d; Peng et al., 2012; Hu et al., 2013a,b; Zhu et al., 2011). Owing to their bright PL, low cytotoxicity, excellent solubility, and biocompatibility, GQDs are particularly eco-friendly and have been demonstrated to be excellent probes for bioimaging (Ju et al., 2014; Wang et al., 2014a,b,c,d,e; Chandra et al., 2014; Nurunnabi et al., 2013a,b; Nahain et al., 2013; Nigam et al., 2014). Liu et al. (2013a,b) demonstrated that nitrogen-doped graphene quantum dots (N-GQDs) were facilely prepared via a one-pot solvothermal approach using dimethylformamide (DMF) as solvent and nitrogen source. N-GQDs exhibited a two-photon absorption cross section as high as 48,000 GM and were demonstrated to be an efficient two-photon fluorescent probe for cellular and deep-tissue imaging. Similarly, a one-step solvothermal method for preparing fluorescent GQDs for labeling and imaging cells was reported by Zhu et al. (2011). Chandra et al. (2014) reported a simple method for reducing the toxicity of GQDs by embedding them in a PEG matrix. The enhanced biocompatibility of polymer-modified

GQDs can be used to reduce the generation of reactive oxygen species, as well as for cell imaging. GQDs could be used for loading cancer drugs by simple physical absorption via $\pi-\pi$ stacking (Nigam et al., 2014; Zheng et al., 2013; Wang et al., 2013a,b,c,d; Chowdhury et al., 2015). By exploiting the excellent optical properties of GQDs, many applications have been devised for bioimaging while delivering drugs. Nigam et al. (2014) applied hyaluronic acid and GQD-functionalized human serum albumin NPs for bioimaging and targeted delivery of gemcitabine to pancreatic cancer. Gemcitabine, the most preferred drug for pancreatic cancer treatment, was encapsulated in albumin NPs. Chen et al. (2014a,b,c) designed a biocompatible pH-responsive drug delivery system by grafting GQDs onto the surface of mesoporous silica nanoparticles (MSNs) via acid-cleavable acetal bonds, which could effectively prevent the leakage of drug molecules at neutral pH and release them at acidic pH (Fig. 4.3).

Jing et al. (2011) reported one-step fabrication of multifunctional core–shell-structured capsules using a coaxial electrospray method to achieve the integration of targeted therapy and bioimaging. A TiO_2 shell suppressed the initial burst release of paclitaxel, whereas Fe_3O_4 and GQDs inside the oil core functioned successfully for magnetic targeting and fluorescence imaging, respectively. Moreover, depending on their intrinsic fluorescence properties, GQDs with ultrasmall sizes and visible fluorescence emission have also been developed in recent years for applications in biomedical imaging. Zheng et al. (2014a,b,c) successfully used GQDs to enhance the nuclease activity of

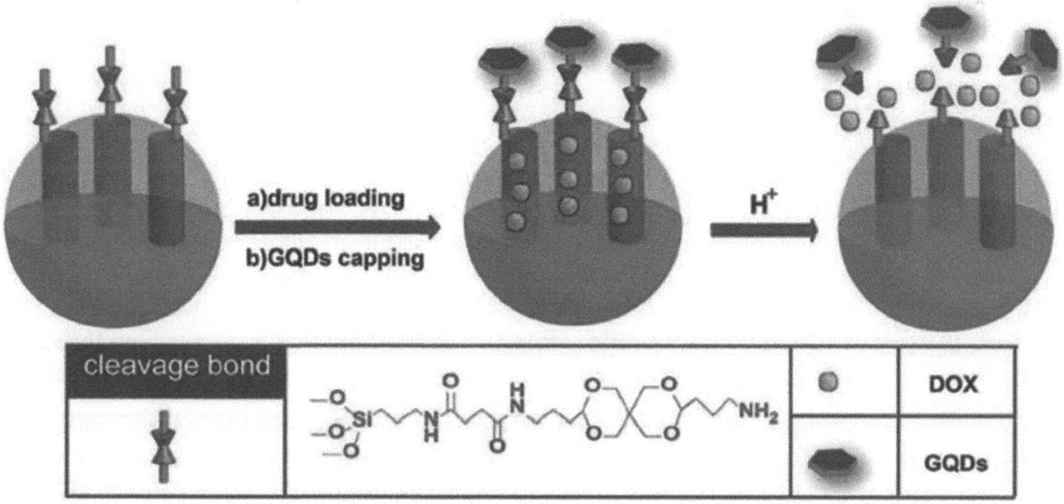

FIGURE 4.3

Schematic representation of pH-triggered release of drug molecules from graphene quantum dot (GQD)-capped mesoporous silica nanoparticles (Chen et al. 2014a,b,c). *DOX*, Doxorubicin. *Copyright by the Royal Society of Chemistry.*

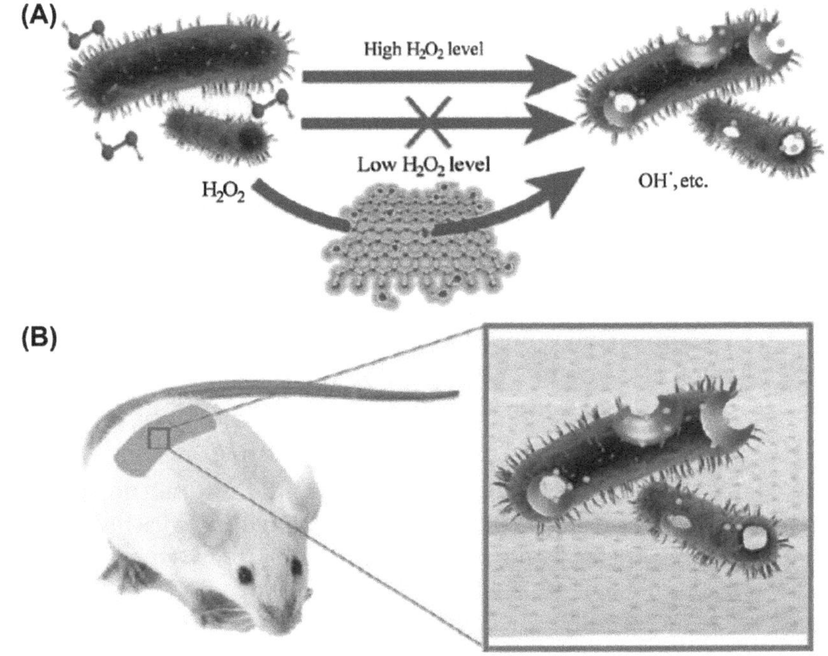

FIGURE 4.4
(A) System designed based on graphene quantum dots (GQDs) and low levels detection of H_2O_2 for antibacterial applications. (B) GQD Band-Aids used in wound disinfection in vivo (Sun et al., 2014). *Copyright by the American Chemical Society.*

copper complexes. Owing to efficient electron transfer from electron-rich GQDs to copper complexes, GQDs promoted the reduction of copper ions and accelerated their reaction with O_2 to form superoxide anions and copper-centered radicals, which then oxidized DNA molecules. Recently, the development of GQDs as antibacterial agents has developed quickly (Ristic et al., 2014). Sun et al. (2014) reported an antibacterial system, which combined GQDs with a low level detection of H_2O_2. They found that the peroxidase-like activity of GQDs originated from their ability to catalyze the decomposition of H_2O_2, generating •OH. As •OH had higher antibacterial activity, the conversion of H_2O_2 into •OH improved the antibacterial performance of H_2O_2, which made it possible to avoid the toxicity of H_2O_2 at high levels in wound disinfection. More importantly, GQDs Band-Aids were prepared that exhibited excellent antibacterial features in vivo when combined with a low concentration of H_2O_2, as illustrated in Fig. 4.4, which indicates that GQDs Band-Aids have promising applications for wound disinfection.

Zhu et al. (2011 and 2012a,b) dispersed graphene oxide in DMF under sonication and then transferred the suspension into Teflon autoclaves and treated them at high temperature for a few hours to get GQDs with a QY of 11%. Tetsuka et al. (2012) used the hydrothermal approach to treat graphene oxide sheets in ammonia solution to get GQDs with a QY between ~19% and 29%. The emission of GQDs can be tuned by controlling the hydrothermal

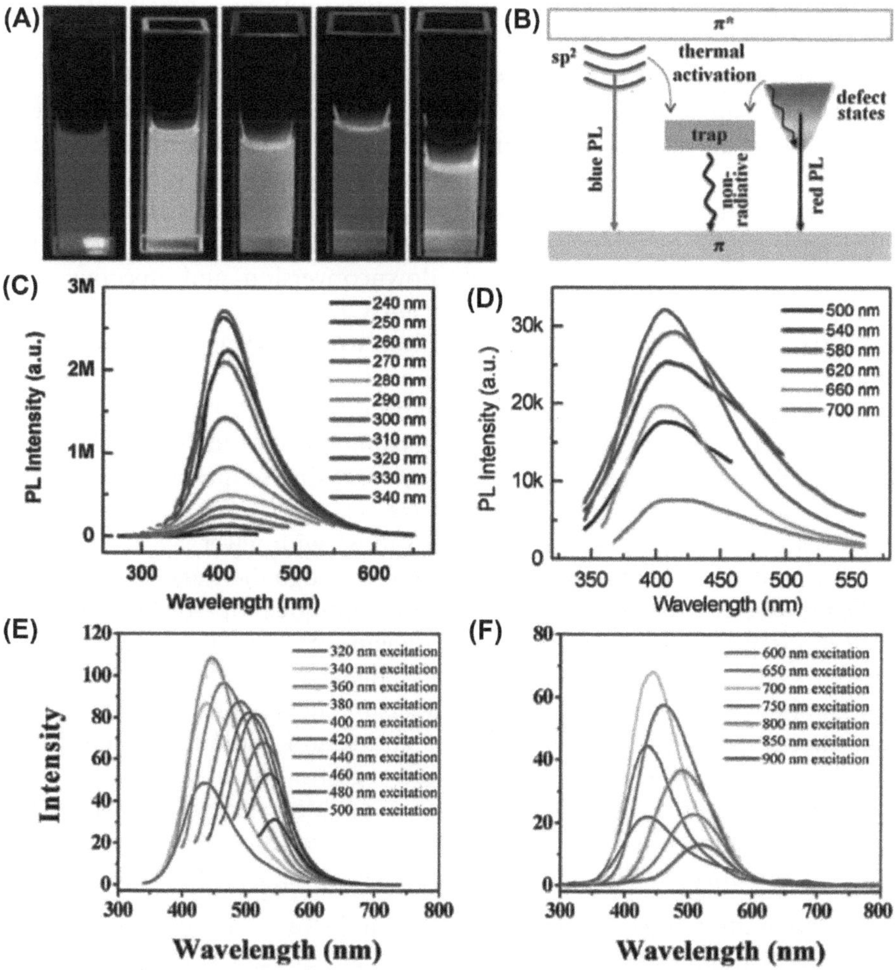

FIGURE 4.5

(A) Image of graphene quantum dot (GQD) solution under UV-light (Tetsuka et al., 2012). (B) Schematic emissions in GQDs (Xu et al., 2013). (C and D) Excitation-independent downconversion and upconversion spectra of GQDs (Zhuo et al., 2012). (E and F) Excitation-dependent downconversion and upconversion spectra of GQDs (Zhu et al. 2012a,b). *PL*, photoluminescence. *(A, E, F) Copyright Wiley-VCH Verlag GmbH & Co. KGaA. Reproduced with permission. (B–D) Copyright by the American Chemical Society.*

temperature (Fig. 4.5A), and the QYs can be further enhanced to ~46% after modification with PEG. Wu et al. (2013) used a one-step pyrolysis of a natural amino acid (i.e., glutamic acid) to prepare fluorescent GQDs with a QY of 54.5%. Dong et al. (2013) used L-cysteine as precursor to prepare S,N-co-doped GQDs with a QY up to 73%.

However, the preparation process significantly influences the optical properties of GQDs. There are two types of emissions in GQDs, i.e., intrinsic state emission and defect state emissions (Fig. 4.5B; Xu et al., 2013). The competition between these two states could be changed during preparation or post surface modification. Zhuo et al. (2012) oxidized graphene in concentrated H_2SO_4 and HNO_3 and then sonicated the mixture and calcinated it at 350°C to remove acid. The resultant fluorescent GQDs did not exhibit excitation-dependent fluorescence (Fig. 4.5C and D; Zhuo et al., 2012). Zhu et al. (2012a,b) prepared green fluorescent GQDs through the hydrothermal approach. The green fluorescence was changed into blue after the GQDs were modified with alkylamines or reduced with $NaBH_4$ (referred to as *m*-GQDs and *r*-GQDs, respectively), while the particle size was similar. The fluorescence shift was attributed to the suppression of nonradiative processes and to the enhanced integrity of the π-conjugated system. These three types of GQDs exhibited strong excitation-dependent downconversion and upconversion emissions, demonstrated by that of the *m*-GQDs in Fig. 4.5E and F (Zhu et al. 2012a,b), which is in contrast to Zhuo's report (Fig. 4.5C and D). In addition to the preparation, the optical properties of GQDs are also influenced by the solution pH, solvent, and concentration (Li et al., 2013a,b,c). Xu et al. studied the fluorescence of GQDs on a substrate at the single particle level. All the GQDs investigated had the same spectral line shapes and peak positions, despite notable differences in particle size and the number of layers. GQDs with more layers were brighter than those with fewer layers, but were associated with shorter fluorescence lifetimes. Although there are some debates on the fluorescence mechanisms of GQDs, their unique properties afford many applications in cellular and deep-tissue imaging. Sun et al. (2008) demonstrated the first bioapplication of nanographene oxide (NGO), i.e., single-layer graphene oxide sheets a few nanometers in lateral width. The PEGylated NGO sheets used were soluble in buffers and serum without agglomeration and showed PL in the visible and infrared regions. These NGO sheets had low background PL in the NIR window. In addition, simple physisorption through π-stacking was used to load the anticancer drug doxorubicin onto NGO functionalized with antibody for selective killing of cancer cells in vitro.

Compared with fluorescent CDs, GQDs can be used for two-photon or multiphoton luminescence imaging (Qian et al., 2012; Liu et al. 2013a,b). Qian et al. (2012) used PEGylated graphene oxide NPs to label HeLa cells and observed that graphene oxide NPs were mainly localized in the mitochondria, endoplasmic reticulum, Golgi apparatus, and lysosomes of HeLa cells with a two-photon scanning microscope. They intravenously injected graphene oxide NPs into mouse bodies from the tail vein and observed their flow, distribution, and clearance in the blood vessels, utilizing a deep-penetrating two-photon imaging technique. These NPs were also injected into the brains of gene transfected mice, and the in vivo two-photon luminescence imaging results showed that graphene oxide NPs were located at 300 μm depth in the brain, demonstrating the advantage of QGDs for deep imaging in tissues. Liu et al. (2013a,b) prepared nitrogen-doped GQDs as efficient two-photon fluorescent probes. These N-GQDs exhibited the highest two-photon absorption cross section (up to 48,000 Göppert-Mayer units) among the

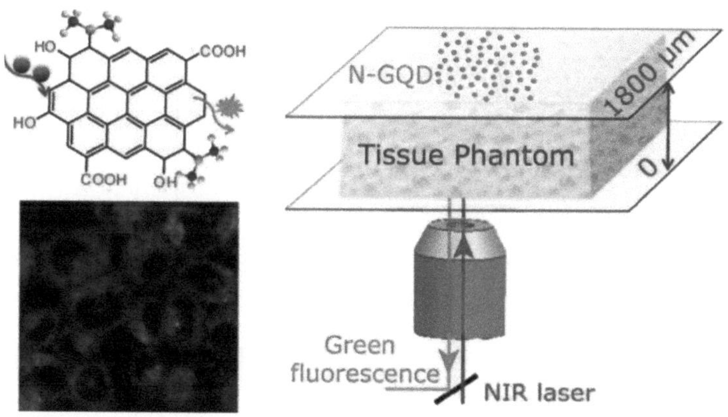

FIGURE 4.6
Nitrogen-doped graphene quantum dots (GQDs) for cellular and deep tissue imaging (Liu et al. 2013a,b). *N-GQD*, nitrogen-doped graphene quantum dot; *NIR*, near-infrared. *Copyright by the American Chemical Society.*

carbon-based materials. They also demonstrated a large imaging depth of 1800 μm by a study of penetration depth in tissue phantom (Fig. 4.6). Table 4.1 lists several works reporting synthetic methods and bioimaging applications of the GQD.

4.2.1.3 Fullerene, Carbon Nanohorns, and Carbon Nanoonions for Bioimaging

There have also been applications of fullerene, CNHs, and CNOs for bioimaging. Liu et al. (2013a,b) fabricated a composite nanofibrous material, which consisted of fullerene NPs and poly(L-lactide), by a simple electrospinning method, which was successfully used as substrates for bioimaging in vitro. Li et al. (2012a,b) synthesized a nanocomposite, which consisted of amino-modified $NaYF_4$:Yb,Er upconversion luminescent NPs and single-walled CNHs, via covalent linkage for the first time. This nanocomposite covalently coupled with rabbit anti-CEA8 antibody was successfully used as a cell-labeling agent for the immunolabeling and imaging of HeLa cells. Sonkar et al. (2012) used water-soluble CNOs as highly fluorescent bioimaging agents for the common food web of two model organisms, unicellular *Escherichia coli* and multicellular *Caenorhabditis elegans*, with no toxic effects.

4.2.2 Bioimaging Probe Using Carbon Nanotubes

CNTs have been widely studied as in vitro and in vivo imaging agents (Liu et al., 2014a,b,c,d; Gong et al., 2013). Ag et al. (2013) demonstrated a simple cell-specific imaging and diagnosis by combining folic acid (FA)-treated carboxyl-functionalized SWCNTs with a copolymer of poly(para-phenylene) (PPP) containing polystyrene (PSt) and poly(e-caprolactone) (PCL) side chains (PPP-g-PSt-PCL) via $\pi-\pi$ stacking reactions. The conjugates bound to FA can specifically target HeLa cells and possess great potential for targeting and

Table 4.1 Lists Several Works Reporting Synthetic Methods and Bioimaging Applications of the Graphene Quantum Dots (Song et al., 2014)

References	Starting Materials	Synthetic Methods	Size (nm)	Quantum Yield	Photoluminescent Color	Cell	Organelle
Zhu et al. (2011)	Graphene oxide and dimethylformamide (DMF)	Solvothermal method	~5.3	11.4%	Green	MC3T3	Cytoplasm
Peng et al. (2012)	Carbon fibers	Acidic exfoliation	1–4		Green	T47D	Cytoplasm
Pan et al. (2012)	Graphene oxide and ammonia	Hydrothermal cutting	3	7.5%	Green	HeLa	Cytoplasm
Zhang et al. (2012a,b,c,d)	Graphite rods and hydrazine	Electrochemical and reduction	5–10	14%	Yellow	PPCs, CPCs	Cytoplasm
Zhang et al. (2012a,b,c,d)	Triiodotriptycene	Organic synthesis	40–50	12.8%	Green	A2780, RAW264.7	Cytoplasm
Hu et al. (2013a,b)	Graphene oxide and ammonia	Hydrothermal treatment	2–6	24.6%	Green	HeLa	Cytoplasm
Dong et al. (2012)	CX-72 carbon black	Chemical oxidation	~15	4.04%	Green	MCF-7	Cytoplasm, nucleus
Liu et al. (2013a,b)	Graphene oxide and DMF	Solvothermal method	~3	31%	Green	HeLa	Membrane, cytoplasm
Zhou et al. (2013)	Polycyclic aromatic hydrocarbon	Carbonization, hydrothermal reduction	5–10	11.7%	Green	MCF-7	Cytoplasm
Sun et al. (2013)	Graphite powder	Oxidizing and etching	2–4	1%	Green/Blue	A549	Cytoplasm
Zhang et al. (2013a,b,c)	Ultrasmall graphite powder	One-pot hydrothermal	1–20	1.1–3.2%	Green/yellow/red	A549	Cytoplasm

imaging studies. El-Sayed et al. (2013) introduced a novel method for in vivo imaging of the biodistribution of recombinant thermostable *Luciola cruciata* luciferase (LcL) labeled with SWCNTs. They demonstrated for the first time that LcL chemically bound to SWCNTs was a powerful tool for in vivo imaging applications of CNTs. Moreover, they also showed that loading CNTs with drugs did not inhibit the chemiluminescence of LcL. Khandare et al. (2012) have synthesized a highly versatile multifunctional nanosystem by covalently assembling Fe_3O_4 NPs, PEG, and fluorescein isothiocyanate (FITC) dye on CNTs for fluorescence bioimaging both in vitro and in vivo. In vitro time kinetic experiments using confocal microscopy demonstrated a higher uptake of the Fe_3O_4-PEG-FITC-CNT nanosystem localized in the perinuclear region of MCF7 cells compared to free FITC. In addition, the CNT nanosystem demonstrated no evidence of toxicity toward cell growth. In vitro surface conjugation of multiple components enhanced cellular uptake for FITC and site-specific targeting ability with no toxicity.

Furthermore, Raman imaging is one of the most promising and powerful bioimaging methods using CNTs. NIR excitation for Raman imaging can minimize auto-fluorescence of biological specimens and/or photobleaching of CNTs. A versatile immunoassay using biotinylated SWCNTs as a Raman label was reported by Bajaj et al. (2014). They used avidin–biotin to link targeting ligands to the label and confocal Raman microscopy to image whole cells. Liu et al. (2014a,b,c,d) reported a flexible nanoplatform based on electrostatic assembly of conjugated polyelectrolytes (CPEs) and carboxylated multiwalled carbon nanotubes. The obtained nanocomposites inherited the intrinsic optical properties of CPEs and the characteristic Raman vibration modes of MWCNTs, providing a fluorescence-Raman dual-imaging method for intracellular tracking and location of MWCNTs in both in vitro and in vivo applications.

CNTs are also widely used for magnetic resonance imaging (MRI). Chen et al. (2010) developed a simple and novel layer-by-layer assembly in combination with a covalent connection strategy for the synthesis of multifunctional magnetic fluorescent nanohybrids based on CNTs and used as multimodal cellular imaging agents for detecting human embryonic kidney 293T cells via MRI and confocal fluorescence imaging with higher intracellular labeling efficiency due to the ability of CNTs to penetrate into cells.

CNTs have great potential for effective drug delivery. The large surface area, availability of multiple functional groups, and hydrophobic nature of CNTs are advantages for loading and delivering drugs, proteins, DNA, and siRNA, including fluorophores, efficiently (Minati et al., 2012; Datir et al., 2012; Wang et al., 2012; Wong et al., 2013; Liu et al. (2013a,b); Siu et al., 2014). Many biological applications of CNTs have been designed that combined drug delivery and bioimaging. Moreover, by modification, CNTs offer more advantages such as higher drug-loading capacity, high internalization into cells, selective targeting and imaging. There are many biological applications of CNTs for bioimaging combined with drug delivery. Anderson et al. (2014) have reported PEG-modified SWCNTs as nanocarriers for delivery of siRNA into pancreatic cancer cells. Positively charged SWCNTs were complexed with siRNAs for targeting the mutant K-Ras gene in PANC-1 cells by electrostatic interaction, thereby promoting gene therapy. High siRNA transfection efficiency mediated by the nanoplex formulation was observed via fluorescence imaging and quantitative flow cytometric analysis.

CNTs also offer excellent photo-to-acoustic conversion efficiency and photo—thermal—acoustic response, which makes them one of the most promising contrast agents for photoacoustic imaging of tumors. Wang et al. (2014a,b,c,d,e) reported that the RGD-conjugated silica-coated gold nanorods on the surface of MWCNTs were successfully used for targeted photoacoustic imaging of gastric cancer cells in vivo. 'Arg-Gly-Asp' (RGD)-conjugated silica-coated gold nanorods/MWCNT probes with good water solubility and low cellular toxicity could target gastric cancer cells in vivo, as well as providing strong photoacoustic imaging in a nude animal model. Zerda et al. (2012a,b) presented a family of novel photoacoustic contrast agents based on binding of small optical dyes to SWCNTs (SWCNT dye). They found that SWCNTs that were coated with either QSY_{21} (SWCNT-QSY) or indocyanine green (SWCNT-ICG) exhibited over 100-fold higher photoacoustic contrast in living animals with high sensitivity compared to plain SWCNTs. Then, conjugated SWCNT dye with cyclic Arg-Gly-Asp peptides to molecularly target $\alpha_v\beta_3$ integrin that is associated with tumor angiogenesis. Intravenous administration of these tumor-targeted imaging agents to tumor-bearing mice provided a significantly stronger photoacoustic signal in the tumor than in mice injected with an untargeted contrast agent. They were also able to spectrally separate the photoacoustic signals of SWCNT-QSY and SWCNT-ICG in living animals that were injected subcutaneously with both particles in the same location, opening the possibility of in vivo multiplexing studies.

4.2.3 Bioimaging Probe Using Graphene/Graphene Oxide

4.2.3.1 Graphene for Bioimaging

Owing to their intrinsic physical, especially optical properties, graphene and its derivatives can be applied in bioimaging. The excellent optical properties of graphene and its derivatives such as visible and NIR PL, characteristic Raman bands, and photoacoustic and photothermal responses (Yang et al., 2013) make them attractive for bioimaging in live cells. Much effort has been devoted to exploiting graphene and its derivatives as fluorescent probes for intracellular imaging in vitro and in vivo (Zhu et al., 2011; Nurunnabi et al., 2013a,b, 2014; Shi et al., 2014; Zhang et al. 2013a,b,c; Chen et al., 2011a,b). Li et al.

(2013a,b,c) developed a method to detect and image DNA in vivo via a GO-Ru hybrid. GO was employed to effectively quench the fluorescence of $Ru(phen)_3Cl_2$ via $\pi-\pi$ interaction and electrostatic interaction, as well as loading and delivering $Ru(phen)_3Cl_2$ into the nuclei of living cells to detect and image DNA, whereas $Ru(phen)_3Cl_2$ alone cannot enter into cells. Liu et al. (2014a,b,c,d) successfully employed GO to deliver propidium iodide for imaging of live cells. Hu et al. (2012) reported that semiconductor QDs with strong fluorescence could be bound to water-soluble polypeptide-modified fluorescence-Raman dual-imaging method to obtain QDs bound to reduced graphene oxide (QD-rGO) nanocomposites which largely retained the fluorescence of QDs because of a suitable nanosized spacer separating QDs and reduced graphene oxide (rGO) and were useful in imaging cells. Wang et al. (2011a,b,c) reported that gold nanoclusters exhibiting NIR PL were anchored on rGO and used for imaging cells. Sheng et al. (2013) reported a portable method for preparing fluorescent nanocomposites incorporating water soluble GO sheets and Zn-doped $AgInS_2$ NPs. Pegylated AIZS-GO nanocomposites could be easily uptaken by NIH/3T3 cells (mouse embryonic fibroblast cell line), whereas no distinct cytotoxicity was observed. Moreover, these as-prepared AIZS-GO-PEG nanocomposites could be used for in vitro cellular imaging of NIH/3T3 cells. Graphene and its derivatives can also be used for MRI. For instance, by growing iron oxide nanoparticles (IONP) on the surface of GO, a large number of groups have successfully fabricated superparamagnetic GO-IONP nanocomposites, which could be employed as anticancer drug carriers as well as contrast agents in MRI (Chen et al., 2011a,b; Yang et al., 2011).

Its large surface area, two-dimensional π-stacked structure, and CNT-like surface chemistry allow graphene and its derivatives to chemically conjugate or physically adsorb a wide range of cargos such as anticancer drugs (Justin and Chen, 2014; Kavitha et al., 2013; Cheng et al., 2013; Ni et al., 2013; Hu et al., 2014; Miao et al., 2013; Shim et al., 2014; Chen et al., 2014a,b,c; Wang et al., 2014a,b,c,d,e; Weaver et al., 2014). By exploiting the excellent optical properties of graphene and its derivatives, many applications have been designed for bioimaging while delivering drugs. Ma et al. (2013) have successfully fabricated gold NPs wrapped in graphene oxide (Au@NGO) by a one-step process. Surface-enhanced Raman scattering from Au@NGO NPs was utilized for intracellular Raman imaging in HeLa cells. Au@NGO NPs could also serve as carriers for delivery of anticancer drugs with sustained intracellular drug release in cancer treatment. Zhang et al. (2011) reported a real-time method for monitoring drug loading and release on GO in a cuvette using rhodamine B (RB) as a model drug, where RB can be loaded on GO with a capacity of 0.5 mg mg^{-1}. The release of RB was pH-sensitive, showing that higher pH values led to weaker hydrophobic forces and hydrogen bonding interactions and therefore a higher release rate. Some biomolecules can also be transported by conjugating them with graphene and its derivatives, and this process is based on fluorescence bioimaging. Zhang et al. (2014) developed a novel multifunctional aptamer-QD-GO nanocomposite via facile decoration of aptamer-labeled CdSe@ZnS QDs on GO nanosheets (Fig. 4.7). The formation of such nanocomposites was based on $\pi-\pi$ stacking between DNA bases on the surfaces of QDs and GO. QDs decorated on the surfaces of GO could serve as fluorescent labeling probes for tracking intracellular transport, whereas GO combined with the aptamer conjugated on the outside of the nanocomposites facilitated targeted drug delivery with enhanced loading capability. Apart from drug delivery, graphene and its derivatives also show great potential in therapy. Owing to the low toxicity, low production costs, and strong optical absorbance in the NIR region of graphene and its derivatives, many applications that combined therapy and bioimaging have been investigated (Tian et al., 2011; Robinson et al., 2011; Li et al., 2014; Akhavan et al., 2012a,b; Zheng et al., 2014a,b,c; Gollavelli and Ling, 2014). Hu et al. (2012) reported a novel strongly fluorescent nontoxic nanocomposite of semiconductor QD-rGO, which served as an imaging agent in the visible light region and a photothermal cancer therapy agent in the NIR region. Remarkably, the heat generated from QD-rGO simultaneously caused an increase in

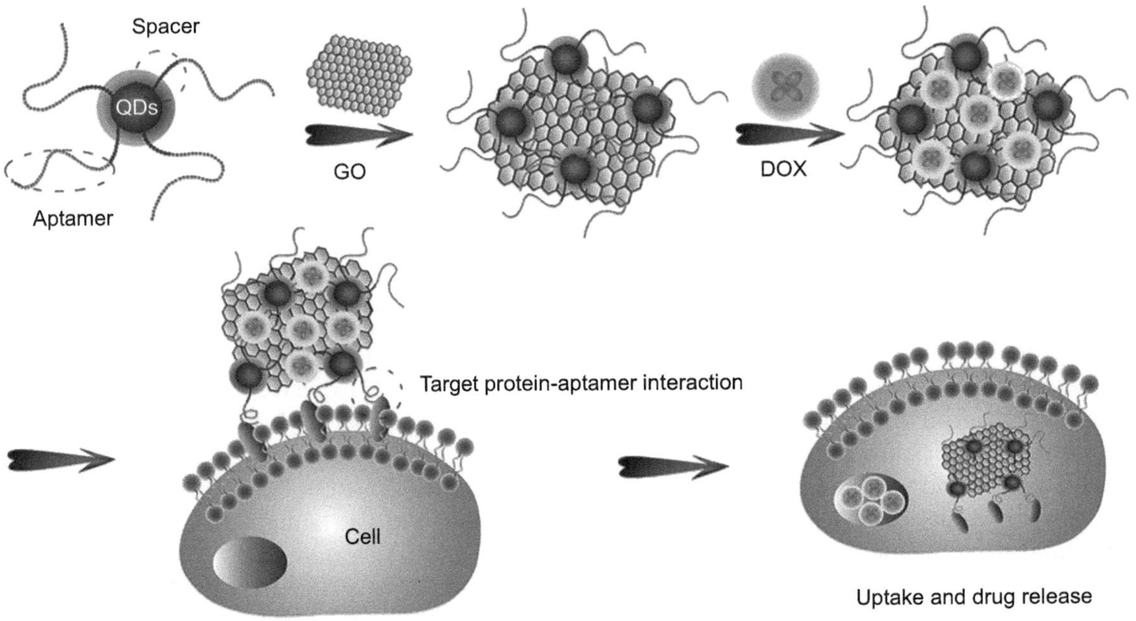

FIGURE 4.7

Schematic presentations of bioimaging and cell-targeted drug delivery using aptamer—QD—GO nanocomposites (Zhang et al., 2014). *GO*, graphene oxide; *QD*, quantum dot. *Copyright by the Royal Society of Chemistry.*

temperature and a marked decrease in the brightness of the QDs, which provided a method for in situ heat/temperature sensing, as well as photothermal therapy. Yang et al. (2010) prepared nanographene sheets (NGS) with a PEG coating by a fluorescent labeling method. Pegylated NGS exhibited highly efficient passive targeting of tumors and relatively low retention in reticuloendothelial systems. Moreover, ultraefficient ablation of tumors was achieved after intravenous administration of NGS and low-power NIR laser irradiation of the tumor, owing to the strong optical absorbance of NGS in the NIR region.

4.2.3.2 Graphene Oxide for Bioimaging

Owing to the excellent biocompatibility, ready cellular uptake, flexible chemical modifications, and unique optical properties, GO has been explored for biological imaging. The protection of DNA from enzymatic digestion by the self-assembly of ssDNA on GO sheets was reported by Tang et al. (2010) (Fig. 4.8). No detectable hydrolysis by deoxyribonuclease (DNase) was observed after ssDNA adsorbed onto the GO surface compared to free ssDNA. The ssDNA was firmly adsorbed on a graphene surface and effectively protected from enzymatic cleavage due to a steric hindrance effect that prevented DNase I from binding to the DNA to initiate enzymatic digestion. This protection effect opens an opportunity for the delivery of DNA/aptamer probes for in vivo

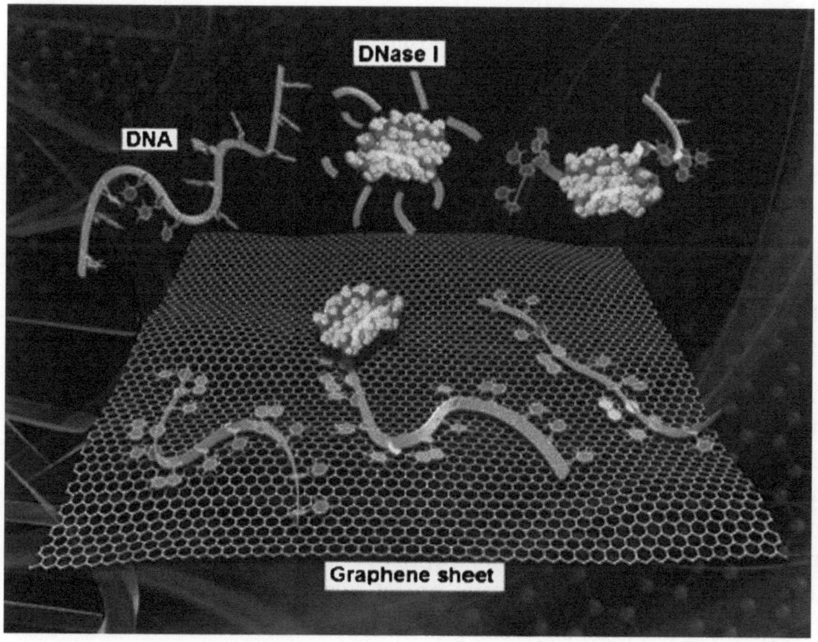

FIGURE 4.8
Schematic illustration of the constraint of DNA molecules on functionalized graphene and its effects. Single-stranded DNA can be effectively constrained on the surface of graphene through adsorption. The enzymes DNase I can digest free DNA but not graphene-bound DNA (Tang et al., 2010). *Copyright Wiley-VCH Verlag GmbH & Co. KGaA. Reproduced with permission.*

imaging of biomolecules in living cells. An aptamer-carboxyfluorescein (FAM)/ graphene oxide nanosheet (GO-nS) nanocomplex was fabricated for the imaging of adenosine-50-triphosphate (ATP) and guanosine-50-triphosphate (GTP) in living cells (Wang et al., 2013a,b,c,d). The aptamer is protected from enzymatic digestion; and the fluorescence of FAM is quenched by absorbing onto the GO surface. After uptake by JB6 cells (Wang et al., 2010a,b,c) or a human breast cancer cell MCF-7 (Wang et al., 2013a,b,c,d), the recognition between the aptamer and the intracellular ATP or GTP induced the conformation change of the aptamer, leading to the restoration of fluorescence of FAM.

Researchers also introduced fluorescent molecules to GO and utilized the functionalized graphene as an in vitro and in vivo imaging probe. It is a challenge to prepare fluorescent graphene because graphene efficiently quenches the fluorescence of fluorophores. Yang et al. (2010) conjugated NGS with Cy7 via a bridge of polyethylene glycol. The in vivo fluorescence imaging reveals a surprisingly high tumor uptake of NGS in xenograft 4T1 murine breast cancer tumor, KB human epidermoid carcinoma tumor and U87MG human glioblastoma tumor mouse models. By using polyacrylic acid and bovine serum albumin as the bridge, fluorescein o-methacrylate (Gollavelli et al., 2012), and CdTe QDs (Chen et al., 2011a,b) were linked to graphene and the complexes showed strong fluorescence in vitro.

Besides fluorescence imaging, MRI is also a powerful in vivo and noninvasive imaging technique that has been widely used in clinical practice (Pohlmann et al., 2013). Superparamagnetic Fe_3O_4 NPs are always utilized as a T2 contrast agent in MRI because they can induce local inhomogeneity of the magnetic field and, therefore, cause a decrease in regional signal intensity because of more rapid dephasing of the precessing spins (shorter T2 relaxation time). To enhance biocompatibility and the cellular MRI, aminodextran-coated Fe_3O_4 NPs were immobilized onto GO (Chen et al., 2011a,b). The composites possessed good physiological stability, low cytotoxicity, and could be efficiently internalized by Hela cells. Compared with isolated Fe_3O_4 NPs, the Fe_3O_4-GO composites exhibited significantly enhanced cellular MRI signal.

PET imaging is a nuclear medical imaging which has the capability to quantitatively measure radioisotope concentrations in vivo with excellent tissue penetration (Zhang et al. 2012a,b,c,d). Hong et al. (2012) conjugated GO with CD105 (a marker for tumor angiogenesis) antibody TRC105 and labeled it with 64Cu for in vivo targeting and PET imaging in a breast cancer mouse model. Serial in vivo PET imaging demonstrated the radiolabeled GO conjugates accumulated in 4T1 murine breast tumor quickly, which demonstrated that TRC105-conjugated GO can be used as the in vivo imaging agent for CD105 targeting (Hong et al., 2012).

4.2.4 Carbon Nanomaterials for Other Optical Imaging

4.2.4.1 Carbon Nanomaterials for Raman Imaging

Raman spectroscopy is a very important and noninvasive tool to analyze vibration in molecules. Raman imaging of SWNTs in live cells was first reported by Heller et al. (2005). DNA-functionalized SWNTs were incubated with 3T3 fibroblast and myoblast stem cells, and Raman spectroscopic mapping was conducted under 785 nm laser excitation, showing high SWNT Raman signals inside cells. It was found that the photostability of SWNTs was many orders of magnitudes better than that of organic fluorescent dyes and NIR QDs.

Targeted in vivo Raman imaging was conducted by Zavaleta et al. (2008). In their experiment, mice bearing U87MG tumors were intravenously injected with PEGylated SWNTs conjugated with an RGD peptide to target integrin $\alpha_v\beta_3$ upregulated on tumor vasculature and tumor cells. Strong SWNT Raman signals were detected from tumors receiving targeting SWNTs but not plain SWNTs without RGD conjugation. SWNTs with different isotope compositions show shifted G-band peaks (Liu et al. 2001), thus can be used as different colors for Raman imaging under a single laser excitation. Multiplexed three-color imaging of tumor cells and tissues with CNTs was achieved by Liu et al. (2008). Pure ^{12}C, pure ^{13}C, and mixed $^{12}C/^{13}C$ SWNTs with different Raman G-band peaks at positions of 1590, 1528, and 1544 cm^{-1}, respectively, were used as three different Raman "colors" for multiplexed Raman imaging. By varying mixed $^{12}C/^{13}C$ ratios during nanotube synthesis, as many as five SWNT Raman "colors" were obtained in subsequent studies by the same group (Liu et al., 2010).

Compared with fluorescence dyes, Raman spectroscopy of SWNTs are resistant to autofluorescence and photobleaching, but its inherently weak effect has limited its application to some extent. The full width at half maximum of the SWNT G peak is very narrow, allowing for high degrees of multiplicity (Liu et al., 2008; Zavaleta et al., 2008). In the Raman imaging, excitation (785 nm) and scattered photons (892–897 nm) are all in the NIR window which allows excellent tissue absorption (Liu et al., 2011) and low autofluorescence background.

4.2.4.2 Carbon Nanomaterials for Transient Absorption Imaging

Transient absorption microscopy is an emerging imaging tool allowing for label-free imaging of endogenous biomolecules such as melanin and hemoglobin in cells and biological tissues (Fu et al., 2007a,b; Min et al., 2009). In this imaging process, a temporally modulated pump laser is used to perturb the electronic states and the distribution of electrons in these states for a certain molecule or material, while a probe laser is simultaneously used to interrogate the perturbed electronic states by detecting the change in optical absorption of the probe laser as a response to the pump laser. SWCNTs as an example, when a pump laser at 707 nm and a probe laser at 885 nm are used, the pump laser wavelength matches the E_{11} transition of metallic SWCNTs (600–800 nm) while the probe laser wavelength matches the E_{22} transition of semiconducting SWCNTs (850–1100 nm) (Wu et al., 2012), this pump–probe combination results in reduced absorption of the probe field (in-phase modulation) for semiconducting SWCNTs owing to the stimulated emission and enhanced absorption of the probe field (antiphase modulation) for metallic SWCNTs due to the broadening of the metallic E_{11} transition and the off resonance probe wavelength (Jung et al., 2010). The phase of the transient absorption signal in the probe field can thus be used as imaging contrast to form images that discriminate different electronic types of SWCNTs by performing 2D raster scan of a field of view, where semiconducting SWCNTs appear with positive contrast and metallic SWCNTs appear with negative contrast.

With this powerful tool of imaging both semiconducting SWCNTs and metallic SWCNTs with microscopic spatial resolution, Tong et al. (2012) have shown that cellular uptake and intracellular trafficking of DNA-functionalized SWCNTs can be imaged with transient absorption microscopy. When the SWCNTs are administered into the mouse via tail vein injection, the circulation of individual SWCNTs can be monitored by intravital transient absorption microscopy with high temporal resolution. Liver tissue slices from the mouse injected with SWCNTs reveal accumulation of SWCNTs mainly in the Kupffer cells, which are labeled with ED-1 antibody. One important advantage of transient absorption microscopy for liver tissue imaging is the negligible interference from tissue autofluorescence, which is a long-lasting, nontrivial problem for conventional fluorescence imaging of liver tissue with immunohistochemical staining (Krawczynski et al., 1992). The transient absorption microscope images reveal little to no background from autofluorescence, owing to the selection of pump/probe wavelengths that specifically match the transition energy of the SWCNT contrast agents and avoid excitation and stimulated emission of the endogenous, autofluorescent species in the tissue (Tong et al., 2012).

Another imaging process is nonfluorescent molecules and materials, i.e., imaging without photon emission. Chen et al. (2013a,b,c) have applied this imaging technique using nanodiamonds in cells, and the cell internalization dynamics can be probed by using a pseudo-3D transient absorption microscopy forming images at two focal planes simultaneously. Although there has not been any report on using transient absorption microscopy for biomedical imaging based on the fullerene NPs with low fluorescence QYs, or pristine, unoxidized graphene sheets without photoluminescent edge states, several works have already shown the proof of concept that these materials can be imaged by transient absorption microscopy. One possible application along this direction could be in situ imaging of graphene-based tissue reinforcing and scaffolding materials (Li et al., 2013a,b,c; Ramanathan et al. 2008) to reveal the local morphological changes and interactions with the surrounding biochemical environment of these tissue implants without additional dye labeling.

4.2.4.3 Near-Infrared Fluorescence Imaging

Semiconducting SWNTs exhibit band gap PL when excited by the NIR light and have been used in fluorescence imaging of biological samples. Cherukuri et al. (2004) found that macrophage cells could actively take up significant quantities of SWNTs by detecting NIR fluorescence signals (>1100 nm) from nanotubes inside macrophage cells. Subsequently, the first report of in vivo NIR imaging of SWNTs in *Drosophila* larvae was achieved by Leeuw et al. (2007). Welsher et al. (2008) reported the use of SWNTs as NIR fluorescent tags for selective probing and imaging of cells. In this study, they first conjugated Rituxan (anti-CD20 antibody) onto biocompatible SWNTs, and incubated these with two types of cells: Raji B-cell lymphoma ($CD20^{+}$) and CEM T-cell lymphoma ($CD20^{-}$). Strong NIR PL of SWNTs could be detected from the positive Raji cells after labeling with the nanotube–antibody conjugate.

Low QY is a major limitation for SWNT-based fluorescence imaging (Welsher et al., 2004). It has been found that the PL QY is closely related to the length and surface coating of nanotubes. Although PEGylated SWNTs exhibit excellent biocompatibility, their QY was significantly lower than that of surfactant-suspended nanotubes, the latter, however, were not biocompatible and thus not suitable for applications in bioimaging. Welsher et al. (2009) revealed that the surface coating of SWNTs presonicated in sodium cholate could be replaced by PEGylated phospholipid (PL-PEG) to gain biocompatibility with retained high QY. It was reported that the QY of the exchanged SWNTs prepared by this method was more than one order of magnitude enhanced over SWNTs directly suspended in PL-PEG. These exchange-SWNT conjugates were then used as contrast agents for in vitro and in vivo NIR imaging with outstanding performance. Hong et al. (2010) observed the first metal enhanced fluorescence of SWNTs. In their experiments, SWNTs modified by gold showed enhanced fluorescence resulting from radioactive lifetime shortening by >10-fold through resonance coupling of SWNT emission to plasmonic modes in the metal. A latest study by the same laboratory further showed that the Au substrate could be used for NIR fluorescence–enhanced (NIR-FE) cellular imaging by using both SWNTs and organic fluorescent labels (Hong et al., 2011). The noble metal nanostructure-enhanced SWNT PL provides a novel approach for improved sensitivity in SWNT-based fluorescence imaging and detection.

4.3 CONCLUSION

In this chapter we have covered the recent progress in the biomedical imaging applications of carbon nanomaterials, which constitute a new class of materials with exciting chemical, optical, and mechanical properties. Although all made of the same chemical element, these nanomaterials with different allotropic forms of carbon exhibit distinct properties and behaviors depending on how the carbon atoms are bonded to form the larger structures on the nanoscale and on the size of the nanostructures. With a library of well-established surface functionalization and passivation methodologies and to use these nanostructures for a myriad of biomedical image applications, carbon nanomaterials have proven themselves to be highly promising fluorescence bioimaging tools. To realize their more significant bioimaging applications in combined diagnosis and therapy, it is important to appropriately control and tune the size, shape, morphology, and surface modifications of carbon nanomaterials to satisfy the requirements for high biocompatibility, long-term stability, and accurate targeting ability in vivo. Current research suggests that carbon nanomaterials have a promising outlook for various biological applications. The CDs and GQDs are surface-modified fluorescent carbon nanomaterials that have small size, distinctive PL properties, low toxicity, and low cost. These advantages offer them great potential for optical imaging and biomedical applications, as they might gradually replace conventional semiconductor QDs in these aspects. The self-assembly of fluorescent chemodosimeter molecules on a GO surface can also solve these problems by taking advantage of the excellent chemical catalysis and nanocarrier functions of GO.

References

Ag, D., et al., 2013. From invisible structures of SWCNTs toward fluorescent and targeting architectures for cell imaging. Biomacromolecules 14, 3532–3541.

Aharonovich, I., et al., 2011. Diamond photonics. Nat. Photonics 5, 397–405.

Akhavan, O., et al., 2012a. The use of a glucose-reduced graphene oxide suspension for photothermal cancer therapy. J. Mater. Chem. 22, 13773–13781.

Akhavan, O., et al., 2012b. Nontoxic concentrations of PEGylated graphene nanoribbons for selective cancer cell imaging and photothermal therapy. J. Mater. Chem. 22, 20626–20633.

Anderson, T., et al., 2014. Pancreatic cancer gene therapy using an siRNA-functionalized single walled carbon nanotubes (SWNTs) nanoplex. Biomater. Sci. 2, 1244–1253.

Bajaj, P., et al., 2014. A carbon nanotube-based Raman-imaging immunoassay for evaluating tumor targeting ligands. Analyst 139, 3069–3076.

Baker, S.N., Baker, G.A., 2010. Luminescent carbon nanodots: emergent nanolights, Angew. Chem. Int. 49, 6727–6744.

Bhunia, S.K., et al., 2013. Carbon nanoparticle-based fluorescent bioimaging probes. Sci. Rep. 3 (1473), 1–7.

Cao, L., et al., 2007. Carbon dots for multiphoton bioimaging. J. Am. Chem. Soc. 129, 11318–11319.

Cao, L., et al., 2011. Reverse Stern-Volmer behavior for luminescence quenching in carbon nanoparticles. Can. J. Chem. 89, 104–109.

Chandra, A., et al., 2014. Mitigating the cytotoxicity of graphene quantum dots and enhancing their applications in bioimaging and drug delivery. ACS Macro Lett. 3, 1064–1068.

Chen, et al., 2004. In vivo near-infrared fluorescence imaging of integrin alphavbeta3 in brain tumor xenografts. Cancer Res. 64 (21), 8009–8014.

Chen, B., et al., 2010. Carbon nanotube-based magnetic-fluorescent nanohybrids as highly efficient contrast agents for multimodal cellular imaging. J. Mater. Chem. 20, 9895–9902.

Chen, W.H., et al., 2011a. Composites of aminodextran-coated Fe_3O_4 nanoparticles and graphene oxide for cellular magnetic resonance imaging. ACS Appl. Mater. Interfaces 3, 4085–4091.

Chen, M.L., et al., 2011b. Conjugation of quantum dots with graphene for fluorescence imaging of live cells. Analyst 136 (20), 4277–4283.

Chen, B.S., et al., 2013a. Large scale synthesis of photoluminescent carbon nanodots and their application for bioimaging. Nanoscale 5, 1967–1971.

Chen, T., et al., 2013b. Optical imaging of non-fluorescent nanodiamonds in live cells using transient absorption microscopy. Nanoscale 5, 4701–4705.

Chen, et al., 2013c. Fluorescent probe for highly selective and sensitive detection of hydrogen sulfide in living cells and cardiac tissues. Analyst 138, 946–995.

Chen, T., et al., 2014a. Graphene quantum dot-capped mesoporous silica nanoparticles through an acid-cleavable acetal bond for intracellular drug delivery and imaging. J. Mater. Chem. B 2, 4979–4982.

Chen, J.Q., et al., 2014b. Restoring basal planes of graphene oxides for highly efficient loading and delivery of β-lapachone. Biomaterials 35, 4986–4995.

Chen, et al., 2014c. One-step reduction and PEGylation of graphene oxide for photothermally controlled drug delivery. Biomaterials 35, 4986–4995.

Cheng, F.F., et al., 2013. Highly dispersible PEGylated graphene/Au composites as gene delivery vector and potential cancer therapeutic agent. J. Mater. Chem. B 1, 4956–4962.

Cheng, L., et al., 2014. Polycation-*b*-Polyzwitterion copolymer grafted luminescent carbon dots as a multifunctional platform for serum-resistant gene delivery and bioimaging. ACS Appl. Mater. Interfaces 6, 20487–20497.

Chernov, A.I., et al., 2013. Optical properties of graphene nanoribbons encapsulated in single-walled carbon nanotubes. ACS Nano 7, 6346–6353.

Cherukuri, P., et al., 2004. Near-infrared fluorescence microscopy of single-walled carbon nanotubes in phagocytic cells. J. Am. Chem. Soc. 126, 15638–15639.

Chowdhury, S.M., et al., 2015. Graphene nanoribbons as a drug delivery agent for lucanthone mediated therapy of glioblastoma multiforme. J. Nanomed. Nanotechnol. 11, 109–118.

Datir, S.R., et al., 2012. Hyaluronate tethered, "smart" multiwalled carbon nanotubes for tumor-targeted delivery of doxorubicin. Bioconjug. Chem. 23, 2201–2213.

De la Zerda, A., et al., 2012a. A brain tumor molecular imaging strategy using a new triple-modality MRI-photoacoustic-Raman nanoparticle. Nat. Med. 18 (5), 829–834.

De la Zerda, A., et al., 2012b. Family of enhanced photoacoustic imaging agents for high-sensitivity and multiplexing studies in living mice. ACS Nano 6, 4694–4701.

Ding, et al., 2013. Functional surface engineering of C-dots for fluorescent biosensing and in vivo bioimaging. Acc. Chem. Res. 47 (1), 20–30.

Dong, Y., et al., 2012. One-step and high yield simultaneous preparation of single- and multi-layer graphene quantum dots from CX-72 carbon black. J. Mater. Chem. 22, 8764–8766.

Dong, Y., et al., 2013. A graphene quantum dot photodynamic therapy agent with high singlet oxygen generation. Angew. Chem. Int. Ed. 52, 7800–7804.

Dresselhaus, M.S., Dresselhaus, G., 2002. Intercalation compounds of graphite. Adv. Phys. 51, 1–186.

El-Sayed, R., et al., 2013. Thermostable luciferase from *Luciola cruciate* for imaging of carbon nanotubes and carbon nanotubes carrying doxorubicin using in vivo imaging system. Nano Lett. 13, 1393–1398.

Fang, Y.X., et al., 2012. Easy synthesis and imaging applications of cross-linked green fluorescent hollow carbon nanoparticles. ACS Nano 6, 400–409.

Fu, D., et al., 2007a. Two-color, two-photon, and excited-state absorption microscopy. J. Biomed. Opt. 12, 054004.

Fu, C.C., et al., 2007b. Characterization and application of single fluorescent nanodiamonds as cellular biomarkers. Proc. Natl. Acad. Sci. U.S.A 104, 727–732.

Gao, X.H., et al., 2005. In vivo molecular and cellular imaging with quantum dots. Curr. Opin. Biotechnol. 16, 63–72.

Geim, A.K., 2009. Graphene: status and prospects. Science 324, 1530–1534.

Gogoi, N., Chowdhury, D., 2014. Novel carbon dot coated alginate beads with superior stability, swelling and pH responsive drug delivery. J. Mater. Chem. B 2, 4089–4099.

Gollavelli, G., Ling, Y.C., 2014. Magnetic and fluorescent graphene for dual modal imaging and single light induced photothermal and photodynamic therapy of cancer cells. Biomaterials 35, 4499–4507.

Gollavelli, G., et al., 2012. Multi-functional graphene as an in vitro and in vivo imaging probe. Biomaterials 33 (8), 2532–2545.

Gong, H., et al., 2013. Carbon nanotubes for biomedical imaging: the recent advances. Adv. Drug Deliv. Rev. 65, 1951–1963.

Guldi, D.M., et al., 2003. First comparative emission assay of single-wall carbon nanotubes-solutions and dispersions. Chem. Commun. 2003, 1130–1131.

Han, M.Y., et al., 2007. Energy band-gap engineering of graphene nanoribbons. Phys. Rev. Lett. 98, 206805.

Hardman, R., 2006. A toxicologic review of quantum dots: toxicity depends on physicochemical and environmental factors. Environ. Health Perspect. 114, 165–172.

He, L., et al., 2014. Carbon nanodots@zeolitic imidazolate framework-8 nanoparticles for simultaneous pH-responsive drug delivery and fluorescence imaging. CrystEngComm 16, 3259–3263.

Heller, D.A., et al., 2005. Single walled carbon nanotube spectroscopy in live cells: towards long term labels and optical sensors. Adv. Mater. 17, 2793–2799.

Hong, G., et al., 2010. Metal-enhanced fluorescence of carbon nanotubes. J. Am. Chem. Soc. 132, 15920–15923.

Hong, G., et al., 2011. Near infrared fluorescence enhanced molecular imaging of live cells on gold substrates. Angew. Chem. Int. Ed. 50, 1–6.

Hong, H., et al., 2012. Graphene platform for hairpin-DNA-based impedimetric genosensing. ACS Nano 6 (3), 2361–2370.

Hu, S.H., et al., 2012. Quantum-dot-tagged reduced graphene oxide nanocomposites for bright fluorescence bioimaging and photothermal therapy monitored in situ. Adv. Mater. 24, 1748–1754.

Hu, C., et al., 2013a. One-step preparation of nitrogen-doped graphene quantum dots from oxidized debris of graphene oxide. J. Mater. Chem. B 1 (1), 39–42.

Hu, et al., 2013b. Graphene quantum dots–carbon nanotube hybrid arrays for supercapacitors. Nanotechnology 24, 195401.

Hu, L., et al., 2014. Multifunctional carbon dots with high quantum yield for imaging and gene delivery. Carbon 2014 (67), 508–513.

Huang, X.L., et al., 2013. Effect of injection routes on the biodistribution, clearance, and tumor uptake of carbon dots. ACS Nano 7, 5684–5693.

Jaiswal, J.K., Simon, S.M., 2004. Potentials and pitfalls of fluorescent quantum dots for biological imaging. Trends Cell Biol. 14, 497–504.

Jing, Y., et al., 2011. Ultrasound-triggered smart drug release from multifunctional core-shell capsules one-step fabricated by coaxial electrospray method. Langmuir 27, 1175–1180.

Ju, J., et al., 2014. Nitrogen-doped graphene quantum dots-based fluorescent probe for the sensitive turn-on detection of glutathione and its cellular imaging. RSC Adv. 2014 (4), 52583–52589.

Jung, Y., et al., 2010. Fast detection of the metallic state of individual single-walled carbon nanotubes using a transient-absorption optical microscope. Phys. Rev. Lett. 105 (217401), 1–4.

Justin, R., Chen, B.Q., 2014. Strong and conductive chitosan–reduced graphene oxide nanocomposites for transdermal drug delivery. J. Mater. Chem. B 2, 3759–3770.

Kavitha, T., et al., 2013. pH-sensitive nanocargo based on smart polymer functionalized graphene oxide for site-specific drug delivery. Phys. Chem. Chem. Phys. 15, 5176–5185.

Khandare, J.J., et al., 2012. PEG-conjugated highly dispersive multifunctional magnetic multi-walled carbon nanotubes for cellular imaging. Nanoscale 4, 837–844.

Kim, J., et al., 2013. Transfection and intracellular trafficking properties of carbon dot-gold nanoparticle molecular assembly conjugated with PEI-pDNA. Biomaterials 34, 7168–7180.

Krawczynski, K., et al., 1992. Hepatitis C virus antigen in hepatocytes: immunomorphologic detection and identification. Gastroenterology 103, 622–629.

Lacerda, L., et al., 2006. Luminescence of functionalized carbon nanotubes as a tool to monitor bundle formation and dissociation in water: the effect of plasmid-DNA complexation. Adv. Funct. Mater. 16, 1839–1846.

Leeuw, T.K., et al., 2007. Single-walled carbon nanotubes in the intact organism: near-IR imaging and biocompatibility studies in *Drosophila*. Nano Lett. 7, 2650–2654.

Li, X., et al., 2011. Preparation of carbon quantum dots with tunable photoluminescence by rapid laser passivation in ordinary organic solvents. Chem. Commun. 47, 932–934.

Li, H., et al., 2012a. Carbon nanodots: synthesis, properties and applications. J. Mater. Chem. 22, 24230–24253.

Li, J., et al., 2012b. Synthesis of $NaYF_4$:Yb,Er/single-walled carbon nanohorns nanocomposite and its application as cells label. Anal. Biochem. 428, 4–6.

Li, H.J., et al., 2013a. Interaction of $Ru(phen)_3Cl_2$ with graphene oxide and its application for DNA detection both in vitro and in vivo. J. Mater. Chem. B 2013 (1), 4146–4151.

Li, N., et al., 2013b. Three-dimensional graphene foam as a biocompatible and conductive scaffold for neural stem cells. Sci. Rep. 3, 1–6, 1604.

Li, L., et al., 2013c. Focusing on luminescent graphene quantum dots: current status and future perspectives. Nanoscale 5, 4015–4039.

Li, J.L., et al., 2014. Graphene oxide nanoparticles for enhanced photothermal cancer cell therapy under the irradiation of a femtosecond laser beam. J. Biomed. Mater. Res. Part A 102, 2181–2188.

Lin, Y., et al., 2005. Visible luminescence of carbon nanotubes and dependence on functionalization. J. Phys. Chem. B 109, 14779–14782.

Liu, L., et al., 2001. Isotope labeling of carbon nanotubes and formation of ^{12}C-^{13}C nanotube junctions. J. Am. Chem. Soc. 123, 11502–11503.

Liu, H., et al., 2007. Fluorescent carbon nanoparticles derived from candle soot. Angew. Chem. Int. Ed. 46, 6473–6475.

Liu, Z., et al., 2008. Multiplexed multicolor Raman imaging of live cells with isotopically modified single walled carbon nanotubes. J. Am. Chem. Soc. 130, 13540–13541.

Liu, Z., et al., 2010. Multiplexed five-color molecular imaging of cancer cells and tumor tissues with carbon nanotube Raman tags in the near-infrared. Nano Res. 3, 222–233.

Liu, Z., et al., 2011. Single-walled carbon nanotubes in biomedical imaging. J. Mater. Chem. 21, 586–598.

Liu, Q., et al., 2013a. Strong two-photon-induced fluorescence from photostable, biocompatible nitrogen-doped graphene quantum dots for cellular and deep-tissue imaging. Nano Lett. 13, 2436–2441.

Liu, W., et al., 2013b. Electrospinning of poly(L-lactide) nanofibers encapsulated with water-soluble fullerenes for bioimaging application. ACS Appl. Mater. Interfaces 5, 680–685.

Liu, F.Y., et al., 2014a. Interaction of propidium iodide with graphene oxide and its application for live cell staining. Carbon 71, 190–195.

Liu, X., et al., 2014b. Design of covalently functionalized carbon nanotubes filled with metal oxide nanoparticles for imaging, therapy, and magnetic manipulation. ACS Nano 8, 11290–11304.

Liu, Y., et al., 2014c. A fluorescence-Raman dual-imaging platform based on complexes of conjugated polymers and carbon nanotubes. Nanoscale 6, 1480–1489.

Liu, et al., 2014d. A graphene-based biosensing platform based on the release of DNA probes and rolling circle amplification. ACS Nano 8 (6), 5564–5573.

Luo, P.G., et al., 2013. Carbon "quantum" dots for optical bioimaging. J. Mater. Chem. B 1, 2116–2127.

Ma, X., et al., 2013. Graphene oxide wrapped gold nanoparticles for intracellular Raman imaging and drug delivery. J. Mater. Chem. B 1, 6495–6500.

Miao, W., et al., 2013. Cholesteryl hyaluronic acid-coated, reduced graphene oxide nanosheets for anti-cancer drug delivery. Biomaterials 34, 9638–9647.

Min, W., et al., 2009. Imaging chromophores with undetectable fluorescence by stimulated emission microscopy. Nature 461, 1105–1109.

Minati, L., et al., 2012. Multifunctional branched gold–carbon nanotube hybrid for cell imaging and drug delivery. Langmuir 28, 15900–15906.

Nahain, A.A., et al., 2013. Target delivery and cell imaging using hyaluronic acid-functionalized graphene quantum dots. Mol. Pharm. 10, 3736–3744.

Nakada, K., et al., 1996. Edge state in graphene ribbons: nanometer size effect and edge shape dependence. Phys. Rev. B 54, 17954–17961.

Ni, Y.N., et al., 2013. Graphene oxide as a nanocarrier for loading and delivery of medicinal drugs and as a biosensor for detection of serum albumin. Mater. Lett. 125, 82–85.

Nigam, P., et al., 2014. Graphene quantum dots conjugated albumin nanoparticles for targeted drug delivery and imaging of pancreatic cancer. J. Mater. Chem. B 2, 3190–3195.

Nurunnabi, M., et al., 2013a. Surface coating of graphene quantum dots using mussel-inspired polydopamine for biomedical optical imaging. ACS Appl. Mater. Interfaces 5, 8246–8253.

Nurunnabi, M., et al., 2013b. Near infra-red photoluminescent graphene nanoparticles greatly expand their use in noninvasive biomedical imaging. Chem. Commun. 49, 5079–5081.

Nurunnabi, M., et al., 2014. Photoluminescent graphene nanoparticles for cancer phototherapy and imaging. ACS Appl. Mater. Interfaces 6, 12413–12421.

Pan, D., et al., 2010. Hydrothermal route for cutting graphene sheets into blue-luminescent graphene quantum dots. Adv. Mater. 22 (6), 734–738.

Pan, D., et al., 2012. Cutting sp^2 clusters in graphene sheets into colloidal graphene quantum dots with strong green fluorescence. J. Mater. Chem. 22, 3314–3318.

Pandey, S., et al., 2013a. Carbon dots functionalized gold nanorod mediated delivery of doxorubicin: tri-functional nano-worms for drug delivery, photothermal therapy and bioimaging. J. Mater. Chem. B 1, 4972–4982.

Pandey, S., et al., 2013b. Cysteamine hydrochloride protected carbon dots as a vehicle for the efficient release of the anti-schizophrenic drug haloperidol. RSC Adv. 3, 26290–26296.

Peng, J., et al., 2012. Graphene quantum dots derived from carbon fibers. Nano Lett. 12 (2), 844–849.

Pohlmann, A., et al., 2013. Linking non-invasive parametric MRI with invasive physiological measurements (MR-PHYSIOL): towards a hybrid and integrated approach for investigation of acute kidney injury in rats. Acta Physiol. 207 (4), 673–689.

Qian, J., et al., 2012. Graphene oxide nanoparticles for enhanced photothermal cancer cell therapy under the irradiation of a femtosecond laser beam. Angew. Chem. Int. Ed. 51, 10570–10575.

Ramanathan, T., et al., 2008. Functionalized graphene sheets for polymer nanocomposites. Nat. Nanotechnol. 3, 327–331.

Ray, S.C., et al., 2009. Fluorescent carbon nanoparticles: synthesis, characterization, and bioimaging application. J. Phys. Chem. C 113, 18546–18551.

Riggs, J.E., et al., 2000. Synthesis and characterization of water soluble single-walled carbon nanotube graft copolymers. J. Am. Chem. Soc. 122, 5879–5880.

Ristic, B.Z., et al., 2014. Photodynamic antibacterial effect of graphene quantum dots. Biomaterials 35, 4428–4435.

Robinson, J.T., et al., 2011. Ultrasmall reduced graphene oxide with high near-infrared absorbance for photothermal therapy. J. Am. Chem. Soc. 133, 6825–6831.

Shen, R., et al., 2012a. Dramatic fluorescence enhancement of bare carbon dots through facile reduction chemistry. Chem. Phys. Chem. 13, 3549–3555.

Shen, J.H., et al., 2012b. Graphene quantum dots: emergent nanolights for bioimaging, sensors, catalysis and photovoltaic devices. Chem. Commun. 48, 3686–3699.

Shen, et al., 2012c. The production of pH-sensitive photoluminescent carbon nanoparticles by the carbonization of polyethylenimine and their use for bioimaging. Carbon 55, 343–349.

Sheng, Y., et al., 2013. Graphene oxide based fluorescent nanocomposites for cellular imaging. J. Mater. Chem. B 1, 512–521.

Shi, S.X., et al., 2014. Surface engineering of graphene-based nanomaterials for biomedical applications. Bioconjug. Chem. 25, 1609–1619.

Shim, G., et al., 2014. Reduced graphene oxide nanosheets coated with an anti-angiogenic anticancer low-molecular-weight heparin derivative for delivery of anticancer drugs. J. Control. Release 189, 80–89.

Siu, K.S., et al., 2014. Development of the novel PEG-PE-based polymer for the reversible attachment of specific ligands to liposomes: synthesis and in vitro characterization. Bioconjug. Chem. 25, 1744–1751.

Son, Y.W., et al., 2006. Energy gaps in graphene nanoribbons. Phys. Rev. Lett. 97, 216803.

Song, Y., et al., 2014. Bioimaging based on fluorescent carbon dots. RSC Adv. 4, 27184–27200. Sonkar, et al., 2012. Carbon nano-onions as nontoxic and high-fluorescence bioimaging agent in food chain—an in vivo study from unicellular *E. coli* to multicellular *C. elegans*. Mater. Express 2 (2), 105–114.

Sun, Y.-P., et al., 2002. Functionalized carbon nanotubes: properties and applications. Acc. Chem. Res. 35, 1096–1104.

Sun, Y.-P., et al., 2006. Quantum-sized carbon dots for bright and colorful photoluminescence. J. Am. Chem. Soc. 128, 7756–7757.

Sun, X.M., et al., 2008. Nano-graphene oxide for cellular imaging and drug delivery. Nano Res. 1, 203–212.

Sun, Y., et al., 2013. Large scale preparation of graphene quantum dots from graphite with tunable fluorescence properties. Phys. Chem. Chem. Phys. 15, 9907–9913.

Sun, H., et al., 2014. Graphene quantum dots-band-aids used for wound disinfection. ACS Nano 8, 6202–6210.

Tang, Z., et al., 2010. Constraint of DNA on functionalized graphene improves its biostability and specificity. Small 6 (11), 1205–1209.

Tetsuka, H., et al., 2012. Optically tunable amino-functionalized graphene quantum dots. Adv. Mater. 24, 5333–5338.

Tian, B., et al., 2011. Photothermally enhanced photodynamic therapy delivered by nano-graphene oxide. ACS Nano 5, 7000–7009.

Tong, L., et al., 2012. Label-free imaging of semiconducting and metallic carbon nanotubes in cells and mice using transient absorption microscopy. Nat. Nanotechnol. 7, 56–61.

Tong, G.S., et al., 2015. Amorphous carbon dots with high two-photon fluorescence for cellular imaging passivated by hyperbranched poly(amino amine). J. Mater. Chem. B 2015 (3), 700–706.

Wang, Y., et al., 2010a. Aptamer/graphene oxide nanocomplex for in situ molecular probing in living cells. J. Am. Chem. Soc. 132 (27), 9274–9276.

Wang, X., et al., 2010b. Bandgap-like strong fluorescence in functionalized carbon nanoparticles. Angew. Chem. Int. Ed. 49, 5310–5314.

Wang, et al., 2010c. One-step synthesis of highly luminescent carbon dots in noncoordinating solvents. Chem. Mater. 22 (16), 4528–4530.

Wang, C.S., et al., 2011a. Gold nanoclusters and graphene nanocomposites for drug delivery and imaging of cancer cells. Angew. Chem. Int. Ed. 50, 11644–11648.

Wang, Y., et al., 2011b. Carbon dots of different composition and surface functionalization: cytotoxicity issues relevant to fluorescence cell imaging. Exp. Biol. Med. 236, 1231–1238.

Wang, et al., 2011c. Microwave assisted one-step green synthesis of cell-permeable multicolor photoluminescent carbon dots without surface passivation reagents. J. Mater. Chem. 21, 2445–2450.

Wang, Y.X., et al., 2012. Ageing process of some pyrogenic silica samples exposed to controlled relative humidities: Part I: kinetic of water sorption and evolution of the surface silanol density. Colloids Surf. B 97, 62–69.

Wang, P., Slipchenko, M.N., Mitchell, J., Yang, C., Potma, E.O., Xu, X.F., Cheng, J.X., 2013a. Far-field imaging of non-fluorescent species with subdiffraction resolution. Nat. Photonics 7, 450–454.

Wang, Y., et al., 2013b. Situ live cell sensing of multiple nucleotides exploiting DNA/RNA aptamers and graphene oxide nanosheets. Anal. Chem. 85 (14), 6775–6782.

Wang, Q.L., et al., 2013c. Hollow luminescent carbon dots for drug delivery. Carbon 59, 192–199.

Wang, Z., et al., 2013d. Synthesis of strongly green-photoluminescent graphene quantum dots for drug carrier. Colloids Surf. B 112, 192–196.

Wang, X.J., et al., 2014a. Multifunctional graphene quantum dots for simultaneous targeted cellular imaging and drug delivery. Colloids Surf. B 122, 638–644.

Wang, C., et al., 2014b. RGD-conjugated silica-coated gold nanorods on the surface of carbon nanotubes for targeted photoacoustic imaging of gastric cancer. Nanoscale Res. Lett. 9 (264), 1–10.

Wang, H., et al., 2014c. Magnetic iron oxide–fluorescent carbon dots integrated nanoparticles for dual-modal imaging, near-infrared light-responsive drug carrier and photothermal therapy. Biomater. Sci. 2, 915–923.

Wang, J., et al., 2014d. Carbon nanodots featuring efficient FRET for two-photon photodynamic cancer therapy with a low fs laser power density. Biomaterials 35, 9372–9381.

Wang, et al., 2014e. Common origin of green luminescence in carbon nanodots and graphene quantum dots, 8/3, 2541–2547.

Weaver, C.L., et al., 2014. Electrically controlled drug delivery from graphene oxide nanocomposite films. ACS Nano 8, 1834–1843.

Welsher, et al., 2004. Efficient isolation and solubilization of pristine single-walled nanotubes in bile salt micelles. Adv. Funct. Mater. 14 (11), 1105–1112.

Welsher, K., et al., 2008. Selective probing and imaging of cells with single walled carbon nanotubes as near-infrared fluorescent molecules. Nano Lett. 8, 586–590.

Welsher, K., et al., 2009. A route to brightly fluorescent carbon nanotubes for near-infrared imaging in mice. Nat. Nanotechnol. 4, 773–780.

Wong, B.S., et al., 2013. Carbon nanotubes for delivery of small molecule drugs. Adv. Drug Deliv. Rev. 65, 1964–2015.

Wu, J., et al., 2012. Short channel field-effect transistors from highly enriched semiconducting carbon nanotubes. Nano Res. 5, 388–394.

Wu, X., et al., 2013. Fabrication of highly fluorescent graphene quantum dots using L-glutamic acid for in vitro/in vivo imaging and sensing. J. Mater. Chem. C 1, 4676–4684.

Xu, Q., et al., 2013. Single-particle spectroscopic measurements of fluorescent graphene quantum dots. ACS Nano 7, 10654–10661.

Yang, S.T., et al., 2009a. Carbon dots for optical imaging in vivo. J. Am. Chem. Soc. 131, 11308–11309.

Yang, S.-Y., et al., 2009b. Carbon dots as nontoxic and high-performance fluorescence imaging agents. J. Phys. Chem. C 113, 18110–18114.

Yang, K., et al., 2010. Graphene in mice: ultrahigh in vivo tumor uptake and efficient photothermal therapy. Nano Lett. 10 (9), 3318–3323.

Yang, X.Y., et al., 2011. Multi-functionalized graphene oxide based anticancer drug-carrier with dual-targeting function and pH-sensitivity. J. Mater. Chem. 21, 3448–3454.

Yang, K., et al., 2013. Nano-graphene in biomedicine: theranostic applications. Chem. Soc. Rev. 42, 530–547.

Zavaleta, C., et al., 2008. Noninvasive Raman spectroscopy in living mice for evaluation of tumor targeting with carbon nanotubes. Nano Lett. 8, 2800–2805.

Zhang, R.Y., et al., 2011. Real time monitoring of the drug release of rhodamine B on graphene oxide. Carbon 49, 1126–1132.

Zhang, M., et al., 2012a. Facile synthesis of water-soluble, highly fluorescent graphene quantum dots as a robust biological label for stem cells. J. Mater. Chem. 22 (15), 7461–7467.

Zhang, Y., et al., 2012b. Graphene: a versatile nanoplatform for biomedical applications. Nanoscale 4 (13), 3833–3842.

Zhang, C., et al., 2012c. Three-dimensional nanographene based on triptycene: synthesis and its application in fluorescence imaging. Org. Lett. 14, 5912–5915.

Zhang, et al., 2012d. Preparation of graphene quantum dots for bioimaging application. J. Nanosci. Nanotechnol. 12 (3), 2924–2928.

Zhang, M.X., et al., 2013a. Graphene oxide based theranostic platform for T_1-weighted magnetic resonance imaging and drug delivery. ACS Appl. Mater. Interfaces 5, 13325–13332.

Zhang, X., et al., 2013b. Size tunable fluorescent nano-graphite oxides: preparation and cell imaging applications. Phys. Chem. Chem. Phys. 15, 19013–19018.

Zhang, et al., 2013c. Highly photoluminescent carbon dots for multicolor patterning, sensors, and bioimaging. Angew. Chem. Int. Ed. 52, 3953–3957.

Zhang, L., et al., 2014. Aptamer-mediated nanocomposites of semiconductor quantum dots and graphene oxide as well as their applications in intracellular imaging and targeted drug delivery. J. Mater. Chem. B 2, 8558–8565.

Zheng, X.T., et al., 2013. Multifunctional graphene quantum dots-conjugated titanate nanoflowers for fluorescence-trackable targeted drug delivery. RSC Adv. 3, 24853–24857.

Zheng, M., et al., 2014a. Integrating oxaliplatin with highly luminescent carbon dots: an unprecedented theranostic agent for personalized medicine. Adv. Mater. 26, 3554–3560.

Zheng, B., et al., 2014b. Electron transfer from graphene quantum dots to the copper complex enhances its nuclease activity. J. Phys. Chem. C 118, 7637–7642.

Zheng, X.W., et al., 2014c. Design of multifunctional FePt/GO nanocomposites for targeting, dual-modal imaging diagnostic and in situ therapeutic potential theranostic platform. RSC Adv. 4, 58489–58494.

Zhou, J., et al., 2007. An electrochemical avenue to blue luminescent nanocrystals from multiwalled carbon nanotubes (MWCNTs). J. Am. Chem. Soc. 129, 744–745.

Zhou, L., et al., 2013. Graphene quantum dots from polycyclic aromatic hydrocarbon for bioimaging and sensing of Fe^{3+} and hydrogen peroxide. Part. Part. Syst. Charact. 30, 1086–1092.

Zhu, S.J., et al., 2011. Strongly green-photoluminescent graphene quantum dots for bioimaging applications. Chem. Commun. 47, 6858–6860.

Zhu, et al., 2012a. Temperature-dependent dual emission from sucrose-derived carbon nanodots: a ratiometric fluorescent thermometer. Angew. Chem. Int. Ed. 51, 7185–7189.

Zhu, S.J., et al., 2012b. Surface chemistry routes to modulate the photoluminescence of graphene quantum dots: from fluorescence mechanism to up-conversion bioimaging applications. Adv. Funct. Mater. 22, 4732–4740.

Zhu, X.H., et al., 2014. Preparation and characterization of the fluorescent carbon dots derived from the lithium-intercalated graphite used for cell imaging. Part. Part. Syst. Charact. 31, 771–777.

Zhuo, S.J., et al., 2012. Upconversion and downconversion fluorescent graphene quantum dots: ultrasonic preparation and photocatalysis. ACS Nano 6, 1059–1064.

FURTHER READING

De La Zerda, A., et al., 2008. Carbon nanotubes as photoacoustic molecular imaging agents in living mice. Nat. Nanotechnol. 3, 557–562.

Ding, C., et al., 2014. Functional surface engineering of C-dots for fluorescent biosensing and in vivo bioimaging. Acc. Chem. Res. 47 (1), 20–30.

Grancini, G., Polli, D., Fazzi, D., Cabanillas-Gonzalez, J., Cerullo, G., Lanzani, G., 2011. Transient absorption imaging of P3HT: PCBM photovoltaic blend: evidence for interfacial charge transfer state. J. Phys. Chem. Lett. 2, 1099–1105.

Liu, P., 2013. Modification strategies for carbon nanotubes as a drug delivery system. Ind. Eng. Chem. Res. 52, 13517–13527.

Liu, Z., et al., 2006. In vivo biodistribution and highly efficient tumour targeting of carbon nanotubes in mice. Nat. Nanotechnol. 2, 47–52.

Zhou, B., et al., 2006. Polymeric nanocomposite films from functionalized vs suspended single-walled carbon nanotubes. Polymer 47, 5323–5329.

CHAPTER 5

Application of Carbon-Based Nanomaterials as Drug and Gene Delivery Carrier

5.1 INTRODUCTION

The emergence of carbon nanomaterials has enabled new possibilities and approaches to simultaneously detecting and treating a multitude of diseases, the most prominent of all being cancer. While nanomaterials may be applied to treating a wide variety of ailments and in innovated uses such as vaccine vectors and the delivery of therapeutics to the immune system. Cancer is one of the most deadly diseases facing humanity, and current treatment methodologies suffer from many deficits such as the inability to completely destroy all cancerous cells as well as the severe toxic side effects of both chemotherapy and radiotherapy (Barreto et al., 2011; Liu et al., 2014). A great deal of the research into the therapeutic applications of carbon nanomaterials focuses on the targeted delivery of anticancer drugs and the selective destruction of cancerous bodies (Bianco et al., 2008). Carbon nanomaterials have many useful electrical and optical properties, yet their size, which ranges from a few nanometers up to a few hundred nanometers, is perfectly tuned for use in both in vitro and in vivo biological settings as their size is comparable to that of biological macromolecules such as proteins, enzymes, and DNA plasmids (Liu et al., 2011a,b). Furthermore, these materials demonstrate facile transport across the cell membrane, presumably through endocytosis (Kam et al., 2006; Hong et al., 2012). The ability to easily cross the cell membrane allows fullerenes, carbon nanotubes (CNTs), graphene, and nanodiamonds to act as transport vehicles for both small molecule-based drugs and biopharmaceuticals, making many exciting and innovative approaches to both drug delivery and gene therapy possible (Bates et al., 2013). The size of carbon nanomaterials offers additional benefits to morbid regions such as tumors. The enhanced permeation and retention (EPR) effect occurs through abnormally leaky and tortuous blood vasculature in conjunction with the poor lymphatic drainage commonly associated with tumors which allows nanomaterials to selectively accumulate within cancerous tissue at much higher concentrations than in the surrounding healthy tissue (Liu et al., 2004; Perr et al., 2007; Schroeder et al., 2012). Taking advantage of this selective tumor uptake, carbon nanomaterials can be applied to the selective destruction of cancerous bodies

CONTENTS

Carbon Nanomaterials for Biological and Medical Applications. http://dx.doi.org/10.1016/B978-0-323-47906-6.00005-9

through photothermal therapy (PTT) by utilizing their high absorbance of near infrared (NIR) light to ablate the tumor or by activating photosensitizing agents piggybacking on the nanomaterial which produce singlet oxygen or reactive oxygen species that in turn kill the surrounding cancer cells.

5.2 BIODISTRIBUTION OF CARBON NANOMATERIALS

Interactions of engineered nanomaterials with environmental interfaces have become a critical aspect of environmental health and safety evaluations. Carbon nanomaterials have emerged at the forefront of nanoscale research and applications due to its unique properties. Although there are concerns associated with the harmful effects of fullerene toward living organisms, the mechanisms of fullerene toxicity are still under debate. A first step toward assessing these mechanisms requires evaluation of the bioaccumulation and biouptake of fullerene through lipid membranes that serve as biological barriers in cells. Evidence suggests that the lipid membrane composition can be a critical factor for assessing bioaccumulation of fullerene (Su et al., 2010; Dellinger et al., 2010). It is also important to understand the mechanism of engineered carbon nanomaterial—based delivery systems to intracellular targets for the internalization of carbon nanomaterials into the live cells, and it has been considered to be decisive. The ability of carbon nanomaterials to be internalized into a model lipid bilayer as a function of their length was explored by molecular dynamics simulations and statistical models, on highly functionalized carbon nanomaterials, which revealed a large propensity of their passive uptake by phospholipid cell membrane. Hydrophobicity and membrane asymmetry are the key factors responsible for the insertion and complete translocation. Data demonstrate that shorter CNTs have a stronger propensity to passively penetrate the bilayer and reach the cytoplasm (Kraszewski et al., 2012). Although much progress has been made in understanding how CNTs traverse the lipid membrane of a given cell type, the details of the proposed mechanisms are still debated. Such considerations are important in that the failure to understand the uptake mechanisms of nanoscale materials and their influence on toxicity could create another level of unpredictability. Internalization to the cell has always been a preferred mechanism of drug delivery. It has been reported that functionalized multiwalled carbon nanotubes (MWCNTs) are uptaken by cells in an energy-independent manner. This holds true for various types of cell lines that undergo phagocytosis and also the ones that do not. Thus, the drug payload is effectively carried across the cell membrane and the same is effectively delivered, as well (Lacerda et al., 2012).

Two major mechanisms have been widely considered: (1) endocytosis/phagocytosis and (2) nanopenetration (Wick et al., 2007; Cherukuri et al., 2004). Endocytosis represents the engulfing of an extracellular particle by the cell, for example, viruses (~100 nm in size), through the creation of a vesicle that is then integrated into the cell. Phagocytosis is similar to endocytosis but usually involves uptake of larger particles, such as bacteria (~1 μm), and is characteristic to a subset of immune cells/phagocytes (e.g., neutrophils, macrophages, dendritic cells). These processes are energy dependent and are hindered at low temperatures and in low ATP environments. Several studies implicate endocytosis/phagocytosis as the cellular uptake mechanism of CNTs (Kam et al., 2006).

Nanopenetration is an energy-independent passive process, where the nanotubes diffuse across the cellular membrane. In this respect, it is similar to passive diffusion of nanoneedles from extracellular to intracellular space (Cai et al., 2006). It was also pointed that CNTs could behave similarly to cell-penetrating peptides (CPPs), which represent polycationic sequences that enhance uptake of proteins into mammalian cells. An interesting study in this context was the passive penetration of water-solubilized fluorescein isothiocyanate-attached CNTs and G-protein–functionalized CNTs into fibroblasts and keratinocytes at 37°C. Such investigations hint that *f*-CNTs that resemble CPPs in morphology and possess an overall charge may more likely penetrate the plasma membrane rather than undergo endocytosis. Further experimental testing is crucial in that the type of functionalization could ultimately determine the precise mechanism. Despite of it, evaluation of the partitioning thermodynamics of fullerene demonstrated that the partitioning mechanism of fullerene is different from that of molecular level chemicals. It is generally acknowledged that molecular level chemicals partition into the hydrophobic center of lipid membranes (i.e., absorption). However, the partitioning mechanism of fullerene is a combination of adsorption on the lipid membrane surface and absorption. Energy-dependent endocytosis is hypothesized to be the main cellular transport mechanism based on an observed temperature dependence of cellular uptake and evidence for saturation of the active sites of transport during cellular uptake of fullerene. Metabolic inhibitors decrease the mass of fullerene taken up by the cells, which supports an active transport mechanism of fullerene through the cell membranes.

Different studies illustrated the significant contribution of plasma membrane translocation in the cellular uptake of CNTs. Multiple internalization pathways may simultaneously operate and determine CNT cellular uptake and trafficking. Furthermore, the balance between the possible mechanisms in operation will be strongly dependent on the type of CNT functionalization (i.e., small organic groups or molecules against macromolecules or biopolymers such as lipids, proteins, or DNA), the physicochemical nature of the CNT dispersions (i.e., individualization against small or large bundles), and the type of cells (i.e., nonphagocytic against phagocytic cells). In general, it can be concluded that there was no single, unique mechanism responsible for CNT cellular uptake, and that chemical functionalization could represent a way to tailor the fate of CNTs by tilting the balance toward specific mechanisms of internalization, cellular processing, and elimination/degradation depending on the desired application (Lacerda et al., 2012).

Mao et al. (2013) investigated the uptake, intracellular distribution, and cellular effects of the collagen-single-walled carbon nanotubes (SWCNTs), using them for culture of bovine articular chondrocytes (BACs). Well-dispersed SWCNTs in aqueous solutions can be obtained through functionalization of SWCNTs with type I collagen. The inherent properties of SWCNTs were retained after collagen functionalization, and the collagen-SWCNT suspension was stable for over 63 days. The collagen-SWCNTs did not show any detrimental cellular effects on the BACs. Cellular uptake of collagen-SWCNTs by BACs was confirmed and up to 10 million SWCNTs were internalized in one cell on an average. The distribution of collagen-SWCNTs in the cells was most prevalent in the perinuclear region. The results suggested that SWCNTs functionalized by collagen should be suitable for applications in biomedicine and biotechnology (Mao et al., 2013).

In comparison to large-sized particles, nanoparticles show a different biodistribution profile, accumulation in tissues/organs, organ specificity, body clearance, and chemical composition. These parameters play an important role in biodistribution studies and in vivo interactions of nanoparticles. Studies carried out so far point at involvement of physical clearance processes (viz., mucociliary movement, epithelial endocytosis, interstitial translocation, lymphatic drainage, blood circulation translocation, and sensory neuron translocation) and chemical clearance processes such as dissolution, leaching, and protein binding. Certain kinds of nanoparticles can pass through the gastrointestinal (GIT) and are rapidly eliminated in feces and in urine, indicating the absorption across the GIT barrier and entry into the systemic circulation. However, some nanoparticulates can accumulate in the liver during the first-pass metabolism. After intravenous administration, nanoparticles get distributed to the colon, lungs, bone marrow, liver, spleen, and the lymphatics. Such distribution is followed by rapid clearance from the systemic circulation, predominantly by action of the liver and spleenic macrophages clearance, and opsonization of nanoparticles depends on size and surface characteristics. Differential opsonization translates into variations in clearance rates and macrophage sequestration of nanoparticles. To increase the passive retention of nanomaterials in systemic circulation, the suppression of opsonization events is necessary at desired sites or anatomical compartments. For example, in case of hydrophobic particles, a coating with polyethylene glycol (PEG) would increase their hydrophilicity, hence increasing the systemic circulation time (Arora et al., 2012). The understanding of the mechanisms involved in the interaction of biological systems with nanomaterials is of interest to both fundamental and applied disciplines. The adsorption of proteins modulates the formation of biofilms onto surfaces, a process important in infections associated with medical implants, in dental caries, and in environmental technologies. A deep understanding of the mechanisms driving the interaction between biological fluids or cell constituents and surfaces is instrumental in designing strategies apt to prevent the toxicity and premature clearance of nanoparticles used in diagnosis and therapy, and for avoiding adverse reactions to materials used as implants or toxic effects that may follow the accidental exposure of organisms to nanomaterials. The physicochemical properties of the surface definitely play a pivotal role in modulating the various possible processes at the interface between biological fluids and solid surfaces; the peculiar features and behavior of biomacromolecules further complicate the picture (Fenoglio et al., 2010).

Quite a few studies are focused on the development of CNT and other carbon nanomaterial–based delivery systems. Attempts to systematically expound the mechanisms of cellular uptake of carbon nanotropes are still rather limited in the presence of different uptake inhibitors. Therapeutic or diagnostic cargos loaded into the nanocarbons showed the release of active molecules directly into the cytoplasm and increased their biological activity along with the therapeutic efficacy (Lacerda et al., 2012; Debouzy et al., 2010).

5.3 APPLICATION OF CARBON NANOMATERIALS IN DRUG DELIVERY

Nanoparticles are very useful to control and manipulate biomacromolecular constructs and supramolecular assemblies that are critical to living cells. The nanoconstructs and assemblies including drugs, proteins, DNA/RNA, viruses, cellular lipid bilayers, cellular receptor sites, and antibody variable regions (decisive for immunology) are involved in events of nanoscale proportions. The nanotherapeutics/diagnostics may lead to deeper understanding of human ills such as cancer, cardiovascular disease, and genetic disorders. Size, shape, surface chemistry, and other nanotechnical dimensions are decisive steps in understanding scientific basis of nanoparticles (Mudshinge et al., 2011). Carbon nanomaterials such as fullerenes, CNTs, and graphenes show wide applicability in drug delivery due to their small size and biological activity. Nanomolecular carbon cages such as fullerenes are used for the delivery of drugs and imaging agents in several functional modes. These are also useful drug vectors or drug delivery scaffolds with noncovalent or covalent linkages between the fullerene and a bioactive moiety (Bolskar et al., 2012). Fullerene and the other carbon nanomaterials have numerous points of attachments, which allow accurate grafting of active chemical groups in 3D orientations.

This attribute is the hallmark of rational drug design, which allows positional control in matching these carbon nanocompounds to biological targets. It is possible to modify some essential pharmacokinetic characteristics of carbon nanotropes and to optimize their therapeutic effect in association with other attributes such as, size, redox potential, and the relative inertness in biological systems (Jain et al., 2005). Because of these peculiar characteristics, fullerenes hold great promise for biomedical applications (Mudshinge et al., 2011; Chawla et al., 2010; Jensen et al., 1996; Bakry et al., 2007; Partha et al., 2009). The biological activities of carbon nanotropes depend upon the core and its chemical modification. Functional groups attached to the core put in more complexity to the behavior of the highly hydrophobic fullerene molecule. Hydrophilic functionalization results into water-soluble fullerenes that are competent to carry drugs and genes for the cellular delivery. Derivatized fullerene binds to the mitochondria by crossing the cell membrane. Moreover, DNA-functionalized fullerenes are capable of entering into the COS-1 cells (fibroblast-like cell line derived from monkey kidney tissue) and offer better efficiency than that of commercial lipid-based vectors. A protective sheath of fullerene reagent is formed around the bound DNA. This increases the lifetime of DNA in endosomes and hence supports their chromosomal incorporation. Amino fullerenes are among the preferred nanomaterials for the attachment of DNA sequences. Detachment of DNA in the cytoplasm is accomplished by loss of the amino groups or loss of the binding ability of amines by transforming them into neutral compounds (Isobe et al., 2006; Nakamura et al., 2003). In tissue culture, fullerene derivatives were employed to enhance the therapeutic efficacy of a lipophilic slow-release drug delivery system (Zakharian et al., 2005). In C60−paclitaxel (PTX) conjugate, modified fullerenes provide a lipophilic slow-release system exhibiting significant anticancer activity in the cell culture. Furthermore, the ability of fullerene derivatives in penetrating through intact skin has broadened their application in cellular drug and gene delivery (Ryman-Rasmussen et al., 2006). A fullerene-based peptide derivative has been synthesized and evaluated for its ability to penetrate through flexed and inflexed skin by Rouse et al. (2007). Development of safe and efficient carriers to transport genes or drugs has attracted interest in the area of direct delivery of drugs and biomolecules into cells through the cell membrane. The cell, endosomal, and nuclear membrane barriers pose major challenges in transporting any compound into the nucleus of an intact cell. Therefore, it is important to understand the mechanism through which carriers enter the cells. Presently, organic cationic compounds, viral carriers, recombinant proteins, and inorganic nanoparticles are the four major categories of drug and gene carriers. Because of good biocompatibility, selective targeted delivery, and controlled release of carried drugs, a huge number of carbon nanotropes are being used in cellular delivery (Azzam et al., 2004; Xu et al., 2005).

While a variety of drug delivery systems have been made available to researchers and medical practitioners (Xie et al., 2013; Ma et al., 2011), carbon nanomaterials offer a promising and efficient alternative for transporting therapeutic molecules. Drug delivery strives to maximize bioavailability at a specific location in the body at the requisite time (Liu et al., 2008a,b,c; Li et al., 2008). The overall dosage of a drug and the associated adverse side effects can both be simultaneously reduced through the utilization of carbon nanomaterials as drug delivery vehicles with specificity to targeted locations in vivo (Bianco et al., 2008). The utilization of carbon nanomaterials can help get a drug to a specific location within the body, and given the predisposition for nanomaterials to accumulate within tumors, the payload drug molecules can be specifically delivered into the cancerous tissue while leaving the normal tissue unaffected by the drug. Through both covalent and noncovalent methods, the ultrahigh surface area of carbon nanomaterials such as graphene, CNTs, and nanohorns can be highly decorated with drugs including the anticancer drugs doxorubicin (DOX), cisplatin, and methotrexate (Barreto et al., 2011; Ajima et al., 2005, 2008; Zhang et al., 2011; Dhar et al., 2008) as well as anti-inflammatory drugs such as dexamethasone (Muralkami et al., 2004). These nanomaterials can also serve to highly increase the water solubility of hydrophobic small molecule drugs and facilitate their release in the often acidic conditions found within tumors (Bianco et al., 2005). Interestingly, the nanomaterial–drug complex can significantly increase the drug's efficacy and produce orders of magnitude increases in overall cytotoxicity. In addition to small molecule drugs, carbon nanomaterials can serve as scaffolding to help shuttle peptides, proteins, and other biologics into cells via endocytosis (Liu et al., 2011a,b).

5.3.1 Drug Delivery Using Carbon Nanoparticles

5.3.1.1 Carbon Quantum Dots in Drug Delivery

Carbon nanoparticles (CNPs) and carbon quantum dots (CQDs) have attractive prospect for medical therapy and bioimaging diagnostics for visual drug distribution and monitoring of their effects (Cheng et al., 2014). Zheng et al. (2014) prepared a multifunctional theranostic agent (CD-Oxa) by the conjugation of an anticancer agent (oxidized oxaliplatin, oxa (IV)-COOH) onto the surface of CDs containing amine groups and that successfully integrates with the optical properties of the CDs and the therapeutic performance of Oxa. The in vitro results indicated that CD-Oxa possesses good biocompatibility, bioimaging function, and anticancer effects. The in vivo results demonstrate that it is possible to follow the distribution of the drug by monitoring the fluorescence signal of CD-Oxa, which helps customize the injection time and dosage of the medicine (Fig. 5.1A).

Kim et al. (2013a,b) coupled CQDs with gold nanoparticles and then conjugated with PEI-pDNA for delivering DNA to cells. The results suggested that

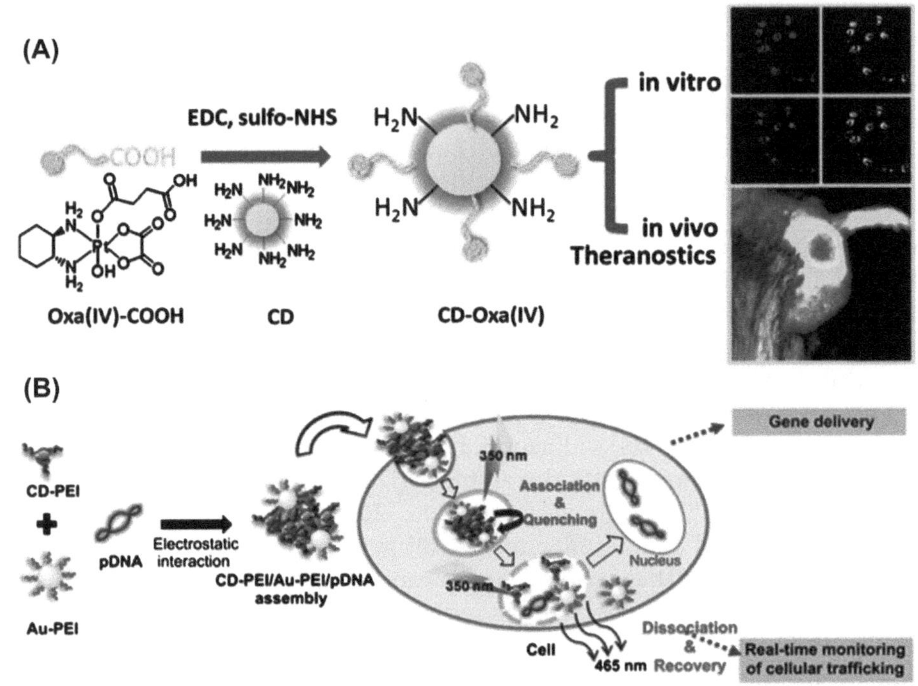

FIGURE 5.1

(A) Synthetic scheme for CD-Oxa and its applications in bioimaging and theranostics (Zheng et al., 2014). (B) A schematic illustration for the gene delivery and real-time monitoring of cellular trafficking utilizing CD-PEI/Au-PEI/pDNA assembled nanohybrids. *(A) Copyright Wiley-VCH Verlag GmbH & Co. KGaA. Reproduced with permission. (B) Adapted with permission from Kim, H., et al., 2013a. Photothermally triggered cytosolic drug delivery via endosome disruption using a functionalized reduced graphene oxide. ACS Nano 7, 6735–6746 and Kim, H., et al., 2013b. Transfection and intracellular trafficking properties of carbon dot-gold nanoparticle molecular assembly conjugated with PEI-pDNA. Biomaterials 34 (29), 7168–7180.*

the assembly entered the cells with the CQDs located in the cell cytoplasm and the released pDNA entered the cell nuclei, achieving significant transfection efficiency (Fig. 5.1B). Pandey et al. (2013a,b) used CQD-functionalized gold nanorods for the delivery of DOX in a multimodality fashion, including drug delivery, PTT, and bioimaging using the same platform. The widely used antipsychotic drug haloperidol (HaLO)-grafted CQDs with cysteamine hydrochloride (CysHCl) as a linker can offer controlled release under physiological conditions for more than 40 h following the Hixson–Crowell model under standardized conditions (Pandey et al., 2013a,b). A broad spectrum antibiotic, ciprofloxacin attached to CQDs with bright green fluorescence can not only pave a way for bioimaging but also provide an efficient new nanocarrier for controlled drug release with high antimicrobial activity under physiological conditions (Thakur et al., 2014).

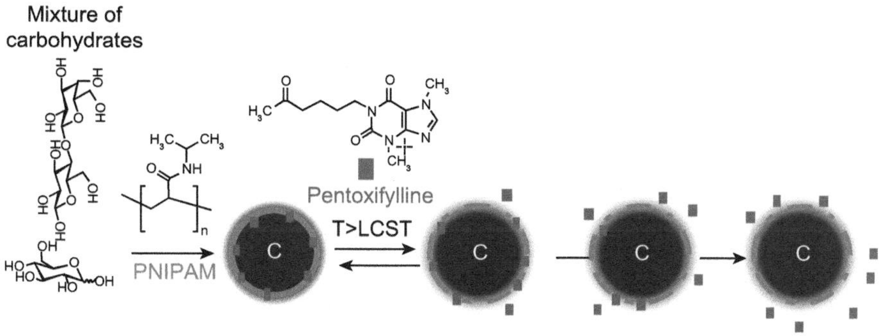

FIGURE 5.2
Luminescent carbon nanoparticle mixture of carbohydrates for controlled drug delivery. Poly(N-isopropylacrylamide) (PNIPAM), a thermoresponsive (heat-sensitive) polymer, was chosen to encapsulate pentoxifylline drug [red (gray in print versions)] molecules. PNIPAM undergoes a phase transition at a higher temperature, releasing the drug molecule in a controlled fashion (Mukherjee et al., 2015).
Copyright Wiley-VCH Verlag GmbH & Co. KGaA. Reproduced with permission.

Mukherjee et al. (2015) used carbohydrates mixture with luminescent CNPs for controlled drug delivery (Fig. 5.2). In this process poly (N-isopropylacrylamide) (PNIPAM), a thermoresponsive polymer, was chosen to encapsulate pentoxifylline drug (red) molecules, and then PNIPAM undergoes a phase transition at a higher temperature, releasing the drug molecule in a controlled fashion.

5.3.1.2 Fullerene in Drug Delivery

A fullerene is composed entirely of carbon in the form of a hollow sphere, ellipsoid, or tube. Spherical fullerenes are also called buckyballs, and cylindrical ones are called CNTs or buckytubes. Fullerenes are similar in structure to the graphite, which is composed of stacked graphene sheets of linked hexagonal rings, additionally they may also contain pentagonal (or sometimes heptagonal) rings to give potentially porous molecules (Holister et al., 2003). Buckyball clusters or buckyballs composed of less than 300 carbon atoms are commonly known as endohedral fullerenes and include the most common fullerene, buckminsterfullerene, C60. Megatubes are larger in diameter than nanotubes and prepared with walls of different thickness which is potentially used for the transport of a variety of molecules of different sizes (Mitchell et al., 2001). "Nanoonions" are spherical particles based on multiple carbon layers surrounding a buckyball core, which are proposed for lubricants (Sano et al., 2001). These properties of fullerenes hold great promise in health and personal care application. The versatile biomedical applications are tabulated in Table 5.1.

Table 5.1 Biomedical Applications of Fullerenes (Mudshinge et al., 2011)

Fullerenes Composition	Application	References
Fullerene (C60)	HIV proteases	Friedman et al. (1993) and Sijbesma et al. (1993)
Fulleropyrrolidines	HIV-1 and HIV-2	Marchesan et al. (2005)
Dendrofullerene 1	HIV-1 replication	Brettreich and Hirsch (1998) and Schuster et al. (2000)
Amino acid derivatives of fullerene C60 (ADF)	HIV and human cytomegalovirus replication	Kotelnikova et al. (2003)
Buckminsterfullerene	Semliki forest virus (Togaviridae) or vesicular stomatitis virus (Rhabdoviridae)	Kaesermann et al. (1997)
Cationic, anionic, and amino acid type fullerene	HIV-reverse transcriptase and hepatitis C virus replication	Mashino et al. (2005)
Fullerene (C60) 34 methyl radicals	Free radicals and oxidative stress	Krusic et al. (1991)
Fullerene (C60)	Liver toxicity and diminished lipid peroxidation	Slater et al. (1985)
C3-fullero-tris-methanodicarboxylic acid	Apoptosis of neuronal cells	Dugan et al. (1997)
Carboxyfullerene	Apoptosis of hepatoma cells	Huang et al. (1998)
Carboxyfullerenes	Neurological disease including Parkinson's disease	Dugan et al. (1997)
Fullerene (C60) with organic cationic compounds, viral carriers, recombinant proteins, and inorganic nanoparticles	Gene transfer	
Metallofullerol	Leukemia and bone cancer	Thrash et al. (1999)

Copyright permission from Mudshinge, S.R., et al., 2011. Nanoparticles: emerging carriers for drug delivery. Saudi Pharm. J. 19 (3), 129–141.

5.3.2 Drug Delivery Using Carbon Nanotubes

Out of all carbon nanomaterials, CNTs are most commonly used because of their unique spectroscopic properties, and because they can be easily functionalized either covalently or noncovalently and decorated with bioactive peptides, small molecule drugs, proteins, and nucleic acids for gene delivery. After attaching anticancer, antiviral, or antibacterial drugs to the CNT, the efficacy of the drug must be preserved after crossing the cell membrane or the drug must be anchored to the nanotube through a cleavable bond that promotes the drug's release once inside the cell (Liu et al., 2008a,b,c). With proper surface functionalizations, CNTs display low levels of cytotoxicity and can easily be transported across the cell membrane, both of which are key factors for a drug delivery vehicle. Carbon nanotubes are most commonly researched and applied to drug delivery, yet graphene, graphene nanosheets, and carbon nanohorns have also been utilized effectively for drug delivery applications (Liu et al., 2011a,b).

Prior to utilizing CNTs for drug delivery, functionalization either covalently or noncovalently is necessary (described in Chapter 4). Appropriate surface functionalization is the key to generating sufficient anchoring sites to load and immobilize the cargo molecules on the CNTs, and imparting stability and cell penetration capability to the CNTs. Both covalently and noncovalently functionalized CNTs favor PEGylation to improve stability in physiological buffers (Liu et al., 2007a,b,c,d; Schipper et al., 2008) in the meantime allowing for easy functionalization of CNTs with terminal functional groups on the PEG chains (Liu et al., 2007a,b,c,d, 2009a,b,c,d). Functionalized CNTs provide two possible locations where the exogenous drugs, biologics, or targeting molecules can be loaded: for molecules with functional groups, they can be easily conjugated to the terminal functional groups on the CNTs or the PEG chains; on the other hand, for hydrophobic molecules without any available functional groups such as DOX and SN38, they can adsorb to the similarly hydrophobic backbone of CNTs through $\pi-\pi$ stacking or hydrophobic interaction between the drug molecules and the CNTs.

CNT-based drug delivery vehicles can be designed so that the drug payload is only released after the nanotube is within the cell. The PL-PEG-functionalized SWCNTs were used as a platform to improve the internalization process of an inert platinum (IV) complex, a prodrug compound developed by the Lippard group, into cancer cells (Feazell et al., 2007). The platinum (IV) complex was tethered onto the end of the PEG chain through an amide bond. After the SWCNT is taken up by the cell through endocytosis, the SWCNTs become engulfed in endosomes where a drop in environmental pH causes the release of the platinum (II) core complex, which kills the cancerous cells. Due to the selective release of platinum (II) in the cell, the cytotoxicity of the platinum (IV)-SWCNT complex increased by over 100-fold when compared to platinum (IV) alone. This effect was only demonstrated in vitro, but as SWCNTs tend to aggregate at high concentrations in tumors this strategy has the potential to be utilized after intravenous injection.

Liu et al. (2008a,b,c) demonstrated successful drug delivery of PTX, a commonly used anticancer drug, with CNTs in vivo. The PTX molecule has an activatable hydroxyl group that can be covalently linked to the amine terminal groups of PEG chains through succinic anhydride, rendering the PTX molecule soluble in aqueous biological environment. Due to the passive uptake of CNTs by cancerous tissue by the EPR effect, the SWCNT–PTX conjugate demonstrated a 10-fold increase in PTX uptake within xenograft murine tumor models compared with Taxol that is the clinical drug formulation of PTX. It is noteworthy that the ester bond through which the PTX molecule is anchored to the surface of SWCNT is cleavable, leading to controlled release of the drug molecules inside the tumor tissue and successful suppression of the tumor growth in a murine model of 4T1 breast cancer (Liu et al., 2008a,b,c). After accumulation of SWCNT–PTX conjugates in the tumor at a much higher concentration than the surrounding tissue, cellular internalization of the SWCNT–PTX complex promoted PTX release due to the lower pH in the local tumor environment. The SWCNTs are accumulated within the RES system during clearance from circulation as has been demonstrated in previous studies (Liu et al., 2007a,b,c,d, 2008a,b,c) yet PTX was found to be excreted rapidly due to cleavage of the ester bond, indicating that residual PTX was rapidly removed, much more quickly than the CNTs, and did not cause an increase in toxicity due to long-term retention in the organs of the RES (Liu et al., 2008a,b,c). This model can be applied to a variety of water-insoluble drugs to boost their tumor accumulation abilities. It was proposed that the use of targeting ligands for the selective targeting of particular cell surface receptors would further contribute to the efficacy of this method. In addition to attaching anticancer drugs covalently through the terminal functional groups on the PL-PEG chains, many anticancer drugs can be loaded onto CNTs through supramolecular chemistry by loading aromatic drug molecules through $\pi-\pi$ stacking on the CNT surface (Fig. 5.3). DOX is a commonly used drug in chemotherapy and can be loaded in extremely high ratios on CNTs of up to 4 g of DOX for every 1 g of SWCNTs (Liu et al., 2007a,b,c,d). In addition to the weight of the PL-PEG coating, the SWCNT–DOX complex contained ~50–60 wt% DOX, in stark contrast to the ~8–10 wt% found in conventional liposomes used to solubilize the drug in its commercial formulation. The SWCNT–DOX conjugate demonstrated excellent stability in buffer and serum, yet in acidic conditions as found

FIGURE 5.3
Representation of the SWCNT–DOX complex (Liu et al., 2009a,b,c,d). *DOX*, doxorubicin. *Copyright Wiley-VCH Verlag GmbH & Co. KGaA. Reproduced with permission.*

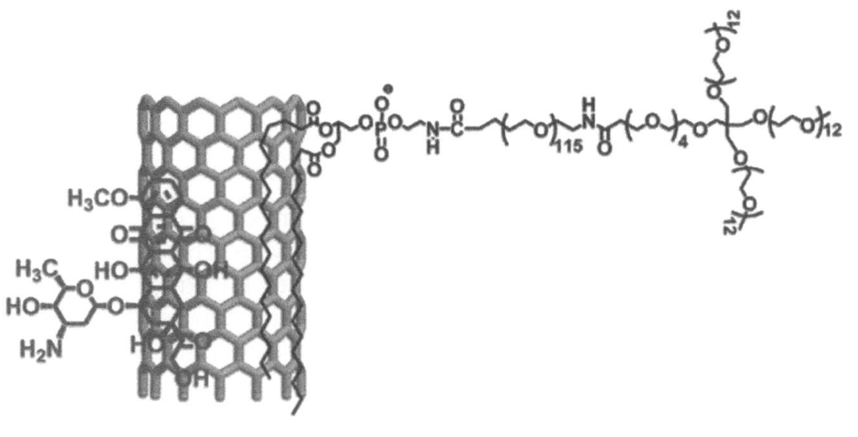

in liposomes, in endosomes, and within tumor microenvironments, DOX demonstrated rapid release from the CNT surface, both of which are desirable properties for in vivo drug carriers. In this case, while the anticancer drug was loaded onto the CNT surface, the terminal functional groups on the PL-PEG chains were functionalized with an RGD peptide to facilitate the selective targeting of cancerous cells in an in vitro setting. DOX-loaded SWCNTs have also demonstrated significantly enhanced in vivo therapeutic efficacy with greatly mitigated toxicity to treated mice in comparison to other formulations of DOX such as the free DOX and DOXIL, which is liposome-solubilized DOX (Liu et al., 2009a,b,c,d). In these in vitro and in vivo studies, CNTs were used; yet a variety of different carbon nanomaterials have shown similar drug loading behaviors. Supramolecular loading through $\pi-\pi$ stacking is not unique to SWCNTs, as MWCNTs, carbon nanohorns, and nanographene oxide (NGO) all exhibit similar drug loading behaviors as well as variable drug release as a function of pH (Liu et al., 2008a,b,c; Sun et al. 2008; Ali-Boucetta et al., 2008; Muralkami, 2006). Furthermore small-molecule anticancer drugs, biological therapeutics can be successfully attached to CNTs for translocation across the cell membrane. Unlike small molecules, these larger biomacromolecules rarely cross cell membranes unaided. SWCNTs have been used as scaffolds for introducing these bioactive peptides to the immune system, and CNTs have been proposed as novel vaccine delivery tools (Pantarotto et al., 2003). Proteins such as the epidermal growth factor receptor inhibitors are routinely attached to CNTs for the targeted delivery of drug payloads to enhance the efficacy of chemotherapy (Bhirde et al., 2009).

5.3.3 Graphene/Graphene Oxide/Reduced Graphene Oxide

5.3.3.1 Graphene for Drug Delivery

Since their discovery as a biosafe material, graphene has been explored as a carrier molecule in drug delivery research (Goenka et al., 2014; Novoselov et al., 2012). The large specific surface area of graphene enhances opportunities for multidrug delivery to the target site from the site of administration. Polymeric modification and conjugation strategies also enhance biocompatibility and circulation times in vivo (Nurunnabi et al., 2013a,b). Several studies have been conducted on the delivery of anticancer drugs, genes, and peptides through graphene derivatives in the last couple of years (Nurunnabi et al., 2013a,b; Dembereldorj et al., 2012; Depan et al., 2011; Yang et al., 2009, 2011, 2012; Xu et al., 2014). Simple physisorption via $\pi-\pi$ stacking can be used for loading many hydrophobic drugs, such as DOX and docetaxel, with antibodies for the selective killing of cancer cells. Owing to its small size, intrinsic optical properties, large specific surface area, low cost, and useful noncovalent interactions with aromatic drug molecules, graphene is a promising new material for drug delivery through the nanocarrier approach. The large specific surface area, $\pi-\pi$ stacking, and electrostatic or hydrophobic interactions of graphene can assist in high drug loading of poorly soluble drugs without compromising potency or efficiency.

Dembereldorj et al. (2012) reported that PEGylated GO loaded DOX via π−π interactions shows promising real-time release of DOX from PEGylated GO at specific loci after an external triggering by glutathione (GSH). Depan et al. (2011) reported that GO loaded with DOX exhibits higher drug release at pH 5.3 due to the reduced interaction between DOX and the drug carrier. GO loaded with DOX shows enhanced cellular toxicity and promising tumor growth inhibition, with almost 66−91% cell death (Yang et al., 2009, 2011, 2012; Xu et al., 2014). Other chemotherapy drugs, such as paclitaxel and methotrexate loaded on GO via π−π stacking and amide bonds, exhibited amazing cancerous effect on lung cancer and breast cancer, which has resulted in inhibition of about 66−90% tumor growth (Xu et al., 2014; Wojtoniszak et al., 2013). When ibuprofen, which is used as an NSAID, was conjugated with chitosan-functionalized GO via amide linkages, the functionalized GO exhibited higher (20%) biocompatibility than GO sheets for CEM and MCF-7 cell lines (Rana et al., 2011). GO loaded with a second generation photosensitizer chlorin e6 (Ce6) resulted in its higher accumulation in tumor cells, leading to a higher photodynamic efficacy upon irradiation (Tian et al., 2011; Huang et al., 2011). Nano GO is another important material for the drug delivery research area. NGO is used as a novel and efficient nanocarrier for the delivery of water-insoluble aromatic anticancer drugs into cells. The approach first conjugated nano GO with PEG to obtain PEG-NGO.

Further, DOX and camptothecin (CPT) analog, SN38 were conjugated by noncovalent π−π interactions. These complexes showed pH-dependent drug release and prominent cytotoxicity levels in HCT-116 cell lines which was 1000-fold potential than CPT (Sun et al. 2008; Liu et al., 2008a,b,c). Kim et al. (2013a,b) reported that NIR radiation, acidic pH, and high intracellular concentrations of GSH favored intracellular cytosolic delivery of DOX. Cells treated with PEG and branched polyethylenimine (BPEI)-functionalized rGO (PEG−BPEI-rGO) nanocarriers exposed to NIR irradiation encouraged endosomal disruption and consequent DOX release, which triggered cellular toxicity. Graphene derivatives have been conjugated with biopolymers, such as gelatin and polyethylene glycol (PEG), as functionalizing agents for drug delivery applications. Gelatin and PEG not only favored the reduction of graphene but also functionalized GNS, facilitating the loading of DOX onto GNS. The GNS−DOX complex also exhibited high toxicity toward U251, 1800, and A-5RT3 cells through endocytosis (Liu et al., 2011a,b; Wang et al., 2013). Poly NIPAM and other polymers have been used with graphene for loading various drugs such as CPT (Pan et al. 2011), methotrexate (An et al., 2013), and 5-fluorouracil (Fan et al., 2013). In the last 10 years, researchers have begun to synthesize smaller graphene derivatives, often referred as graphene quantum dots (GQDs). These GQDs exhibit intrinsic fluorescence and are also used for theranostic purposes. Synthesized GQDs with different color emissions and loaded with anticancer drugs such as DOX were reported to have an image-guided higher therapeutic efficacy of about 55–90% cell growth inhibition (Table 5.2) (Chen et al., 2014; Abdullah Al et al., 2013; Wang et al., 2014a,b). Maity et al. (2014) also synthesized carbohydrate-coated and folate-functionalized colloidal graphene for nanocarrier for wide variety of hydrophobic and hydrophilic drugs. These

Table 5.2 The Application of Different Graphene Derivatives for Drug Delivery

Graphene Derivatives	Drug	Application	References
GO	DOX	In vitro: A549 cells In vivo: Cg-Foxn1nu/CrljOri nude mice Results: released 15–20% increase	Dembereldorj et al. (2012)
GO	DOX	In vitro: drug release Results: higher drug release at pH 5.3	Depan et al. (2011)
GO	DOX	In vitro: HeLa cells and OCT-1 mouse osteoblasts In vivo: BALB/c nude mice Results: 71% tumor growth inhibition	Yang et al. (2012)
GO	DOX	Results: pH-triggered controlled magnetic behavior	Yang et al. (2009)
GO	DOX	In vitro: SK3 cells Results: enhanced cellular toxicity	Yang et al. (2011)
GO	Paclitaxel	In vitro: A549 and MCF-7 cells Results: 90%	Xu et al. (2014)
GO		Methotrexate In vitro: MCF-7 cells Results: 66.1%	Wojtoniszak et al. (2013)
GO	Ibuprofen	In vitro: CEM and MCF-7 cells Results: about 95%	Rana et al. (2011)
GO	Ce6	In vitro: KB cells Results: 98%	Tian et al. (2011)
GO	Ce6	In vitro: MGC803 cells Results: 90%	Huang et al. (2011)
Nano GO	DOX	In vitro: CEM.NK T-cells and Raji B-cells Results: 80% cell growth inhibition	Sun et al. (2008)
Nano GO	SN38	In vitro: HCT-116 cells Results: 80% cell growth inhibition	Liu et al. (2008a,b,c)
Reduced GO	DOX	In vitro: PC-3 and HeLa cells Results: 80% cell growth inhibition	Kim et al. (2013a,b)

Continued

Table 5.2 The Application of Different Graphene Derivatives for Drug Delivery continued

Graphene Derivatives	Drug	Application	References
Graphene nanosheet	DOX	In vitro: MCF-7 cells Results: the cytotoxicity enhanced gradually	Liu et al. (2011a,b)
Graphene nanosheet	DOX	In vitro: U251 and 1800 cells Results: 55%	Wang et al. (2013)
Graphene nanosheet	DOX	In vitro: A-5RT3 cells Results: 90%	Pan et al. (2011)
Graphene nanosheet	Methotrexate	In vitro: A549 cells Results: 70.2%	An et al. (2013)
Graphene nanosheet	5FU	In vitro: HepG2 cells Results: 72% growth inhibition	Fan et al. (2013)
Graphene quantum dot	DOX	In vitro: A549 cells Results: 95%	Chen et al. (2014)
Graphene quantum dot	DOX	In vitro: A549 cells In vivo: BALB/c mice Results: 60%	Abdullah Al et al. (2013)
Graphene quantum dot	DOX	In vitro: HeLa, A549, and HEK293A cells	Wang et al. (2014a,b)

DOX, *doxorubicin;* GO, *graphene oxide.*
Copyright permission from the Royal Society of Chemistry.

colloidal graphene is loaded with paclitaxol, CPT, DOX, curcumin and used for their targeted delivery to cancer cells. Maity et al. (2014) demonstrated that these drugs deliver into target cells and offer an enhanced therapeutic performance. The reported approach can be extended to the cellular delivery of other hydrophobic and hydrophilic drugs and the simultaneous delivery of multiple drugs.

5.3.3.2 Graphene Oxide in Drug Delivery

GO, produced by vigorous oxidation of graphite by Hummers method (1958), is an ideal nanocarrier for efficient drug and gene delivery. GO used for drug delivery is usually one to three layers (1–2 nm thick) with size ranging from a few nanometers to several hundred nanometers (Loh et al., 2010; Kovtyukhova et al., 1999; Sun et al., 2008). The unique structural features, such as large and planar sp^2 hybridized carbon domain, high-specific surface area (2630 $m^2\ g^{-1}$), and enriched oxygen-containing groups, render GO excellent biocompatibility, physiological solubility and stability, and capability of loading of drugs or genes via chemical conjugation or physisorption approaches. Moreover, the reactive COOH and OH groups GO bears facilitate conjugation with various systems, such as polymers (Shan et al., 2009), biomolecules (biotargeting ligand (Sun et al. 2008), DNA (Lei et al., 2011), protein (Zhang et al., 2010a,b,c, 2011; Lee et al., 2011), quantum dots (Dong et al., 2010), Fe_3O_4 nanoparticles (Chen et al., 2011a,b), and others (Shen et al., 2010) imparting GO with multifunctionalities and multimodalities for diverse biological and medical applications.

Inspired by the ideas for carbon nanotube-based drug delivery (Liu et al., 2011a,b), Dai et al. explored NGO for the first time as a novel and efficient nanocarrier for delivery of water-insoluble aromatic anticancer drugs into cells (Liu et al., 2008a,b,c). In their approach, NGO was first conjugated with an amine-terminated six armed PEG molecule, followed by loading of a water-insoluble anticancer drug, SN38, onto NGO surface by simple noncovalent adsorption via $\pi-\pi$ stacking. The PEG-functionalized NGO loaded with SN38 exhibited high cytotoxicity for HCT-116 cells is 1000-fold more potent than CPT-11. In another work, the same group studied targeted delivery of chemical drugs into cells by using a Rituxan (CD20+ antibody)-conjugated NGO–PEG (Sun et al. 2008) (Fig. 5.4). It was further demonstrated that the drug releases from the GO surface was pH dependent, suggesting the possibility of pH-controlled drug release. The pH-sensitive drug release behavior from many different GO-based drug delivery systems was also studied later by Yang et al. (2008a,b,c), Bai et al. (2010), Depan et al. (2011), Zhang et al. (2010a,b,c). Apart from pH-activated drug release, Pan et al. (2011) developed a thermoresponsive drug delivery cargo, PNIPAM-grafted graphene sheets.

Combined use of multiple drugs is a widely adopted clinical practice in cancer therapy to overcome drug resistance of cancer cells (Andersson et al., 1999; Gavrilov et al., 2005). However, few reports on nanomaterial-based drug delivery systems for controlled loading and delivery of multiple drugs can be found in the literature due to technical difficulties. Zhang et al. (2010a,b,c) explored the feasibility of GO as a nanocarrier for controlled loading and targeted delivery of mixed chemical drugs. Rana et al. (2011) reported the delivery of an antiinflammatory drug, ibuprofen, by using a chitosan-grafted GO. In this case, the loading rate of ibuprofen on the GO sheet was determined to be 9.7%. Furthermore, the work demonstrates that controlled drug release can be achieved by adjustment of pH value.

To enhance the anticancer effect, Yang et al. (2011) designed and prepared a magnetic- and bio-dual targeting drug delivery cargo based on GO-Fe_3O_4 nanoparticle hybrid. The in vitro experiments indicated specific targeting of the multifunctional drug carriers by SK3 human breast cancer cells. Clearly, in vivo study is desired to demonstrate the performance of this external magnetic field–guided and biotargeted drug delivery system.

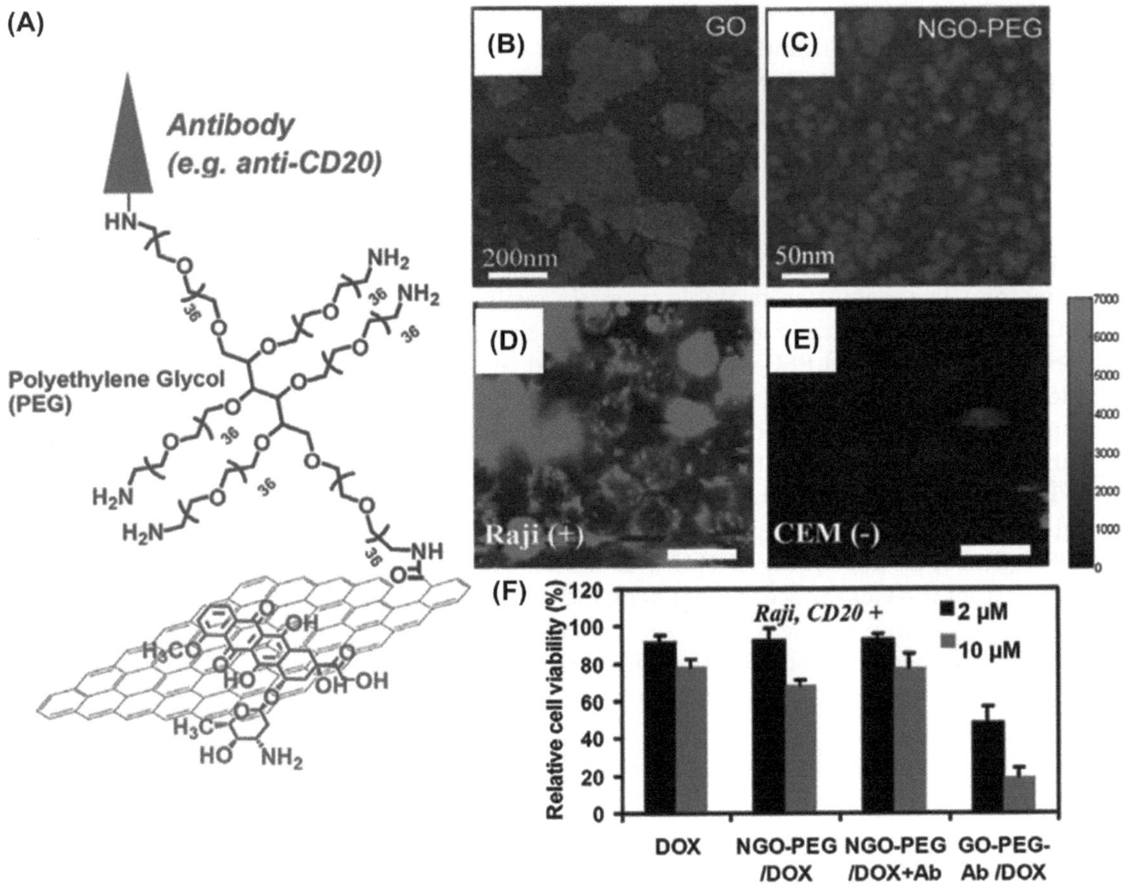

FIGURE 5.4

Nanographene oxide (NGO) for target cell imaging and drug delivery. (A) A schematic illustration of doxorubicin (DOX) loading onto NGO-PEG-Rituxan via π-stacking. Atomic force microscopy images of as-prepared (B) graphene oxide (GO) and (C) NGO-PEG. Near-infrared (NIR) fluorescence image of (d) CD20-positive Raji B-cells and (E) CD20-negative CEM cells treated with the NGO-PEG-Rituxan (anti-CD20 antibody) conjugate. Scale bar shows intensity of total NIR emission in the range of 1100–2200 nm under the 785 nm excitation. Scale bar = 25 μm. (F) In vitro toxicity test at 2 μM and 10 μM DOX concentrations showing that Rituxan conjugation selectively enhanced DOX delivery into Raji B-cells by comparing NGO-PEG-Rituxan/DOX with free DOX; NGO-PEG/DOX; and the mixture of DOX, Rituxan, and NGO-PEG (Sun et al., 2008). *PEG*, polyethylene glycol. *Reproduced with kind permission from Springer Science @ Business Media and Tsinghua Press.*

5.4 GENE DELIVERY USING CARBON NANOMATERIALS

In a similar fashion to drug delivery, carbon nanomaterials, in particular CNTs and graphene, are frequently researched and applied to gene delivery applications. Gene therapy as an efficient delivery system allows the transfer and expression of the therapeutic gene in the target organ or tissue. Gene therapy technique can be divided into two major groups: viral gene delivery (or viral vector) and nonviral gene delivery (or nonviral vector). In first group, the genes are carried by a virus into the cells due to its ability to enter the cells (Tran et al., 2009; Ji et al., 2010; Singh et al., 2005). Although viral gene delivery achieves high levels of gene expression, this group of gene delivery has some disadvantages, for example, viral vectors can be immunogenic or induce inflammation that renders transgene expression transient and can have oncogenic effects (Tran et al., 2009; Ji et al., 2010; Singh et al., 2005). Since nonviral vectors can overcome some of these concerns, it might be more attractive. In addition, since these vectors are typically assembled in cell-free systems from well-defined components, in comparison with viral vectors they have considerable manufacturing and safety advantages. To date, numerous nonviral delivery systems have been established to efficiently deliver short interfering RNA (siRNA), including chemical modification of siRNA, cationic polymers, liposomes, cationic lipids, cell-penetrating peptide, and so on. However, the efficiency of gene expression mediated by these vectors is lower than viral vectors, due to their poor capability in reaching and crossing the nuclear membrane; so to achieve therapeutically relevant levels of gene expression improvements, nonviral vectors are still required (Ji et al., 2010; Singh et al., 2005; Elfinger et al., 2008). The advantages and disadvantages of viral and nonviral gene delivery are shown in Table 5.3.

Gene therapy and RNA interference (RNAi) are two types of therapies: The first one (gain-of-function) removes the need for subsequent drug administration, while the second one (loss-of function) confers highly specific gene silencing

Table 5.3 Comparison of Viral and Nonviral Gene Delivery

Gene Therapy Technique	Advantages	Disadvantages	References
Viral gene delivery (viral vector)	▪ High delivery efficacy ▪ High levels of gene expression	▪ Immunogenic and inflammation responses of host tissue ▪ Can have oncogenic effects	Tran, P.A. et al. (2009) and Liu, S. et al. (2009b)
Nonviral gene delivery (nonviral vector)	▪ Manufacturing ▪ Safety	Gene transfer was not high enough	
Carbon nanotube (CNT) as a nonviral vector	▪ Can mediate endocytosis ▪ Are biodegradable ▪ Can be noncytotoxicafter functionalization	▪ Cytotoxicity of pristine CNTs ▪ Insolubility of pristine CNTs	Daniel, D. et al. (2007), Tagmatarchis, N. et al. (2004), and Liu, S. et al. (2009b)

(Goldberg et al., 2007). The advantage of RNAi in comparison with other gene therapeutic strategies lies in its high affinity and specificity to their target sites. Promising results have been achieved with siRNA in animal models and several clinical trials are underway (Cheung et al., 2010). To generate protein expression, DNA payloads should first enter the cell, escape the endosome, dissociate from their carrier, pass the nuclear membrane, and are transcribed by host machinery (Goldberg et al., 2007). Knocking down expression of genes by means of siRNA is a promising advance for the treatment of various disorders, but the efficacy of this therapeutic strategy is restricted by low capacity of siRNA to pass cell membranes, avoid protease degradation, and the rapid excretion via kidney owing to the size of siRNA: these molecules are quite small and thus are quickly excreted through urine when administrated into the blood stream, even though the siRNA molecules remain stable (Cheung et al., 2010). A variety of transfection agents have been applied to overcome these problems, but the majority of these vectors lead to serious cytotoxic side effects (Goldberg et al., 2007; Herrero et al., 2009).

A tremendous amount of interest has been given to the treatment of various diseases by introducing nucleic acids which can repair, replace, regulate, and add or delete a certain genetic target responsible for a particular disease given the advent of understanding of genetic processes and molecular biology (Bates et al., 2013).

Carbon nanomaterials can be covalently or noncovalently functionalized to contain positive charges to bind and deliver the negatively charged DNA and siRNA (Feng et al., 2011; Kam et al. 2005; Pantarotto et al., 2004; Zhou et al., 2012). One benefit of using carbon nanomaterials for gene therapy is their low intracellular toxicity even at very high concentrations, which is in contrast to commercial gene transfection agents. While overall transfection rates are on par with or superior to commercial gene transfection agents such as lipofectamine 2000, new approaches using carbon nanomaterial–based transporters for gene delivery have significantly decreased the required dosage and dramatically increased transfection success rate. Other carbon-based transfection agents such as carbon dots (Hu et al., 2014; Liu et al., 2012) and nanodiamonds (Zhang et al., 2009; Chen et al., 2010) also show promise as novel gene therapy agents as well.

5.4.1 Carbon Nanotubes for Gene Delivery

The application of CNTs as gene carriers in gene delivery has been considered quite promising. Gene therapy involves not only the gene-based treatment for cancers but also that for the infectious diseases by introducing genetic materials. It is generally believed that the tumor formation is the result of the gene alterations, and gene therapy aims to correct them. For all of the gene-based therapeutic strategies, efficient and safe gene delivery systems have become imperative to develop, especially the gene vectors because it is relatively easy to obtain corresponding genes. There have been two subcategories of gene vectors including many viral and nonviral vectors. Viral vectors have been modified to eliminate their toxicity and maintain their high gene transfer capability. However, their limited capacity for transgenic materials and safety, particularly immunogenicity, has compelled researchers to increasingly shift attention upon nonviral vectors as an alternative. Nonviral vectors are mainly based on cationic polymers. It is just recent thing that CNTs emerge as DNA carriers owing to their unique physical, chemical, and biological properties (El-Aneed et al., 2004).

Charged hybrid DNA/SWCNT complexes can be obtained by sonicating the suspensions composed of single-stranded DNA and CNTs. The aromatic nucleobases are believed to bind to the graphene side walls through $\pi-\pi$ stacking effects. DNA molecules can be confined and oriented by CNTs that acted as scaffolds, which were wrapped around by DNA macromolecules. Moreover, there are different interaction energies with nanotubes for different nucleobases. The sugar and phosphate groups remain at the periphery relative to the bases, playing the roles of enhancing the dispersibility of CNTs. The spontaneous wrapping of DNA around nanotubes has been also confirmed by atomic force microscopy and spectroscopic studies. Such a system can definitely prove useful in gene delivery (Albertorio et al., 2009). Several experimental studies in this area are discussed in the next section.

5.4.2 Short Interfering RNA Delivery Using Carbon Nanotubes

The SWCNTs modified with lipopolymers and attached siRNA were used to transfect human cells to silence the expression of HIV-specific cell surface receptors. The SWCNTs covalently functionalized with small groups containing distal ammonium functions is the second approach used to form electrostatic complexes with the negatively charged siRNA (Herrero et al., 2009). Also it was reported that the gene silencing effect by cleavable nanotube–siRNA conjugates was more enhanced about 20%, in comparison with lipofectamine/siRNA complexes and noncleavable nanotube–-siRNA conjugates in vitro (Jeong et al., 2009). Yang et al. prepared positively charged SWCNTs (SWCNTs+) using 1,6-diaminohexane and used SWCNTs+ to form complexes with siRNA. The siRNA–SWCNT+ complexes then were preferentially taken up by splenic CD11c+ dendritic cells; CD11b+ cells; and also Gr-1+CD11b+ cells comprising dendritic cells, macrophages, and other myeloid cells to silence the targeting gene (Yang et al., 2006). Liu et al. (2007a,b,c,d) functionalized SWCNTs and made them water soluble by the strong adsorption of phospholipids grafted onto amine-terminated PEG and then applied it for the delivery

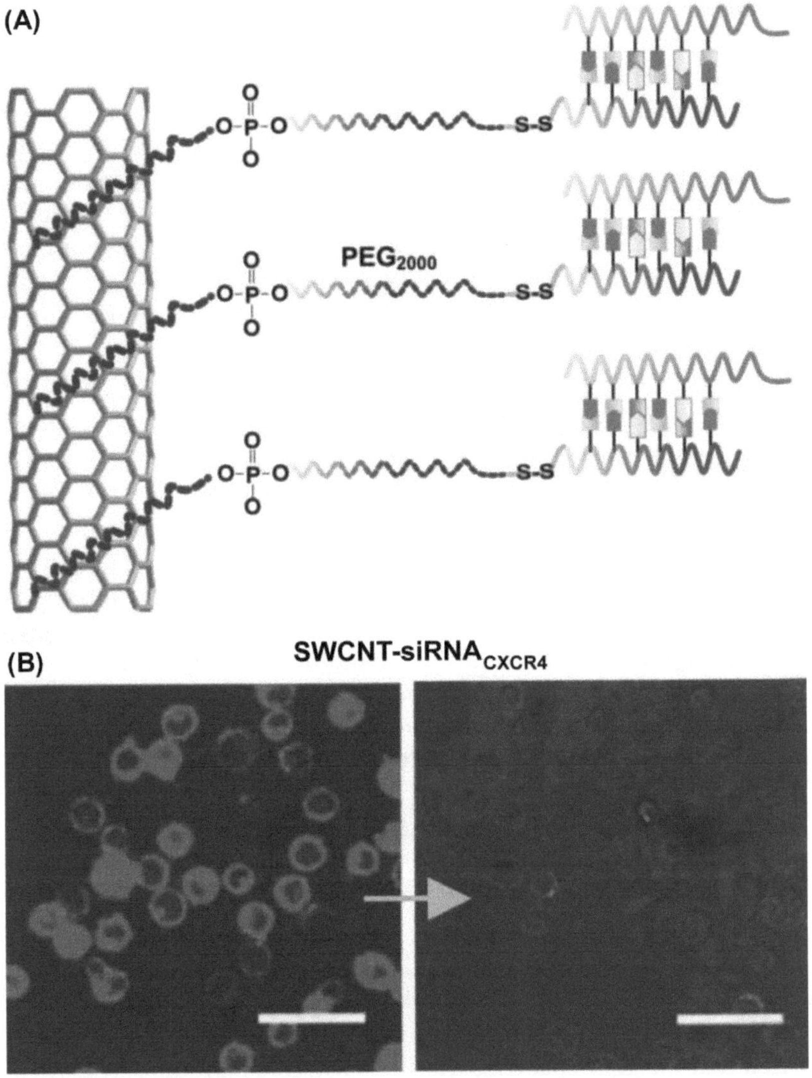

FIGURE 5.5
(A) Functionalization of SWCNTs with PL-PEG2000-NH2 (PL = phospholipid) for the conjugation of thiol—siRNA through disulfide linkages for siRNA delivery into human T cells. (B) Confocal images of untreated cells (left) and SWCNT—siRNA CXCR4 treated cells (right) after PE-anti CXCR4 staining (Liu et al., 2007a,b,c,d). *siRNA*, short interfering RNA. *Copyright Wiley-VCH Verlag GmbH & Co. KGaA. Reproduced with permission.*

of siRNA into human T cells and primary cells (Fig. 5.5). They showed that nanotubes are capable of siRNA delivery to afford efficient RNAi of CXCR4 and CD4 receptors on human T cells and peripheral blood mononuclear cells (Liu et al., 2007a,b,c,d).

The use of phospholipid-functionalized SWCNTs through cleavable disulphide linkage was explored by Kam et al. (2005) to attach a variety of biological molecules. With this functionalization, the transporting, releasing, and nuclear translocation of DNA oligo-nucleotides in mammalian cells with SWCNT transporters was demonstrated. Kam et al. (2005), showed highly efficient delivery of siRNA by SWCNTs and more potent RNAi functionality than a generally used transfection agent, lipofectamine. In another studies, McCarroll et al. (2010) functionalized SWCNTs with lipids and natural amino acid–based dendrimers and complexed to siRNA and investigated silence of targeted gene in mice by delivering siRNA at doses of $<1\ \mathrm{mg\ kg^{-1}}$. They demonstrated that the injection of this complex at $0.96\ \mathrm{mg\ kg^{-1}}$ in mice showed an endogenous gene for apoliproprotein B, which was silenced in liver, lead to decrease in plasma levels of both apoliproprotein B and total plasma cholesterol. A nearly 80% of the RNA trigger molecules assembled with natural amino acid–based dendrimers was cleared from the body, after injection (48 h) indicating that the nanotubes did not cause siRNA aggregation or inhibit biodegradation and drug clearance in in vivo.

Synthesis, characterization, and cell uptake of a series of different-generation alkylated dendron MWCNT derivatives had been reported by Herrero et al. (2009). A fluorescently labeled, noncoding siRNA sequence showed that internalization of the alkylated Dendron MWCNT-siRNA complexes was improved for the later Dendron generations. The study demonstrated that dendron-functionalized CNTs are very efficient carriers of siRNA for the improvement of novel gene silencing and knockdown therapies (Herrero et al., 2009). Bartholomeusz et al. (2009) verified a new method, for delivering siRNA into cancer cells by noncovalently complexing unmodified siRNA with pristine SWCNTs. They showed that siRNA can be used to solubilize pristine SWCNTs and that noncovalent SWCNT/siRNA complexes can transfect cancer cells and effectively silence a targeted gene in cell culture and also in tumors both in vivo and in vitro (Bartholomeusz et al., 2009). Podesta et al. (2009) studied the treatment of a human carcinoma xenograft model using the delivery of a proprietary toxic siRNA sequence complexed with both cationic liposome delivery systems and with amino-functionalized MWCNTs $(\mathrm{MWCNT-NH_3^+})$. A human lung carcinoma tumor model in nude mice was established and treated, based on the hypothesis that functionalized CNTs can be applied to efficiently deliver siRNA directly to the cytoplasm and obtain a specific effect resulting in delayed tumor growth, thus considerably increasing survival (Podesta et al., 2009). This study clearly shows advantages of CNTs for the intracellular delivery of therapeutic agents in vivo in comparison with liposomes (Podesta et al., 2009).

The application of synthetic siRNA provides a potent tool to silence target genes with a high degree of specificity and is a novel approach for targeted cancer therapy and deserves further investigation for utilizing as a therapeutic approach against various tumor types.

5.4.3 DNA Delivery Using Carbon Nanotubes

Introduction of exogenous DNA into mammalian cells signifies a powerful method for manipulating signal transduction. However, the techniques are restricted by low transduction efficiency and low cell possibility after transduction (Cai et al., 2005). DNA could be encapsulated into CNTs in a water solute environment through an enormously rapid dynamic interaction process, provided that the tube size exceeds certain significant value. Either vander Waals or hydrophobic forces are found to be essential for the insertion process, with the former playing a more dominant role on DNA–CNT interaction (Gao et al., 2003). Ammonium-functionalized CNTs were examined for their ability to form supramolecular complexes with nucleic acids via electrostatic interactions. Many cationic systems are being studied for nucleic acid delivery to cells. The successful transfection was proven by showing that the cells produced the protein encoded by the DNA (Yang et al., 2007a,b,c; Kohli et al., 2005). CNTs spearing was used to attain tremendously high transfection efficiency with high postinduction cell viability in hard-to-transfect cells, as well as B cells and primary neurons. This technique uses nickel-embedded nanotubes to penetrate cell membranes via magnetic field driving, thus delivering macromolecules that are immobilized on the nanotubes. This method proposes some benefits, as well as increased control, decreased cytotoxicity, and greater efficiency than standard transfection reagents (Goldberg et al., 2007). Cai et al. successfully attained gene delivery into dividing and nondividing cells using the nanotube spearing technique. The nondividing cells, primary B cells and neurons, are notoriously hard to transfect and had been successfully transduced only by viral vectors so far. Their nonviral method, nanotube spearing, yielded transduction efficiency equal to that of the viral methods. Their techniques also benefit the in vitro gene delivery for overexpression and knockdown in a variety of cells (Cai et al., 2005). Ahmed et al. (2009) used cationic glycol polymer–functionalized SWCNTs as an in vitro gene transfer agent. The presence of net positive charge of the complex generates a "proton sponge" effect thereby leading to the release of DNA in cytosol. By increasing the net positive charge of the complex, it is possible to increase the transfection ability of the vehicle, regardless of the kind of cationic glycol polymer applied for the surface functionalization of SWCNTs. However, optimization of charge ratio and structural characterization should be used for the development of CNT-DNA gene delivery systems. Various features, including complex size, surface charge, DNA topology, and degree of condensation, take part in determining which charge ratio will be most favorable for gene transfer.

For gene delivery, Wu et al. (2008) developed SWCNTs as a carrier for single-stranded DNA probe delivery. Their studies showed that a SWCNT-modified DNA probe, which targets a specific mRNA inside living cells, has improved self-delivery capability and intracellular biostability in comparison with free DNA probes (Wu et al., 2008). To increase the loading efficiency of SWCNTs, the positively charged SWCNTs are functionalized with fluorescein-labeled double strands DNA (dsDNA). Yang et al. (2008a,b,c) applied folic acid–modified phospholipids to encapsulate the SWCNTs loaded with fluorescein-labeled dsDNA to produce the tumor cell—targeted functionalized SWCNTs. These multifunctionalized SWCNTs can target mouse and mammalian tumor cells efficiently in which folate receptors overexpress. Because of the similarity between the structure of dsDNA chains and siRNA, the multifunctionalized SWCNTs can be considered as good vectors for tumor-associated siRNA—targeted therapy (Yang et al., 2008a,b,c). Jung et al. (2010) fabricated SWCNTs dotted with Au nanocrystals (Au-SWCNTs) and then immobilized DNA probes on Au-SWCNTs by the conjugation of DNA functionalized at the 3′ end with a thiol group with Au dots of SWCNTs and followed by hybridization of complementary oligonucleotides. They revealed that Au-SWCNTs can be applied as a multipurpose gene delivery carrier in gene-based disease therapy (Fig. 5.6).

5.4.4 Plasmid DNA Delivery Using Carbon Nanotubes

pDNA was the first type of nucleic acid to be successfully transfected by CNTs in vitro. Pantarotto et al. demonstrated that CNT-mediated delivery of pDNA led to expression of marker proteins such as β-galactosidase in Chinese hamster ovary cells at approximately 10 times higher levels than naked pDNA alone. Furthermore, their group found that the overall charge ratio (+/−) of CNT-NH_3^+/pDNA complexes was an important factor determining the level of gene expression. Furthermore, various types of surface-modified CNTs with a range of different surface modifications have been used (Gao et al., 2006), cationic glycopolymers, PEI (Inoue et al., 2008), PAMAM hybrids (Qin et al., 2011) and ethylenediamine (Karmakar et al., 2011). Different studies have reported successful in vitro transfection of pDNA; however, there is very limited evidence to date of in vivo use of this technology. Delivery of pDNA via nanoparticle CNT (NP-CNT) hybrids has recently been reported in an in vivo canine model of restenosis (Paul et al., 2012).

Similar to the majority of nanoparticle-based delivery systems, the hurdles that CNT-mediated delivery of pDNA needs to overcome are (1) endocytosis; (2) escape from the endosomal vesicle; (3) release of the DNA from the CNT; (4) translocation into the nucleus; (5) DNA transcription; and (6) protein translation (Chen et al., 2012).

New strategies for increasing the efficiency of the delivery of exogenous DNA into cells using CNTs are poised to have significant boosts in the efficacy of gene therapy. An innovative approach involves utilizing ferromagnetic nickel—embedded nanotubes coated with a DNA plasmid that are driven into cells through a variable magnetic field (Cai et al., 2005). This new approach, termed "nanotube spearing," takes advantage of the ferromagnetic nickel catalyst particles that are enclosed in the tips of CNTs after their growth using a plasma-enhanced chemical vapour deposition (CVD) method. The residual nickel catalyst particles

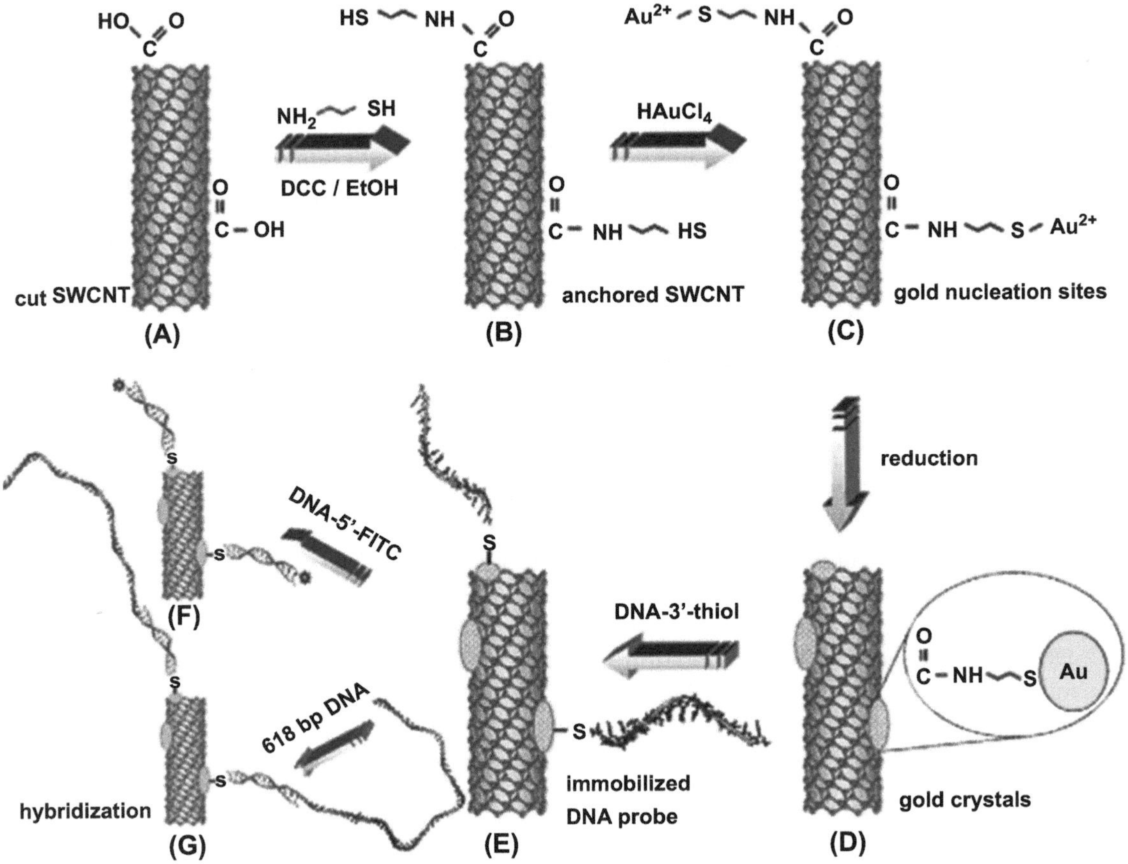

FIGURE 5.6

Overall schemes for fabrication of Au-SWCNTs, and the immobilization and hybridization of DNA. (A) A cut SWCNT with carboxylic acid groups, (B) aminoethanethiols anchored by amide linkages to an SWCNT, (C) Au nucleation sites created in the thiol groups of SWCNTs, (D) Au nanocrystals fabricated by a two-phase redox reaction with the hydrogen tetrachloroaurate, (E) DNA probes immobilized onto the Au nanocrystals of SWCNTs, (F) hybridization of fluorescein isothiocyanate-labeled complementary DNA, and (G) hybridization of a long-chain oligonucleotide with a sequence complementary to that of probe DNA. *Reprinted with permission from Jung, D.H., et al., 2010. Fabrication of single-walled carbon nanotubes dotted with Au nanocrystals: potential DNA delivery nanocarriers. Carbon 48 (4), 1070–1078.*

allow nanotubes to respond to external magnetic agitation. By tuning the magnetic fields surrounding the cells, these CNTs carrying a genetic payload can be used to "spear" the cells and deliver their contents directly into the nucleus. Carbon nanotubes and graphene oxide are naturally taken up through endosomal or lysosomal pathways, and unfortunately many plasmids are hydrolyzed and thus destroyed prior to releasing their genetic payload. Utilizing nanotube spearing, DNA can be directly delivered to the cell nucleus, contributing to the significant boost in efficiency as observed in this transfection method for DNA plasmids containing the sequence of enhanced green fluorescent protein. Using CNTs alone requires concentrations of ~1–5 μM, yet through nanotube spearing facilitated by the external magnetic field; a low concentration of merely ~100 fM is required, thus increasing the molecular shuttling efficiency by 10^7-fold (Cai et al., 2005).

5.4.5 Micro-RNA Delivery Using Carbon Nanotubes

The term miRNA refers to a small noncoding RNA molecule that functions in transcriptional and posttranscriptional regulation of gene expression (Chen et al., 2007). miRNAs function via basepairing with complementary sequences within mRNA molecules, usually resulting in gene silencing via translational repression or target degradation (Kusenda et al., 2006). Currently, little is known about the ability of CNTs to deliver miRNA sequences in an intracellular manner. There is one report describing the use of "unzipped" MWCNTs as miRNA delivery vectors (Dong et al., 2011). It was reported that PEI-grafted graphene nanoribbon (PEI-g-GNR), formed by longitudinally unzipping MWCNT, can effectively transfer a gene probe into cells. The transfer process was monitored for the specific recognition of "locked nucleic acid molecular beacons" (LNA-m-MB) by miRNA in the cytoplasm, which led to their hybridization and induction of a strong fluorescent signal due to the separation of the fluorescent dye from the quencher attached to the LNA-m-MB. The PEI-g-GNR nanocarrier showed little cytotoxicity and could protect the LNA-m-MB from nuclease digestion or single-strand-binding -protein (SSB) interaction. The proton sponge effect of PEI led to relatively high transfection efficiency via endosomolysis, which allowed the sensitive detection of the recognized target miRNA.

There is still incomplete knowledge about the molecular mechanisms and specificity of regulation of gene expression by miRNAs, so the lack of published studies on CNT-mediated miRNA delivery for gene therapy applications is not surprising, although their clinical application is promising (Fig. 5.8) (Broderick et al., 2011).

5.4.6 Graphene in Gene Delivery

Graphene-mediated gene delivery is another emerging area of research that has been considered by research groups who are focusing on nonviral gene delivery. The majority of the scientific studies in this area have reported that graphene derivatives could be used as a promising carrier with high gene packing densities due to their large surface areas. Some reports also demonstrated that the graphene derivatives could overcome many barriers and increase gene accumulation through the targeting of specific cells, resulting in increased gene transfection. However, modifications of graphene derivatives with polymers are required to render cationic surface properties and facilitate electrostatic interactions with anionic oligonucleotides. Mostly polyethylenimine

(PEI), PEG, and poly (sodium 4-styrenesulfonates) have been considered for gene delivery through graphene derivatives (Wang et al., 2014a,b; Chen et al., 2011a,b; Ren et al., 2012). Zhi et al. (2013) reported that layer-by-layer assembled GO carrying miR-21 simultaneously targeted siRNA and adriamycin and was able to overcome multidrug resistance. The polymer associated with GO significantly enhanced cellular uptake in MCF-7 cells. Paul et al. (2014) applied GO-based injectable hydrogels to host angiogenic genes and demonstrated the hydrogels to be a potential cardiac implant for vasculogenesis. This invention can be widely used in tissue engineering research for generating blood vessels to circulate blood and nutrients for the cells located inside the hydrogel. The thermogenesis properties of rGO have been properly implicated by Kim et al. (2014), where they showed that light sensitive rGO can be used to escape or overcome the barriers of current gene therapeutic strategies. The translocation of a gene from cellular membrane to the nucleus is a hurdle due to the endosomal barrier between the routes. Though many strategies have been taken into consideration to overcome this barrier, progress is far away from expectations. This report has suggested that the heat generated through the irradiation of rGO helps to overcome endosomal escape, thus enhancing gene expression (Fig. 5.7).

5.4.7 Carbon Dots for Gene Delivery

Carbon dots, another member of the carbon nanomaterial family, can have diameters lower than 10 nm that facilitate a high level of cellular uptake. Liu et al. (2012) manufactured carbon dots in a one-step microwave synthesis and conjunction with PEI, which not only acted as a surface passivation agent to enhance the fluorescence of carbon dots, but also played an important role in electrostatically loading DNA on the surface of carbon dots. Nanodiamonds have also been intensively studied as a nanocarrier of nucleic acids. Zhang et al. (2009) reported PEI-functionalized nanodiamonds with high transfection efficiency of plasmid DNA and without the high toxicity inherent to the free PEI molecules in delivery vehicles. Chen et al. (2010) also reported on siRNA delivery with PEI-functionalized nanodiamonds with enhanced efficacy. Alhaddad et al. (2011) have compared the siRNA delivery efficacy using different cationic polymers such as PEI and poly(allylamine hydrochloride) (PAH) as the electrostatic "glue" to bind siRNA on the nanodiamonds and then concluded that the PAH-coated nanodiamonds have a higher affinity to siRNA; however, PAH-coated nanodiamonds are much slower in releasing the loaded siRNA and thus have a lower transfection efficacy than the PEI-coated counterparts. The intrinsic fluorescence of carbon dots and nanodiamonds makes these two carbon nanomaterials particularly useful as multifunctional gene delivery vectors that allow for real-time tracking of their locations inside live cells through microscopic fluorescence imaging.

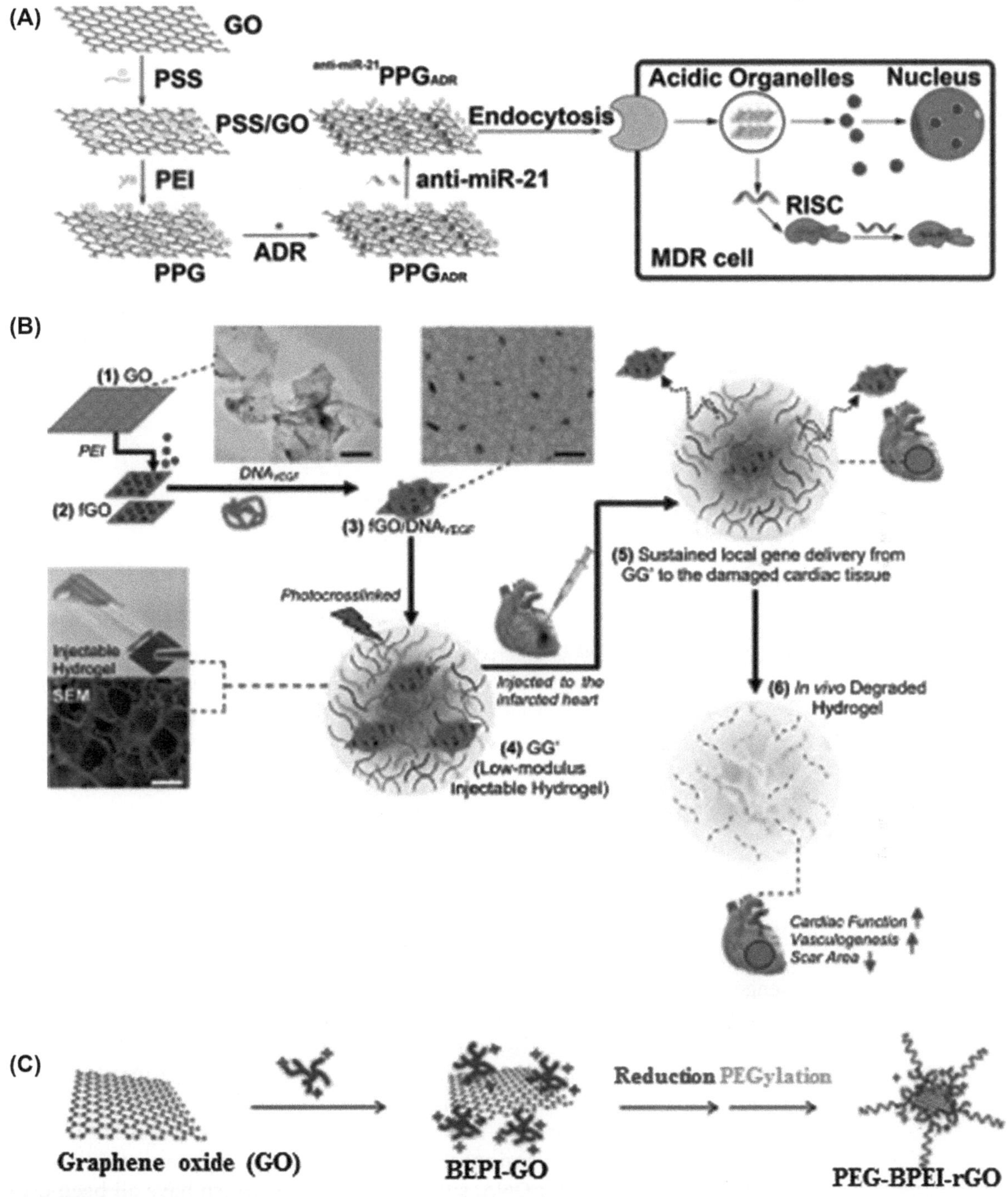

FIGURE 5.7

(A) Schematic of the fabrication of polyethylenimine (PEI)/poly(-sodium 4-styrenesulfonates) (PSS)/graphene oxide and multidrug resistance reversion. (B) The scheme represents the preparation of injectable hydrogel incorporating with plasmid DNA (pDNA) and graphene oxide (GO) for acute myocardial infarction therapy (Paul et al., 2014). (C) Synthesis scheme of PEG—BPEI-rGO nanocomposite. BPEI-rGO was synthesized from BPEI-GO. To enhance the colloidal stability of BPEI-GO and BPEI-rGO, polyethylene glycol monomethyl ether (mPEG) was conjugated with 1,10-carbonyldiimidazole coupling (Kim et al., 2014). *BPEI*, branched polyethylenimine; *PEG*, polyethylene glycol. *(A) Adopted with permission from Wu, S.Y., et al., 2015. (B) Copyright by the American Chemical Society. (C) Copyright Wiley-VCH Verlag GmbH & Co. KGaA. Reproduced with permission.*

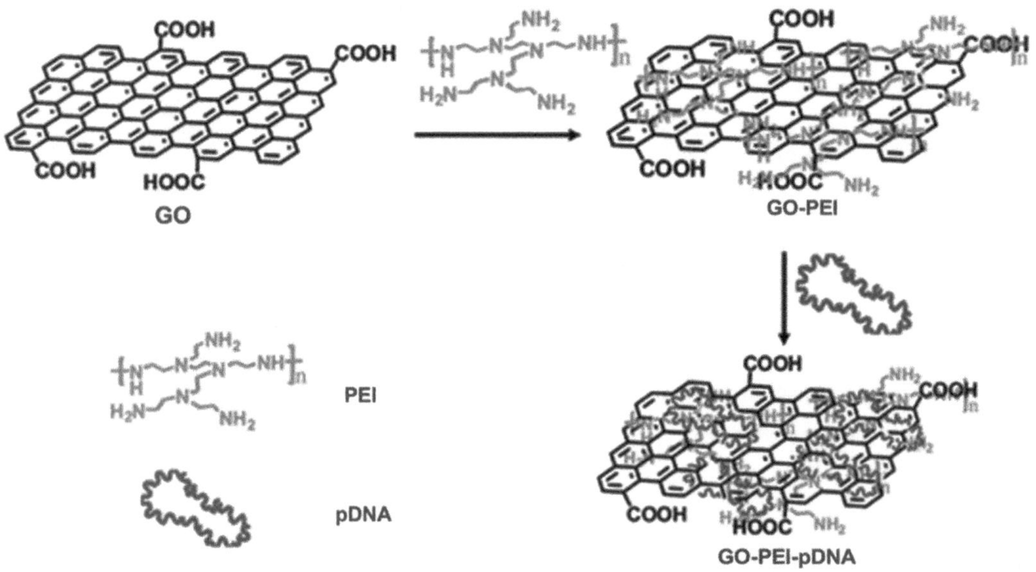

FIGURE 5.8

Schematic illustration showing the synthesis of GO—PEI—DNA complexes via a layer-by-layer assembly process. First, graphene oxide (GO) was noncovalently functionalized by polyethylenimine (PEI) polymers, forming positively charged GO—PEI complexes. Next, negatively charged pDNA was loaded on the GO—PEI complexes also by electrostatic interactions (Feng et al., 2011). *Copyright permission from the Royal Society of Chemistry.*

5.4.8 Graphene Oxide in Gene Delivery

Gene therapy is a novel and promising approach to treat various diseases caused by genetic disorders, including cystic fibrosis, Parkinson's disease, and cancer (Yang et al., 2007a,b,c). Successful gene therapy requires a gene vector that protects DNA from nuclease degradation and facilitates cellular uptake of DNA with high transfection efficiency (Naldini et al., 1996). The major challenge facing the development of gene therapy is lack of efficient and safe gene vectors (Mintzer et al., 2009). A great deal of research effort has been directed toward treating diseases by introducing genetic material (nucleic acids) to replace, regulate, repair, add, or delete a specific genetic target that is responsible for the manifestation of a disease. The therapeutic effects accomplished by successful gene therapy have led to the next generation of disease-modifying medical interventions through a wide range of therapeutically active nucleic acids, including pDNA, siRNA, and miRNA, which have all been used to either overexpress a protein or inhibit its expression. Gene delivery is not only the most important but also the most problematic aspect of gene therapy. Gene delivery systems can be viral vectors or nonviral vector systems (Berindan-Neagoe et al., 2009). Viral vectors (retrovirus, lentivirus, herpes virus simplex, RNA virus, and adenovirus) can achieve high transfection efficiencies. However, the application of viral vectors in the clinical setting is hindered by concerns that they can actually cause diseases including cancer; the limited capacity of genetic payload; systemic instability; poor capability to target specific cell populations, and the expense of production (Atkinson et al., 2007). The alternative, nonviral delivery vectors exhibit advantages over viral vectors in terms of relative safety; they are chemically controllable with reduced toxicity (Zhang et al., 2007) and the ability to deliver genes without any size limitation. However, nonviral gene delivery methods have not been very successful clinically compared to viral vectors due to serious limitations, including incapacity to transverse nuclear membrane, low transfection efficiencies and poor transgene expression (Wang et al., 2012). A variety of nanovectors have been explored as possible gene therapy vehicles, including peptides, liposomes, polymers, dendrimers, micro- and nanoparticles, CNTs, and many others (Lo et al., 2008; Al-Jamal et al., 2011). However, in vivo therapeutic delivery of nucleic acids is challenging for a number of reasons, including lack of stability against endogenous enzymes and inherently poor ability to cross cellular membranes.

5.5 CONCLUSION

Carbon nanomaterials have provided an excellent technology platform in the field of drug delivery. Many achievements have been evidenced in delivering drugs such as anticancer, neurodegenerative disorders, antimycobacterials, antiinflammatory, topical agents, biomolecules, etc. However, there are several limitations associated with the design and characterization of these nanomaterials which include cellular toxicity; ability to achieve optimal drug loading on the nanotropes, to control drug release, delivery, and prediction of exact pharmacokinetic profile. Moreover, carbon nanotropes are considered as a novel paradigm in drug delivery which not only improves the ability to deliver drugs with wide systemic and topical applications but also the ability to control stability, solubility and drug carrying capacity. Thus, it can be concluded that the emergence of nanotherapeutics/nanoformulations provide the pharmaceutical scientists with a tool which, when developed, could revolutionize drug delivery research, and the carbon nanotropes promises to be an integral part of it. The current research on delivery of drugs, genes, we expect, will extend to that of proteins, growth factors, and other biomolecules, either single or combined use of the aforementioned biomolecules, for cancer and other disease therapies. Given its structural features and exceptional physicochemical properties, design and construction of GO-based theranostic platform with multifunctionalities and multimodalities is a new direction to pursue. Definitely, these goals can only be reached by the joint efforts from chemistry, biomedicine, materials sciences, and nanotechnology. To this end, development of suitable chemical synthesis and functionalization approaches for precise control over size, size distribution, morphology, structural defects, and oxygen-containing groups of GO is urgently needed as this is closely correlated to the performance of the GO-based nanomaterials for biomedical applications and the safety issues as well. For GO-based biosensing based on FRET principle, tuning electronic property of GO by controllable modification and reduction of originally prepared GO, and development of techniques to integrate GO into practical devices having high sensitivity, selectivity with acceptable reproducibility, reliability, and low cost remain a big challenge. As for GO-based bioimaging, although the GQDs have distinct advantages over II–IV QDs in their intrinsic biocompatibility, safety, and facile functionalization, the weak fluorescence intensity (with quantum yield about 10%) and broad emission band (with bandwidth beyond 100 nm) are certainly major obstacles for their use in biodetection and labeling, so more efforts must be paid to prepare GQDs with a good control of size and size distribution, surface defects, and their functionalization to improve fluorescence quantum yield and other important properties. The research on graphene-/GO-based scaffold materials for cell culture is a relatively new direction that deserves special attention. The study in this field so far demonstrated that graphene and GO are able to accelerate the growth, differentiation, and proliferation of stem cells, and therefore hold great promise in tissue engineering, regenerative medicine, and other biomedical fields. We have discussed the rapid and exciting advances in the biomedical applications of graphene with focus on drug delivery and bioimaging made in the last few years. This trend, we believe, will continue and even speed up in the years to come.

References

Abdullah Al, N., et al., 2013. Target delivery and cell imaging using hyaluronic acid-functionalized graphene quantum dots. Mol. Pharm. 10, 3736–3744.

Ahmed, M., et al., 2009. Cationic glyco-functionalized single-walled carbon nanotubes as efficient gene delivery vehicles. Bioconjug. Chem. 20 (11), 2017–2022.

Ajima, K., et al., 2005. Carbon nanohorns as anticancer drug carriers. Mol. Pharm. 2, 475–480.

Ajima, K., et al., 2008. Enhancement of in vivo anticancer effects of cisplatin by incorporation inside single-wall carbon nanohorns. ACS Nano 2, 2057–2064.

Albertorio, F., et al., 2009. Base dependent DNA-carbon nanotube interactions: activation enthalpies and assembly-disassembly control. Nanotechnology 20, 395101.

Alhaddad, A., et al., 2011. Nanodiamond as a vector for siRNA delivery to Ewing sarcoma cells. Small 7, 3087–3095.

Ali-Boucetta, H., et al., 2008. Multiwalled carbon nanotube-doxorubicin supramolecular complexes for cancer therapeutics. Chem. Commun. 459–461.

Al-Jamal, K.T., et al., 2011. Functional motor recovery from brain ischemic insult by carbon nanotube-mediated siRNA silencing. Proc. Natl. Acad. Sci. U.S.A. 108 (27), 10952–10957.

An, J., et al., 2013. Synthesis of a biocompatible gelatin functionalized graphene nanosheets and its application for drug delivery. Mater. Sci. Eng. C 33, 2827–2837.

Andersson, M., et al., 1999. Doxorubicin versus methotrexate both combined with cyclophosphamide, 5-fluorouracil and tamoxifen in postmenopausal patients with advanced breast cancer-a randomised study with more than 10 years follow-up from the Danish breast cancer cooperative group. Eur. J. Cancer 35 (1), 39–46.

Arora, S., et al., 2012. Nanotoxicology and *in vitro* studies: the need of the hour. Toxicol. Appl. Pharmacol. 258, 151–165.

Atkinson, et al., 2007. Membrane type matrix metalloproteinases (MMPs) show differential expression in non-small cell lung cancer (NSCLC) compared to normal lung: correlation of MMP-14 mRNA expression and proteolytic activity. Eur. J. Cancer 43, 1764–1771.

Azzam, T., et al., 2004. Current developments in gene transfection agents. Curr. Drug Deliv. 1 (2), 165–193.

Bai, H., et al., 2010. A pH-sensitive graphene oxide composite hydrogel. Chem. Commun. 46 (14), 2376–2378.

Bakry, R., et al., 2007. Medicinal applications of fullerenes. Int. J. Nanomed. 2, 639–649.

Barreto, J.A., et al., 2011. Nanomaterials: applications in cancer imaging and therapy. Adv. Mater. 23, H18–H40.

Bartholomeusz, G., et al., 2009. In vivo therapeutic silencing of hypoxia-inducible factor 1 alpha (HIF-1alpha) using single-walled carbon nanotubes noncovalently coated with siRNA. Nano Res. 2 (4), 279–291.

Bates, K., et al., 2013. Carbon nanotubes as vectors for gene therapy: past achievements, present challenges and future goals. Adv. Drug Deliv. Rev. 65, 2023–2033.

Berindan-Neagoe, I., et al., 2009. p53 gene therapy using RNA interference. J. Buon 14, S51–S59.

Bhirde, A.A., et al., 2009. Targeted killing of cancer cells in vivo and in vitro with EGF directed carbon nanotube-based drug delivery. ACS Nano 3, 307–316.

Bianco, A., et al., 2005. Applications of carbon nanotubes in drug delivery. Curr. Opin. Chem. Biol. 9, 674–679.

Bianco, A., et al., 2008. Opportunities and challenges of carbon-based nanomaterials for cancer therapy. Expert Opin. Drug Deliv. 5, 331–342.

Bolskar, R.D., 2012. Fullerenes for drug delivery. In: Bhushan, B. (Ed.), Encyclopedia of Nanotechnology. Springer, Amsterdam, The Neatherlands, pp. 898–911.

Brettreich, M., Hirsch, A., 1998. A highly water-soluble dendro[60]fullerene. Tetrahedron Lett. 39 (18), 2731–2734.

Broderick, J.A., et al., 2011. MicroRNA therapeutics. Gene Ther. 18 (12), 1104–1110.

Cai, D., et al., 2005. Highly efficient molecular delivery into mammalian cells using carbon nanotube spearing. Nat. Methods 2 (6), 449–454.

Cai, et al., 2006. Peptide-labeled near-infrared quantum dots for imaging tumor vasculature in living subjects. Nano Lett. 6, 669–676.

Chawla, P., et al., 2010. Fullerenes: from carbon to nanomedicine. Mini. Rev. Med. Chem. 10, 662–677.

Chen, K., et al., 2007. The evolution of gene regulation by transcription factors and microRNAs. Nat. Rev. Genet. 8 (2), 93–103.

Chen, M., et al., 2010. Nanodiamond vectors functionalized with polyethylenimine for siRNA delivery. J. Phys. Chem. Lett. 1, 3167–3171.

Chen, B., et al., 2011a. Polyethylenimine-functionalized graphene oxide as an efficient gene delivery vector. J. Mater. Chem. 21, 7736–7741.

Chen, W.H., et al., 2011b. Composites of aminodextran-coated Fe_3O_4 nanoparticles and graphene oxide for cellular magnetic resonance imaging. ACS Appl. Mater. Interfaces 3 (10), 4085–4091.

Chen, H., Ma, X., Li, Z., et al., 2012. Functionalization of single-walled carbon nanotubes enables efficient intracellular delivery of siRNA targeting MDM2 to inhibit breast cancer cells growth. Biomed. Pharmacother. 66 (5), 334–338.

Chen, T., et al., 2014. Graphene quantum dot-capped mesoporous silica nanoparticles through an acid-cleavable acetal bond for intracellular drug delivery and imaging. J. Mater. Chem. B 2, 4979–4982.

Cheng, et al., 2014. C_8-structured carbon quantum dots: synthesis, blue and green double luminescence, and origins of surface defects. Carbon 79, 165–174.

Cherukuri, P., et al., 2004. Near-infrared fluorescence microscopy of single-walled carbon nanotubes in phagocytic cells. J. Am. Chem. Soc. 126, 15638–15639.

Cheung, W., et al., 2010. DNA and carbon nanotubes as medicine. Adv. Drug Deliv. Rev. 62, 633–649.

Daniel, D., et al., 2007. A review of DNA functionalized/grafted carbon nanotubes and their characterization. Sens. Actuator B 122, 672–682.

Debouzy, J.C., et al., 2010. Hydrophobic double walled carbon nanotubes interaction with phopholipidic model membranes: (1)h-, (2)h-, (31)p nmr and esr study. Environ. Toxicol. Pharmacol. 30, 147–152.

Dellinger, A., et al., 2010. Uptake and distribution of fullerenes in human mast cells. Nanomedicine 6, 575–582.

Dembereldorj, U., et al., 2012. A spatiotemporal anticancer drug release platform of PEGylated graphene oxide triggered by glutathione *in vitro* and *in vivo*. J. Mater. Chem. 22, 23845–23851.

Depan, D., et al., 2011. Controlled release of drug from folate-decorated and graphene mediated drug delivery system: synthesis, loading efficiency, and drug release response. Mater. Sci. Eng. C 31 (7), 1305–1312.

Dhar, S., et al., 2008. Targeted single-wall carbon nanotube-mediated Pt (IV) prodrug delivery using folate as a homing device. J. Am. Chem. Soc. 130, 11467–11476.

Dong, H.F., et al., 2010. Fluorescence resonance energy transfer between quantum dots and graphene oxide for sensing biomolecules. Anal. Chem. 82 (13), 5511–5517.

Dong, H., et al., 2011. The use of polyethylenimine-grafted graphene nanoribbon for cellular delivery of locked nucleic acid modified molecular beacon for recognition of microRNA. Biomaterials 32 (15), 3875–3882.

Dugan, L.L., et al., 1997. Carboxyfullerenes as neuroprotective agents. Proc. Natl. Acad. Sci. U.S.A. 94 (17), 9434–9439.

El-Aneed, et al., 2004. An overview of current delivery system in cancer gene therapy. J. Control Release 94 (1), 1–14.

Elfinger, M., et al., 2008. Nanocarriers for gene delivery-polymer structure, targeting ligands and controlled-release devices. Curr. Nanosci. 4, 322–353.

Fan, X., et al., 2013. The preparation and drug delivery of a graphene-carbon nanotube-Fe_3O_4 nanoparticle hybrid. J. Mater. Chem. B 1, 2658–2664.

Feazell, R.P., 2007. Soluble single-walled carbon nanotubes as longboat delivery systems for platinum (IV) anticancer drug design. J. Am. Chem. Soc. 129, 8438–8439.

Feng, L.Z., et al., 2011. Graphene based gene transfection. Nanoscale 3, 1252–1257.

Fenoglio, et al., 2010. Progranulin gene variability increases the risk for primary progressive multiple sclerosis in males. Genes Immun. 11, 497–503.

Friedman, et al., 1993. Childhood conscientiousness and longevity: health behaviors and cause of death. J. Personality Soc. Psychol. 68 (4), 696–703.

Gao, H., et al., 2003. Spontaneous insertion of DNA oligonucleotides into carbon nanotubes. Nano Lett. 3 (4), 471–473.

Gao, L., et al., 2006. Carbon nanotube delivery of the GFP gene into mammalian cells. Chem. Bio Chem. 7 (2), 239–242.

Gavrilov, V., et al., 2005. The combined treatment of 1,25-dihydroxyvitamin D3 and a non-steroid anti-inflammatory drug is highly effective in suppressing prostate cancer cell line (LNCaP) growth. Anticancer Res. 25 (5), 3425–3429.

Goenka, S., et al., 2014. Graphene-based nanomaterials for drug delivery and tissue engineering. J. Control. Release 173, 75–88.

Goldberg, M., et al., 2007. Nanostructured materials for applications in drug delivery and tissue engineering. J. Biomat. Sci-Polym. E 18 (3), 241–268.

Herrero, M.A., et al., 2009. Synthesis and characterization of a carbon nanotube-dendron series for efficient siRNA delivery. J. Am. Chem. Soc. 131 (28), 9843–9848.

Holister, P., et al., 2003. Nanoparticles, technology white papers nr. 3. Cientifica 1–12.

Hong, G.S., et al., 2012. Three-dimensional imaging of single nanotube molecule endocytosis on plasmonic substrates. Nat. Commun. 3, 700, 1–9.

Hu, L.M., et al., 2014. Multifunctional carbon dots with high quantum yield for imaging and gene delivery. Carbon 67, 508–513.

Huang, Y.L., et al., 1998. Blockage of apoptotic signaling of transforming growth factor-β in human hepatoma cells by carboxyfullerene. Eur. J. Biochem. 254, 38–43.

Huang, P., et al., 2011. Folic acid-conjugated graphene oxide loaded with photosensitizers for targeting photodynamic therapy. Theranostics 1, 240–250.

Hummers, W.S., Offeman, R.E., 1958. Preparation of graphitic oxide. J. Am. Chem. Soc. 80, 1339.

Inoue, Y., et al., 2008. Site-specific gene transfer with high efficiency onto a carbon nanotube-loaded electrode. J. R. Soc. Interface 5 (25), 909–918.

Isobe, H., et al., 2006. Nonviral gene delivery by tetraamino fullerene. Mol. Pharm. 3, 124–134.

Jain, K.K., 2005. The role of nanobiotechnology in drug discovery. Drug Discov. Today 10, 1435–1442.

Jensen, A.W., et al., 1996. Biological applications of fullerenes. Bioorg. Med. Chem. 4, 767–779.

Jeong, J.H., et al., 2009. siRNA conjugate delivery systems. Bioconjug. Chem. 20, 5–14.

Ji, S.R., et al., 2010. Carbon nanotubes in cancer diagnosis and therapy. Biochim. Biophys. Acta 1806, 29–35.

Jung, D.H., et al., 2010. Fabrication of single-walled carbon nanotubes dotted with Au nanocrystals: potential DNA delivery nanocarriers. Carbon 48 (4), 1070–1078.

Kaesermann, F., et al., 1997. Photodynamic inactivation of enveloped viruses by buckminsterfullerene. Antivir. Res. 34 (1), 65–70.

Kam, N.W., et al., 2005. Functionalization of carbon nanotubes via cleavable disulfide bonds for efficient intracellular delivery of siRNA and potent gene silencing. J. Am. Chem. Soc. 127 (36), 12492–12493.

Kam, N.W.S., et al., 2006. Carbon nanotubes as intracellular transporters for proteins and DNA: an investigation of the uptake mechanism and pathway. Angew. Chem. Int. Ed. 45, 577–581.

Karmakar, A., et al., 2011. Ethylenediamine functionalized-single-walled nanotube (f-SWNT)-assisted in vitro delivery of the oncogene suppressor p53 gene to breast cancer MCF-7 cells. Int. J. Nanomedicine 6, 1045–1055.

Kim, H., Kim, W.J., 2014. Photothermally controlled gene delivery by reduced graphene oxide–polyethylenimine nanocomposite. Small 10, 117–126.

Kim, H., et al., 2013a. Photothermally triggered cytosolic drug delivery *via* endosome disruption using a functionalized reduced graphene oxide. ACS Nano 7, 6735–6746.

Kim, H., et al., 2013b. Transfection and intracellular trafficking properties of carbon dot-gold nanoparticle molecular assembly conjugated with PEI-pDNA. Biomaterials 34 (29), 7168–7180.

Kohli, P., et al., 2005. Smart nanotubes for biotechnology. Curr. Pharm. Biotechnol. 6, 35–47.

Kotelnikova, R.A., et al., 2003. Nanobionics of pharmacologically active derivatives of fullerene C_{60}. J. Nanoparticle Res. 5, 561–566.

Kovtyukhova, N.I., et al., 1999. Layer-by-layer assembly of ultrathin composite films from micron-sized graphite oxide sheets and polycations. Chem. Mater. 11, 771–778.

Kraszewski, S., et al., 2012. How long a functionalized carbon nanotube can passively penetrate a lipid membrane. Carbon 50, 5301–5308.

Krusic, P.J., et al., 1991. Radical reactions of C_{60}. Science 254 (5035), 1183–1185.

Kusenda, B., et al., 2006. MicroRNA biogenesis, functionality and cancer relevance. Biomed. Pap. Med. Fac. Univ. Palacky. Olomouc Czech Repub. 150 (2), 205–215.

Lacerda, L., et al., 2012. Translocation mechanisms of chemically functionalised carbon nanotubes across plasma membranes. Biomaterials 33, 3334–3343.

Lee, D.Y., et al., 2011. Blood compatible graphene/heparin conjugate through noncovalent chemistry. Biomacromolecules 12 (2), 336–341.

Lei, H.Z., et al., 2011. Adsorption of double-stranded DNA to graphene oxide preventing enzymatic digestion. Nanoscale 3, 3888–3892.

Li, S.S., et al., 2008. Applications of carbon nanotubes in drug and gene delivery. Prog. Chem. 20, 1798–1803.

Liu, S.R., et al., 2004. Clinical features and mismatch repair gene mutation screening in Chinese patients with hereditary nonpolyposis colorectal carcinoma. World J. Gastroenterol 10 (18), 2647–2651.

Liu, Y., et al., 2007a. Nanomedicine for drug delivery and imaging: a promising avenue for cancer therapy and diagnosis using targeted functional nanoparticles. Int. J. Cancer 120, 2527–2537.

Liu, Z., et al., 2007b. In vivo biodistribution and highly efficient tumour targeting of carbon nanotubes in mice. Nat. Nanotechnol. 2, 47–52.

Liu, Z., et al., 2007c. siRNA delivery into human T cells and primary cells with carbon-nanotube transporters. Angew. Chem. Int. Ed. Engl. 46 (12), 2023–2027.

Liu, Z., et al., 2007d. Supramolecular chemistry on water-soluble carbon nanotubes for drug loading and delivery. ACS Nano 1, 50–56.

Liu, Z., et al., 2008a. Drug delivery with carbon nanotubes for in vivo cancer treatment. Cancer Res. 68, 6652–6660.

Liu, Z., et al., 2008b. Circulation and long-term fate of functionalized, biocompatible single walled carbon nanotubes in mice probed by Raman spectroscopy. Proc. Natl. Acad. Sci. U.S.A. 105, 1410–1415.

Liu, Z., et al., 2008c. PEGylated nanographene oxide for delivery of water-insoluble cancer drugs. J. Am. Chem. Soc. 130 (33), 10876–10877.

Liu, Z., et al., 2009a. Carbon nanotubes in biology and medicine: in vitro and in vivo detection, imaging and drug delivery. Nano Res. 2 (2), 85–120.

Liu, S., et al., 2009b. Sharper and faster "nano darts" kill more bacteria: a study of antibacterial activity of individually dispersed pristine single-walled carbon nanotube. ACS Nano 3 (12), 3891–3902.

Liu, Z., et al., 2009c. Preparation of carbon nanotube bio-conjugates for biomedical applications. Nat. Protoc. 4, 1372–1382.

Liu, Z., et al., 2009d. Supramolecular stacking of doxorubicin on carbon nanotubes for in vivo. Cancer Ther. Angew. Chem. Int. Ed. 48, 7668–7672.

Liu, K., et al., 2011a. Green and facile synthesis of highly biocompatible graphene nanosheets and its application for cellular imaging and drug delivery. J. Mater. Chem. 21, 12034–12040.

Liu, Z., et al., 2011b. Carbon materials for drug delivery & cancer therapy. Mater. Today 14 (7–8), 316–323.

Liu, C.J., et al., 2012. Nano-carrier for gene delivery and bioimaging based on carbon dots with PEI-passivation enhanced fluorescence. Biomaterials 33, 3604–3613.

Liu, Y., et al., 2014. Harnessing the collective properties of nanoparticle ensembles for cancer theranostics. Nano Res. 7, 1719–1730.

Lo, S.L., et al., 2008. An endosomolytic Tat peptide produced by incorporation of histidine and cysteine residues as a nonviral vector for DNA transfection. Biomaterials 29 (15), 2408–2414.

Loh, K.P., et al., 2010. Graphene oxide as a chemically tunable platform for optical applications. Nat. Chem. 2, 1015–1024.

Ma, D., et al., 2011. Tunable supramolecular hydrogel for in situ encapsulation and sustained release of bioactive lysozyme. J. Colloid Interface Sci. 359, 399–406.

Maity, A.R., et al., 2014. Carbohydrate coated, folate functionalized colloidal graphene as a nano-carrier for both hydrophobic and hydrophilic drugs. Nanoscale 6, 2752–2758.

Mao, H., et al., 2013. Uptake and intracellular distribution of collagen-functionalized single-walled carbon nanotubes. Biomaterials 34, 2472–2479.

Marchesan, S., et al., 2005. Anti-HIV properties of cationic fullerene derivatives. Bioorg. Med. Chem. Lett. 15 (15), 3615–3618.

Mashino, T., et al., 2005. Human immunodeficiency virus-reverse transcriptase inhibition and hepatitis C virus RNA-dependent RNA polymerase inhibition activities of fullerene derivatives. Bioorg. Med. Chem. Lett. 15 (4), 1107–1109.

McCarroll, J., et al., 2010. Nanotubes functionalized with lipids and natural amino acid dendrimers: a new strategy to create nanomaterials for delivering systemic RNAi. Bioconjug. Chem. 21 (1), 56–63.

Mintzer, M.A., Simanek, E.E., 2009. Nonviral vectors for gene delivery. Chem. Rev. 109 (2), 259–302.

Mitchell, D.R., et al., 2001. The synthesis of megatubes: new dimensions in carbon materials. Inorg. Chem. 40 (12), 2751–2755.

Mudshinge, S.R., et al., 2011. Nanoparticles: emerging carriers for drug delivery. Saudi Pharm. J. 19 (3), 129–141.

Mukherjee, P., et al., 2015. Tunable luminescent carbon nanospheres with well-defined nanoscale chemistry for synchronized imaging and therapy. Small 11 (36), 4691–4703.

Muralkami, T., et al., 2004. Drug-loaded carbon nanohorns: adsorption and release of dexamethasone in vitro. Mol. Pharm. 1, 399–405.

Muralkami, T., et al., 2006. Solubilization of single-wall carbon nanohorns using a PEG-doxorubicin conjugate. Mol. Pharm. 3, 407–414.

Nakamura, E., et al., 2003. Functionalized fullerenes in water. The first 10 years of their chemistry, biology, and nanoscience. Acc. Chem. Res. 36, 807–815.

Naldini, L., et al., 1996. In vivo gene delivery and stable transduction of nondividing cells by a lentiviral vector. Science 272, 263–267.

Novoselov, K.S., et al., 2012. A roadmap for graphene. Nature 490, 192–200.

Nurunnabi, M., et al., 2013a. In vivo biodistribution and toxicology of carboxylated graphene quantum dots. ACS Nano 7, 6858–6867.

Nurunnabi, et al., 2013b. Surface coating of graphene quantum dots using mussel-inspired polydopamine for biomedical optical imaging. ACS Appl. Mater. Interfaces 5 (16), 8246–8253.

Pan, Y.Z., et al., 2011. Water-soluble poly(N-isopropylacrylamide) -graphene sheets synthesized via click chemistry for drug delivery. Adv. Funct. Mater. 21 (14), 2754–2763.

Pandey, S., et al., 2013a. Carbon dots functionalized gold nanorod mediated delivery of doxorubicin: tri-functional nano-worms for drug delivery, photothermal therapy and bioimaging. J. Mater. Chem. B 1, 4972–4982.

Pandey, S., et al., 2013b. Cysteamine hydrochloride protected carbon dots as a vehicle for the efficient release of the anti-schizophrenic drug haloperidol. RSC Adv. 3, 26290–26296.

Pantarotto, D., et al., 2003. Immunization with peptide-functionalized carbon nanotubes enhances virus-specific neutralizing antibody responses. Chem. Biol. 10, 961–966.

Pantarotto, D., et al., 2004. Functionalized carbon nanotubes for plasmid DNA gene delivery. Angew. Chem. Int. Ed. 43, 5242–5246.

Partha, R., et al., 2009. Biomedical applications of functionalized fullerene-based nanomaterials. Int. J. Nanomed. 4, 261–275.

Paul, A., et al., 2012. The attenuation of restenosis following arterial gene transfer using carbon nanotube coated stent incorporating TAT/DNAAng1+ Vegf nanoparticles. Biomaterials 33 (30), 7655–7664.

Paul, A., et al., 2014. Injectable graphene oxide/hydrogel-based angiogenic gene delivery system for vasculogenesis and cardiac repair. ACS Nano 8, 8050–8562.

Peer, D., et al., 2007. Nanocarriers as an emerging platform for cancer therapy. Nat. Nanotechnol. 2, 751–760.

Podesta, J.E., et al., 2009. Antitumor activity and prolonged survival by carbon-nanotube-mediated therapeutic siRNA silencing in a human lung xenograft model. Small 5 (10), 1176–1185.

Qin, W., et al., 2011. Improved GFP gene transfection mediated by polyamidoamine dendrimer functionalized multi-walled carbon nanotubes with high biocompatibility. Colloids Surf. B Biointerfaces 84 (1), 206–213.

Rana, V.K., et al., 2011. Synthesis and drug-delivery behavior of chitosan-functionalized graphene oxide hybrid nanosheets. Macromol. Mater. Eng. 296 (2), 131–140.

Ren, T., et al., 2012. Engineered polyethylenimine/graphene oxide nanocomposite for nuclear localized gene delivery. Polym. Chem. 3, 2561–2569.

Rouse, J.G., et al., 2007. Effects of mechanical flexion on the penetration of fullerene amino acid-derivatized peptide nanoparticles through skin. Nano Lett. 7, 155–160.

Ryman-Rasmussen, J.P., et al., 2006. Penetration of intact skin by quantum dots with diverse physicochemical properties. Toxicol. Sci. 91, 159–165.

Sano, N., et al., 2001. Synthesis of carbon 'onions' in water. Nature 414 (6863), 506–507.

Schipper, M.L., 2008. A pilot toxicology study of single-walled carbon nanotubes in a small sample of mice. Nat. Nanotechnol. 3, 216–221.

Schroeder, A., et al., 2012. Treating metastatic cancer with nanotechnology. Nat. Rev. Cancer 12, 39–50.

Schuster, D.I., et al., 2000. Evaluation of the anti-HIV potency of a water-soluble dendrimeric fullerene. Proc. Electrochem. Soc. 9, 267–270.

Shan, C.S., et al., 2009. Water-soluble graphene covalently functionalized by biocompatible poly-L-lysine. Langmuir 25 (20), 12030–12033.

Shen, J.F., et al., 2010. Facile synthesis and application of Ag-chemically converted graphene nanocomposite. Nano Res. 3 (5), 339–349.

Sijbesma, R., et al., 1993. Synthesis of a fullerene derivative for the inhibition of HIV enzymes. J. Am. Chem. Soc. 115 (15), 6510–6512.

Singh, R., et al., 2005. Binding and condensation of plasmid DNA onto functionalized carbon nanotubes: toward the construction of nanotube-based gene delivery vectors. J. Am. Chem. Soc. 127 (12), 4388–4396.

Slater, T.F., et al., 1985. Carbon tetrachloride toxicity as a model for studying free-radical mediated liver injury. Philos. Trans. R. Soc. Lond. B Biol. Sci. 311 (1152), 633–645.

Su, Y., et al., 2010. Cellular uptake and cytotoxic evaluation of fullerenol in different cell lines. Toxicology 269, 155–159.

Sun, X.M., et al., 2008. Nano-graphene oxide for cellular imaging and drug delivery. Nano Res. 1, 203–212.

Tagmatarchis, N., et al., 2004. Functionalization of carbon nanotubes via 1, 3-dipolar cycloadditions. J. Mater. Chem. 14, 437–439.

Thakur, M., et al., 2014. Antibiotic conjugated fluorescent carbon dots as a theranostic agent for controlled drug release, bio imaging, and enhanced antimicrobial activity. J. Drug Deliv. 2014, 1–9, 282193.

Thrash, T.P., et al., 1999. Toward fullerene-based radiopharmaceuticals: high-yield neutron activation of endohedral ^{165}Ho metallofullerenes. Chem. Phys. Lett. 308 (3–4), 329–336.

Tian, B., et al., 2011. Photothermally enhanced photodynamic therapy delivered by nano-graphene oxide. ACS Nano 5, 7000–7009.

Tran, P.A., et al., 2009. Carbon nanofibers and carbon nanotubes in regenerative medicine. Adv. Drug Deliv. Rev. 61 (12), 1097–1114.

Wang, T., et al., 2012. Design of multifunctional non-viral gene vectors to overcome physiological barriers: dilemmas and strategies. Int. J. Pharm. 427 (1), 3–20.

Wang, Y., et al., 2013. Multifunctional mesoporous silica-coated graphene nanosheet used for chemo-photothermal synergistic targeted therapy of glioma. J. Am. Chem. Soc. 135, 4799–4804.

Wang, X., et al., 2014a. Multifunctional graphene quantum dots for simultaneous targeted cellular imaging and drug delivery. Colloids Surf. B 122, 638–644.

Wang, Y., et al., 2014b. MRI-visualized, dual-targeting, combined tumor therapy using magnetic graphene-based mesoporous silica. Small 10, 109–116.

Wick, P., et al., 2007. The degree and kind of agglomeration affect carbon nanotube cytotoxicity. Toxicol. Lett. 168, 121–131.

Wojtoniszak, M., et al., 2013. Covalent conjugation of graphene oxide with methotrexate and its antitumor activity. Chem. Phys. Lett. 568–569, 151–156.

Wu, Y., et al., 2008. Carbon nanotubes protect DNA strands during cellular delivery. ACS Nano 2 (10), 2023–2028.

Xie, X., et al., 2013. Nanostraw-electroporation system for highly efficient intracellular delivery and transfection. ACS Nano 7, 4351–4358.

Xu, Z.P., et al., 2005. Inorganic nanoparticles as carriers for efficient cellular delivery. Chem. Eng. Sci. 61, 1027–1040.

Xu, Z., et al., 2014. Covalent functionalization of graphene oxide with biocompatible poly(ethylene glycol) for delivery of paclitaxel. ACS Appl. Mater. Interfaces 6, 17268–17276.

Yang, R., et al., 2006. Single-walled carbon nanotubes mediated in vivo and *in vitro* delivery of siRNA into antigen-presenting cells. Gene Ther. 13 (24), 1714–1723.

Yang, et al., 2007a. Carbon nanotubes for biological and biomedical applications. Nanotechnology 18, 412001:1–12.

Yang, J., et al., 2007b. Oxygen adsorption by carbon nanotubes and its application in radiotherapy. IET Nanobiotechnol 1 (1), 10–14.

Yang, Z.R., et al., 2007c. Recent developments in the use of adenoviruses and immunotoxins in cancer gene therapy. Cancer Gene Ther. 14, 599–615.

Yang, et al., 2008a. Histidylated cationic polyorganophosphazene/DNA self-assembled nanoparticles for gene delivery. Int. J. Pharm. 353, 277–282.

Yang, X., et al., 2008b. Multifunctionalized single-walled carbon nanotubes as tumor cell targeting biological transporters. J. Nanopart. Res. 10, 815–822.

Yang, X.Y., et al., 2008c. High-efficiency loading and controlled release of doxorubicin hydrochloride on graphene oxide. J. Phys. Chem. C 112 (45), 17554–17558.

Yang, X., et al., 2009. Superparamagnetic graphene oxide–Fe_3O_4 nanoparticles hybrid for controlled targeted drug carriers. J. Mater. Chem. 19, 2710–2714.

Yang, X., et al., 2011. Multi-functionalized graphene oxide based anticancer drug-carrier with dual-targeting function and pH-sensitivity. J. Mater. Chem. 21, 3448–3454.

Yang, Y., et al., 2012. Construction of a graphene oxide based noncovalent multiple nanosupramolecular assembly as a scaffold for drug delivery. Chemistry 18, 4208–4215.

Zakharian, T.Y., 2005. A fullerene-paclitaxel chemotherapeutic: synthesis, characterization, and study of biological activity in tissue culture. J. Am. Chem. Soc. 127, 12508–12509.

Zhang, L.W., et al., 2007. Biological interactions of functionalized single-wall carbon nanotubes in human epidermal keratinocytes. Int. J. Toxicol. 26 (2), 103–113.

Zhang, X.Q., et al., 2009. Polymer-functionalized nano-diamond platforms as vehicles for gene delivery. ACS Nano 3, 2609–2616.

Zhang, F., et al., 2010a. Horseradish peroxidase immobilized on graphene oxide: physical properties and applications in phenolic compound removal. Phys. Chem. C 114 (18), 8469–8473.

Zhang, J.L., et al., 2010b. Graphene oxide as a matrix for enzyme immobilization. Langmuir 6 (9), 6083–6085.

Zhang, L.M., et al., 2010c. Functional graphene oxide as a nanocarrier for controlled loading and targeted delivery of mixed anticancer drugs. Small 6 (4), 537–544.

Zhang, L.M., et al., 2011. Enhanced chemotherapy efficacy by sequential delivery of siRNA and anticancer drugs using PEI-grafted graphene oxide. Small 7 (4), 460–464.

Zhang, M., et al., 2014a. Potential application of nanocarbons as a drug delivery system. Carbon 69, 642.

Zheng, M., et al., 2014b. Integrating oxaliplatin with highly luminescent carbon dots: an unprecedented theranostic agent for personalized medicine. Adv. Mater. 26, 3554–3560.

Zhi, F., et al., 2013. Functionalized graphene oxide mediated adriamycin delivery and miR-21 gene silencing to overcome tumor multidrug resistance in vitro. PLoS One 8 (3), 1–8.

Zhou, X., et al., 2012. Ultra-small graphene oxide functionalized with polyethylenimine (PEI) for very efficient gene delivery in cell and zebrafish embryos. Nano Res. 5, 703–709.

FURTHER READING

Atkinson, H., et al., 2010. Delivering the goods: viral and non-viral gene therapy systems and the inherent limits on cargo DNA and internal sequences. Genetica 138 (5), 485–498.

Bao, H.Q., et al., 2011. Chitosan-functionalized graphene oxide as a nanocarrier for drug and gene delivery. Small 7 (11), 1569–1578.

Chavva, S.R., et al., 2014. Theranostic graphene oxide for prostate cancer detection and treatment. Part. Part. Syst. Charact. 31, 1252–1259.

Cheng, Z., et al., 2012. Multifunctional nanoparticles: cost versus benefit of adding targeting and imaging capabilities. Science 338, 903–910.

El-Aneed, A., 2008. Current strategies in cancer gene therapy. Eur. J. Pharmacol. (1–3), 1–8.

Fenoglio, I., et al., 2011. Multiple aspects of the interaction of bio-macromolecules with inorganic surfaces. Adv. Drug Deliv. Rev. 63, 1186–1209.

Fielding, R.M., et al., 1998. Altered tissue distribution and elimination of amikacin encapsulated in unilamellar, low-clearance liposomes (MiKasome). Pharm. Res. 15 (11), 1775–1781.

Firme 3rd, C.P., Bandaru, P.R., 2010. Toxicity issues in the application of carbon nanotubes to biological systems. Nanomedicine 6, 245–256.

Ge, J., et al., 2014. A graphene quantum dot photodynamic therapy agent with high singlet oxygen generation. Nat. Commun. 5, 4596, 1–8.

Jia, N., et al., 2007. Intracellular delivery of quantum dots tagged antisense oligodeoxynucleotides by functionalized multiwalled carbon nanotubes. Nano Lett. 7 (10), 2976–2980.

Nurunnabi, M., et al., 2014. Photoluminescent graphene nanoparticles for cancer phototherapy and imaging. ACS Appl. Mater. Interfaces 6, 12413–12421.

Yu, J.G., et al., 2008. Progress in the research of carbon nanotubes as drug carriers, 43 (10), 985–991.

CHAPTER 6

Toxicology and Biosafety of Carbon Nanomaterials

6.1 INTRODUCTION

Nanomaterials, with sizes ranging from 0.1 to 100 nm, are similar to many biomacromolecules and organelles. When the size of a material decreases to nanometer level, small size effect, surface effect, quantum size effect, and quantum tunneling effect will play significant roles and enable nanomaterials with new properties, such as the increasing hardness and strength, super paramagnetism, strong surface adsorption capacity, chemical reactivity, and quantum properties. Nowadays, various kinds of nanomaterials have been used in commercial and industrial products. ZnO and TiO_2 nanoparticles (NP) are widely used on cellulosic fabrics for UV-protection purpose (Farouk et al., 2013; El-Hady et al., 2013). A lot of cosmetics and skin products contain ZnO and TiO_2 NPs as sunscreening active ingredients (Schmid et al., 2008; Nohynek et al., 2008), together with nanosilver, gold, and other inorganic nano-structured materials to gain antimicrobial activity (Selvam et al., 2012; Hebeish et al., 2013; Dastjerdi et al., 2010). Nanoscale CeO_2 can serve as fuel additives and reduce fuel consumption, CO_2 emissions, and particulate emissions (Park et al., 2008; Cassee et al., 2011). Al_2O_3 NPs are widely used in plastic, rubber, and ceramics to reinforce and toughen the products. Nanotechnology and nanomaterials are also very promising in improving the diagnosis, treatment, and monitoring of many diseases. Studies have shown that incorporation of nanotechnology in medicine increases solubility, stability, targeting, biocompatibility, permeability, and controllability of drugs and vaccines (Fröhlich et al., 2012; You et al., 2012).

CONTENTS

Despite the increasing interest and effort in the development and application of nanomaterials, there are increasing concerns about their potential toxicity. In fact, the number of nanomedicines being approved for clinical use is limited (Eifler and Thaxton, 2011; Ventola et al., 2012). A lot of uncertainties remain as how NPs interact with biomacromolecules after entering into the human body; how they accumulate, degrade, and finally leave the body; and how different organs, tissues, and the circulation system are affected. There are reports that nanomaterials can cause damage to cells by generating reactive

Carbon Nanomaterials for Biological and Medical Applications. http://dx.doi.org/10.1016/B978-0-323-47906-6.00006-0

oxygen species (ROS), which will result in DNA damage, lipid peroxidation, and protein denaturation (Zhang et al., 2005; Lanone et al., 2006; Hoshino et al., 2011). In addition, proteins and nucleic acids may stick to NPs on their entry into the cells (Arvizo et al., 2012; Lai et al., 2012; Tenzer et al., 2011; Zhang et al., 2011a,b,c; Walczyk et al., 2010), interfering metabolic pathways and ultimately leading to cell death.

A number of in vivo and in vitro studies have been performed by different researchers to assess the potential toxicity of nanomaterials. At cellular level, methods that are generally used include cell morphology analysis, tissue staining, cell viability assay, ROS analysis, nitric oxide (NO) assay, LDH or MTT assay, cell cycle analysis, and cytokine (IF8, IL-6, TNF-α, etc.) release test. These assays are carried out to explore how the size, shape, and surface properties of NPs affect cell viability and their in vivo distribution, cellular uptake, subcellular location, metabolism, and degradation (Fröhlich et al., 2012; Tenzer et al., 2011; Kumar et al., 2012; Albanese et al., 2012; Kim et al., 2013; Karakoti et al., 2006). In vivo studies showed that nanomaterials can accumulate in many organs, for instance, the liver, spleen, lung, or kidney, depending on their characteristics and how they are administrated (Almeida et al., 2011). Oxidative stresses are detected in cells exposed to carbon nanomaterials (Shvedova et al., 2012; Haniu et al., 2011). Genomic and proteomic technologies have also been applied to study the biological effects and toxicity of nanomaterials (Larguinho et al., 2012). These high-throughput analyses provide a more comprehensive view of how molecules in the biological network respond to the presence of NPs. Genomic instability (Li et al., 2011; de Lima et al., 2012; Oesch et al., 2012), inflammatory response (Ganguly et al., 2011; Teeguarden et al., 2011), apoptosis (Horie et al., 2012), and protein phosphorylation (Ge et al., 2011; Prasad et al., 2013) are found to be affected by exposure to nanomaterials.

With the increasing interest in nanomaterials and concern of their biological effects and safety, more efforts are being invested in these fields of studies. It is promising that in the near future a systematic and accurate method to evaluate the safety of NPs will be developed and consents on by fields of nanomaterials. This will in turn improve the regulations and standards of nanomaterials preparation, processing, characterization, and application. Despite the efforts on the studies of nanomaterial safety, there remain a lot of problems. Firstly, these studies are not systematic; most studies use only one cell line or one tissue of the animal models. The methods reported are different and therefore no comparisons can be made among different materials and studies. Secondly, most studies report observations on cellular or tissue level; there are still not enough studies on how nanomaterials interact with biomolecules, and finally, the mechanism leads to toxicity or biological effects. The third concern is that among these studies there are contradictory results in the evaluation of toxicity of the same kind of nanomaterials. One example is that nano-Au was observed to cause both oxidative stress (Pan et al., 2012) and antioxidative stress (Menchón et al., 2012; Barathmanikanth et al., 2010). And this happens to CeO_2 (Hosseini et al., 2013) and carbon nanotubes (Herzog et al., 2009) too. Factors including different concentrations, administration procedures, fabrication and processing of NPs, and other details can all lead to inconsistent results. As this is a new area and many nanomaterials are newly engineered, standardized preparation and characterization methods of many nanomaterials are not available. With the development of nanomaterials and nanotechnology, there is a great need to set up regulations and standards based on comprehensive studies on biosafety and bioeffects of nanomaterials. It is critical to develop a systematic and standardized method that is accurate and repeatable for the analysis of the biological effects and safety of nanomaterials at tissue, cellular, and molecular levels. The method should start with a comprehensive and detailed characterization of each nanomaterial before in vitro/vivo analysis, which includes dimensions, compositions, purity, surface charges, etc. The method should be applicable to the analysis of most nanomaterials. In addition, a series of dosage and exposure time to nanomaterials should be designed to monitor the effects. This will allow comparisons of toxicity of different nanomaterials to the same cells and tissues and therefore facilitate the standardized regulations for nanomaterial safety.

6.2 TOXICITY AND BIOSAFETY OF CARBON NANOMATERIALS

6.2.1 Toxicity and Biosafety of Carbon Nanoparticles

Determining toxicity of carbon NPs is quite complicated. The surface coating, the presence of metal catalysts and/or graphite, the exposure to UV radiation, the dispersion properties, and the tendency to deposit as aggregates, due to high Van der Waals forces, influence greatly the behavior of these particles. Furthermore, species differences complicate research because some species, such as rats, are more sensitive to particles, making it difficult to extrapolate the results to humans.

6.2.1.1 Toxicity of Carbon Dots and Nanodiamonds

Carbon dots, on the other hand, can be nontoxic even without any additional surface modification by capping agents. Tao et al. (2012) have found that the as-made, bare carbon dots from different carbon precursors do not exhibit any appreciable cytotoxicity to the human kidney embryonic 293T cells until the incubation concentration reaches as high as 0.5 mg mL^{-1}. This result of low toxicity to cells has been confirmed by another independent study on the same 293T cell line but using carbon dots synthesized from a different electro-oxidation approach (Zhao et al., 2008). A similar level of inhibitory concentration on a different cell line, human hepatocellular carcinoma Hep G2 cells, has been found by Ray et al. (2009), with unaffected cell proliferation and viability until the incubation concentration of bare carbon dots is increased to >0.5 mg mL^{-1}. In contrast to bare, nonfunctionalized pristine graphene that induces considerable cytotoxicity by triggering proapoptotic signaling pathways, the bare carbon dots cause much lower toxicity to cells possibly because of the significantly smaller size, higher degree of oxidation, and greater water solubility. Interestingly, when carbon dots are surface-passivated with H_2N-PEG-NH_2, the cytotoxic effect becomes even greater than that with the bare carbon dots, evidenced by a lower incubation concentration of 0.2 mg mL^{-1} that starts causing appreciable cell death, as reported by Yang et al. (2009). The authors have attributed the observed cytotoxicity of polyethylene glycol (PEG) (1.5 kDa) amine-functionalized carbon dots to the surface-passivating agent PEG (1.5 kDa) amine, which has been found to have a greater cytotoxicity effect by itself (Yang et al., 2009). These studies have suggested carbon dots as nontoxic, biocompatible, and eco-friendly alternatives to semiconductor quantum dots for a variety of fluorescence imaging applications (Baker et al., 2010).

Nanodiamonds have been proposed to have a lower chance of inducing cellular oxidative stress than has been found with other carbon nanomaterials (Manus et al., 2010). Yu et al. (2005) have found minimum loss of cell viability up to a nanodiamond incubation concentration of 0.4 mg mL^{-1} and have proposed that the chemical inertness and lack of toxic impurities are the two main reasons for the observed low cytotoxicity of fluorescent nanodiamonds. A more systematic cytotoxicity study of nanodiamonds with diameters of 2–10 nm carried out by Schrand et al. (2007) has revealed nanodiamonds are nontoxic to many cell types including neuroblastoma cells, macrophages, keratinocytes, and rat pheochromocytoma PC-12 cells. For both raw nanodiamonds and carboxyl group–functionalized nanodiamonds, the mitochondrial functions of the tested cells after exposure to nanodiamonds have been evaluated by the MTT assay to reveal negligible damage to the mitochondria, while the amount of ROS inside the tested cells is not significantly increased on addition of nanodiamonds to the cell culture medium. It is also found that neuroblastoma cells can grow on nanodiamond-COOH–coated collagen substrate without appreciable morphological changes in comparison to bare collagen substrate. All these results have indicated the lack of cytotoxicity and high biocompatibility of nanodiamonds for cell fluorescent labeling and cell culture (Schrand et al., 2007).

6.2.1.2 Toxicity of Fullerenes

Being an efficient light sensitizer and singlet oxygen ($^1O_2$) producer under exposure to light, fullerene and its derivatives can be highly toxic to living systems in the presence of oxygen molecules, because of the damage to many biologically crucial molecules including DNA, proteins, and lipids through the light-triggered formation of ROS such as singlet oxygen and reduced oxygen radicals ($\cdot O_2^-$, $\cdot OH$, etc.). This photochemical reaction lays the foundation of photodynamic therapy and antibiotic/antiviral activities of fullerenes (Yamakoshi et al., 2003); however, in the meantime it induces nontrivial toxicity concerns to normal cells and tissues as well.

Pristine, unfunctionalized fullerene, has been found to have a higher degree of cytotoxicity than the covalently functionalized fullerenes, such as C_{60} fullerenols with different surface coverages of hydroxyl groups and methanofullerenes. The observed cytotoxicity is attributed to light-induced oxidative damage to cell membranes such as peroxidation of lipid bilayers, evidenced by the generation of superoxide anions for water-soluble fullerenes (Sayes et al., 2005). However, this ability to cause oxidative damage decreases as the fullerene cage becomes more functionalized and fully derivatized, explaining why the covalently modified fullerenes have less toxic effect to cells than the pristine fullerenes (Sayes et al., 2004). In contrast, noncovalently functionalized fullerenes with certain capping agents can be even more toxic because of the formation of a charge-transfer complex between the fullerene core and the solubilizing coating molecules. Such capping agents include polyvinylpyrrolidone, which is known to form a charge-transfer complex with the highest known equilibrium constant so far (Ungurenasu et al., 2000). The charge-transfer structure enhances the photosensitizing effect that generates more ROS on exposure to ambient light. It has been found that the PVP-solubilized C_{60} solution has obvious harmful effect to mouse embryos in both in vitro and in vivo settings. A statistically significant inhibitory effect on cell differentiation and proliferation has been revealed for mouse midbrain cell culture incubated in the presence of C_{60}–PVP NPs (Tsuchiya et al., 1996).

Most of the studies performed on the toxicity of C_{60} fullerenes demonstrated that these molecules are not cytotoxic toward human and animal cells in vitro, and their acute toxicity against animal tissues in vivo is quite low. But it is actually not known if the chronic exposure to these NPs could affect, at some extent, the biological environment in which they could be retained for years. Yamago et al. (1995) showed that water-soluble C_{60} fullerenes administered orally to rats were distributed rapidly to various tissues, and most of the material was retained in the body after 1 week and was also able to penetrate the blood–brain barrier (BBB). Acute toxicity was found to be quite low, but no additional data were given about chronic toxicity. Moussa et al. (1995, 1997) showed that C_{60} fullerenes did not influence cultured human leukocytes and did not stimulate the secretion of inflammatory molecules by these cells in vitro. Baierl et al. (1996) observed that C_{60} fullerenes incubated with alveolar macrophages and macrophage-like cells were not cytotoxic and had no or very little influence on the formation of ROS. More recently C_{60} fullerenes and their derivatives were shown to inhibit NO production, a key molecule capable of generating inflammation, by all three NO synthase isoforms (Wolff et al., 2000, 2001). Huczko et al. (1999) showed that water-suspended C_{60} fullerenes did not induce any skin irritation when applied in dermatological testing of skin sensitivity. Surface modifications and chemistry can change the behavior of these molecules and make them cytotoxic. It has been demonstrated that the cytotoxicity induced by C_{60} fullerenes is mainly due to the surface modifications of these molecules either by irradiation or by derivatization or functionalization. Rancan et al. (2002) reported that fullerene derivatives become toxic only when irradiated with UVA or UVB light and that the cell death is mainly caused by cell membrane damage and is UV-dose dependent. Sayes et al. (2005) observed that water-soluble C_{60} fullerene aggregates are cytotoxic to human dermal fibroblasts, human liver carcinoma cells, and neuronal human astrocytes. They disrupt the normal cellular function through lipid peroxidation. Their cytotoxicity is a function of surface derivatization. They can generate superoxide anions that can be responsible for membrane damage and subsequent cell death. So based on these studies, it seems likely that only modified fullerenes could be cytotoxic.

6.2.2 Toxicity and Biosafety of Carbon Nanotubes

The cytotoxicity of carbon nanotubes highly depends on the method of surface functionalization and properties of surface capping molecules. When single-walled carbon nanotubes (SWCNTs) and multiwalled carbon nanotubes (MWCNTs) are not functionalized with any surfactants and used as is, they tend to form tight aggregates and bundles in spite of intensive sonication and cause significant cytotoxicity to alveolar macrophages after incubation for merely 6 h (Jia et al., 2005). An increasing level of ROS indicating a higher degree of oxidative stress along with the activation of nuclear transcription factor-κB (NF-κB) as a response to harmful cellular stimuli has been found in human keratinocytes, which are treated with pristine SWCNTs dissolved in dimethylformamide without any surface functionalization (Manna et al., 2005).

The in vitro cytotoxicity can be significantly mitigated by surface functionalization. When SWCNTs are oxidized on the sidewalls and covalently functionalized with different molecules including fluorescein and poly(ethylene oxide)-linked biotin, they exhibit very little toxicity to HL60 cells, evidenced by no appreciable cell death. This finding suggests oxidized SWCNTs with surface carboxyl groups are nontoxic by themselves (Kam et al., 2004). In a follow-up work, Kam et al. (2005a,b) found that oxidized SWCNTs with proteins nonspecifically adsorbed on the nanotube sidewalls (i.e., the SWCNT–protein bioconjugates) do not show any inhibition to the proliferation of the treated HeLa cells up to 5 days after the initial cell exposure to the SWCNT–protein conjugates, as evidenced in the CellTiter 96 MTS assay results that exhibit little deviation of the formazan absorbance in the treated group compared to that in the control, untreated group. Besides these two examples of covalently oxidized and functionalized SWCNTs, Cherukuri et al. (2004) have shown that SWCNTs noncovalently functionalized with Pluronic surfactant, which is a nonionic surfactant, do not cause appreciable cytotoxicity by showing similar cell confluence, adhesion, and morphology to the control group cultured in the absence of SWCNTs. Besides the Pluronic surfactant, it was observed that SWCNTs noncovalently wrapped with DNA do not lead to any adverse effect on cell proliferation and adhesion either (Kam et al., 2005a,b). A more quantitative in vitro toxicity study has suggested a very high half-maximal inhibitory concentration (IC50) of 0.1778 $g\,L^{-1}$ for DSPE-mPEG (5 kDa)–coated SWCNTs on human dermal microvascular endothelial cells, derived from a sigmoidal fitting to the plotted cell viability curve at increasing SWCNT incubation concentrations (Hong et al., 2012). SWCNTs noncovalently wrapped with a branched PEG polymer, C18-PMH-mPEG, are also found nontoxic at a high incubation concentration, and an IC50 value of 0.136 $g\,L^{-1}$ on 4T1 murine breast cancer cells has been determined for this specific formulation of PEGylated SWCNTs (Kosuge et al., 2012).

However, both SWCNTs and MWCNTs have been used as nanocarriers for parenteral drug and gene delivery, and recently as targeted cancer treatment. The safety of CNT and other carbon nanotropes is still debatable because of the lack of systematic and complete toxicity evaluation. Some common cellular and tissue toxicity noticed with these carbon nanomaterials have been summarized in Table 6.1. The size of aggregated CNT is thought to be a primary concern for toxicity. Recently, studies have indicated that CNT responses are similar to the carcinogenic responses of asbestos fibers when injected into the peritoneal cavity (Poland et al., 2008). Because of the high aspect ratio of CNT (>100), it is expected that CNT would behave as biopersistent fibers in vivo. Studies have implicated size (aggregation), CNT length, and manufacturing impurities as sources for potential toxicity in vivo (Lacerda et al., 2006).

Pristine (nonfunctionalized) CNTs are inherently hydrophobic; therefore, aggregation is expected and observed in vivo. For injection, pristine CNTs are suspended in biocompatible surfactants such as Tween 80 or Pluronic F108. Several studies have been conducted on the in vivo distribution of intravenously injected pristine SWCNTs. Primarily, accumulation of CNTs was determined to be not only in the liver, but also in the spleen and lungs. No acute toxicity was observed in any tissue up to 24 h. Accumulation in the liver was suggested to be due to rapid surfactant displacement followed by opsonization

Table 6.1 Compilation of Cellular and Tissue Toxicity Studies of Pristine or Functionalized Carbon Nanomaterials

Nanotube	Biological System	Dosage	Toxicity
Plasmid DNA-SWCNT and plasmid DNA-MWCNT	f-CNTs: HeLa cell lines in vitro	10 mg mL^{-1}	50% survival of HeLa cells
Fluorescein isothiocyanate-SWCNT and fluorescein isothiocyanate-MWCNT	f-SWCNT and f-MWCNT: HeLa cell lines in vitro	5–10 mg mL^{-1}	50% survival of HeLa cells
Pristine SWCNT	SWCNT: mesothelioma cell line MSTO-211H in vitro	7.5 μg mL^{-1} water	10% decrease in cell proliferation and activity
Ammonium chloride-SWCNT, and poly(ethylene glycol)-SWCNT	Macrophages, B and T lymphocytes from BALB/c mice spleen and lymph nodes in vitro	10 μg mL^{-1} water	5% decrease in viability of B lymphocytes, but no adverse effects on T lymphocytes and macrophages
RNA-polymer SWCNT conjugate	MCF-7 breast cancer cells in vitro	1 mg mL^{-1}	No significant cell damage
[111In] DTPA-SWCNT and [111In] DTPA-MWCNT	Intravenous injection, systemic, female BALB/c mice in vivo	20 μg μL^{-1} PBS	No acute toxicity after single 200 μL dose
Pristine MWCNT	Human T lymphocytes in vitro	40 μg mL^{-1}	Should have no toxicity on human T lymphocytes
Pristine SWCNT	Intravenous injection, systemic, rabbit in vivo	7.5 mL of 20 μg kg^{-1} body mass	No toxicity
125I-SWCNT (OH)	Intraperitoneal, intravenous, subcutaneous, in male Kunming mice (KM) mice in vivo	1.5 μg/mouse	Accumulate in bone, but good biocompatibility
Glucosamine-MWCNT	Intraperitoneally into female KM in vivo	300 μL single dose, suspension concentration unknown	Good biocompatibility
pEGFP-c1 plasmid DNA-SWCNT	Mouse B cells and cortical neurons in vitro	0.1 pM/10 mL serum-free medium	~10% of cells were no longer viable
6-Aminohexanoic acid–derivatized SWCNT	Human epidermal keratinocytes (HEK) in vitro	Multiple tests from 0.00000005 to 0.05 mg mL^{-1}	Highest concentration that can interact with HEKs without toxicity, 0.000005 mg mL^{-1} for 24 h
DNA-Cy3 (fluorescent label)-SWCNT	HeLa cell line in vitro	2.5–5 mg L^{-1} water	No toxic effects, after six pulses of 10 s, 808-nm laser radiations at 1.4 W cm^2
Streptavidin-SWCNT	HL60 and Jurkat cells in vitro	0.025 mg mL^{-1}	No adverse effects
SWCNTs dispersed in DMEM with 5% (vol vol^{-1}) fetal bovine serum	Human epithelial-like HeLa cells in vitro	100 μg mL^{-1}	No effect on growth

CNT, *carbon nanotube;* MWCNT, *multiwalled carbon nanotube;* SWCNT, *single-walled carbon nanotube.*
Copyright permission from Firme III., C.P., et al., 2010. Toxicity issues in the application of carbon nanotubes to biological systems. Nanomedicine 6, 245–256.

of serum proteins. Distribution studies were followed up by looking at serum biomarkers of damage. Additionally, at markers for oxidative stress (glutathione and malondialdehyde) in liver and lung samples postdose, elevated levels of lactose dehydrogenase and alanine aminotransferase were concluded to be due to hepatic injury from accumulation in the liver. The study also found an increase in malondialdehyde and a decrease in glutathione in liver and lung samples (at 1.0 mg/mouse), which were indicative of increased levels of oxidative stress. Although no acute toxicity was determined histologically up to 90 days postdose, biomarkers indicating hepatic injury due to oxidative stress were further investigated (Yang et al., 2008). Studies conducted longer than 90 days may exhibit more toxicity damage over time, similar to onset of damage from asbestos fibers in lungs, which could take years. ROS formation and cytotoxic effects induced by aggregates and accumulation have been observed in other studies of CNT (Wick et al., 2007). Reduction of CNT accumulation and aggregation is achieved by functionalization (Wang et al., 2004). Rapid distribution and renal clearance from almost all tissues was observed 1–3 days after injection of functionalized CNTs, whereas pristine CNTs were determined at high levels in reticuloendothelial system tissues up to 1 month after injection. No comprehensive examination of functionalized CNT degradation-induced toxicity has been reported, but these are important questions as functionalized CNTs have more promise as therapeutic and diagnostic tools.

Along with CNT aggregation as a function of improper macrophage uptake, the length of CNTs has been implicated as a source of improper macrophage translocation. Some studies focused on the effect of length on CNT toxicity by injecting MWCNTs I.P. and observing carcinogenic mechanisms in the abdominal cavity and on the diaphragm. The longer-length (=20 μm) CNTs resulted in an inflammatory response within 24 h, with consequent granuloma 7 days after injection. These longer lengths may cause carcinogenic effects such as mesotheliomas, if longer studies were conducted. In the same study, shorter-length MWCNTs injected I.P. were effectively taken up by macrophages with efficient phagocytosis. In terms of CNT for therapeutics, it can be deduced that SWCNTs may be more favorable than MWCNTs from a toxicology standpoint because of smaller size and length resulting in less aggregation and better uptake by macrophages. It has been further suggested that MWCNTs were recognized and interacted with macrophage receptors on the plasma membrane and can rupture the membrane causing cytotoxicity and damage to the macrophage (Hirano et al., 2008).

Methods of CNT fabrication and purification have been shown to increase toxic effects of CNTs. Arc-discharge, laser ablation, chemical vapor deposition, and high-pressure carbon monoxide synthesis are methods commonly used to fabricate CNT. These methods are often performed in the presence of metal catalysts with CNT growth occurring on solid supports. Impurities such as residual metal or supports and amorphous carbon in the final formulation have been thought to induce oxidative stress. On the other hand, extensive purification and treatment will also cause degradation of the nanotubes (Cheng et al., 2011). Therefore, a fine balance between synthetic methods and purification must be achieved to fabricate highly purified CNT for injection. Low clearance and accumulation of CNTs in vivo warrants studies to determine chronic exposure toxicity. Radiolabeled isotopes were commonly used to evaluate the in vivo performance of CNT. However, because of isotopic decay and degradation of the label from the CNT, radiolabeling suffers from short study timeframes. Therefore, estimation and evaluation of chronic exposure toxicity may be possible based on evaluation of CNT intrinsic properties.

Identification of pharmacological and toxicological profiles is of critical importance for the use of NPs as drug carriers in nanomedicine and for the biosafety evaluation of environmental NPs in nanotoxicology. Lysosomes have been considered the pharmacological target organelles for SWCNTs and mitochondria as the target organelles for their cytotoxicity. The gastrointestinally absorbed SWCNTs were lysosomotropic, but also entered mitochondria at large doses. Genes encoding phosphoinositide-3-kinase and lysosomal-associated membrane protein 2 were found to be involved in such an organelle preference. SWCNT administration resulted in collapse of mitochondrial membrane potentials, giving rise to overproduction of ROS leading to damage of mitochondria, which was followed by lysosomal and cellular injury (Yang et al., 2010).

Owing to their special properties, NPs have the capacity to bypass the BBB. The toxic effects of NPs on central nervous system (CNS) function are still lacking, and their interactions with the cells and tissues in CNS are poorly understood. However, because of the large surface area, some of the NPs may cause neurotoxicity after entering into the brain. Using a fish model, Oberdorster (2004) investigated the toxicity of fullerene NPs on the brain of bass through the evaluation of oxyradical-induced lipid and protein damage as well as total glutathione levels. Significant lipid peroxidation was found in the brains of largemouth bass after 48 h of exposure to 0.5 ppm uncoated C_{60} (30–100 nm). Therefore, the fullerene NPs could cause cell damage in the brains of fish (Oberdorster et al., 2004). In cases where CNTs have a toxic interaction with cells, the mechanisms of toxicity are coming into focus. Results suggest CNTs may cause harm to cells by activating many pathways at once, mostly involving DNA damage (Pacurari et al., 2008). In one study, mesothelial cells exposed to SWCNTs, at concentrations $\sim 25\ \mu g\ cm^{-2}$, activated DNA recovery along with changes in the cell cycle and generation of apoptotic signals. Another approach showed that most cells incubated with CNTs halt at the G1 phase of the cell cycle (Jacobsen et al., 2008). It was also observed that

CNT/DNA interaction was the preferred route of toxicity in a 3 h incubation study with 96 μg SWCNT cm^{-2}, which induced DNA damage (through micronucleus generation) in lung fibroblasts (Uhrich et al., 1999). It should be possible, through the observation of specific toxic events that result from incubations with different types of f-CNTs, to test for functional groups that reduce the severity of such events. In conclusion, toxicity concerns of general CNT use in vivo have stemmed from observed toxicity after delivery to the lungs. Current studies have shown that once in the bloodstream, intrinsic properties, propensity to aggregate, and slow clearance can lead to oxidative stress especially in the liver, lungs, and spleen, ultimately resulting in inflammation. More importantly, the length of CNTs has been shown to result in inefficient phagocytosis and damage to macrophages. Because of less aggregation and shorter lengths, SWCNTs are better suited for in vivo applications over MWCNTs. Functionalization of SWCNTs further reduces aggregation; however, more thorough research on functionalized CNT toxicity is required. Moreover, methodically conducted experiments that mimic chronic exposure to CNTs will be more productive in the determination of CNT nanotoxicity (Aillon et al., 2009).

6.2.2.1 In Vitro Toxicological Studies of Carbon Nanotubes

Some in vitro cytotoxicity assessments of water-dispersible SWCNTs on A549 cells, a human lung cell line, confirmed that there was no intracellular localization of SWCNT in A549 cells and demonstrated that SWCNT could induce an indirect cytotoxicity by alteration of cell culture medium, which caused a false-positive toxic effect (Chang et al., 2011; Firme et al., 2010; Casey et al., 2008). Dumortier et al. (2006) observed that water-soluble SWCNTs marked with fluorescein were nontoxic to cultures of mouse B and T lymphocytes and macrophages and preserved the function of these immune cells. Other studies aimed to evaluate the potential toxicity and the general mechanism involved in two diameter kinds of water-soluble MWCNT-induced cytotoxicity in C6 rat glioma cell line (Han et al., 2012). Results demonstrated that smaller-sized MWCNTs seemed to be more toxic than larger ones. MWCNT-induced cytotoxicity in C6 rat glioma cells was probably due to the increased oxidative stress. However, pristine, water-insoluble CNTs have been found to be highly toxic in vitro to many different types of cells, including human keratinocytes, rat brain neuronal cells, human embryonic kidney cells, and human lung cancer cells (Yang et al., 2012; Shvedova et al., 2009; Fisher et al., 2012; Ursini et al., 2012). It is notified that these insoluble pristine CNTs cannot be used as carriers for drug and gene delivery in therapeutics, but are only found in the workplace of CNT production.

6.2.2.2 In Vivo Toxicological Studies of Carbon Nanotubes

According to the interesting review article about CNT toxicity published by Yang et al. (2012), many in vivo toxicological assessments have been performed by IV or SC injections and gastrointestinal exposure with functionalized or dispersed SWCNTs and/or MWCNTs in different animals (rats, mice). The available safety data collectively indicate that CNTs are of low toxicity via various exposure pathways for biomedical applications. CNTs induced meaningful toxicity only when a very high dosage (60 mg kg^{-1}) under PEG-MWCNTs form was administrated in mice (Yang et al., 2012; Zhang et al., 2010a,b). The toxicity of SWCNTs is closely related to the oxidative stress despite the administration routes (Folkmann et al., 2009). On the other hand, when CNTs were used as tissue engineering materials for cell growth by implanting subcutaneously, they exhibited very good biocompatibility and did not raise any serious toxicity, except very limited inflammation (Fraczek et al., 2008). However, Folkmann et al. (2009) reported that SWCNTs can induce oxidative damages to DNA in mice after oral gavage, and Fraczek et al. (2008) found that implanted SWCNTs and MWCNTs induced inflammation. In contrast, other reports on the toxicity of CNTs to the skin suggested that CNTs were biocompatible to the skin with good biocompatibility after subcutaneously planting (Yang et al., 2012). For the carcinogenicity evaluations, Takahashi et al. (2006) also observed that MWCNTs injected into the subcutaneous tissue of rasH2 mice did not develop neoplasms. However, other scientists found that CNTs introduced into the abdominal cavity of mice showed asbestos-like pathogenicity (Shvedova et al., 2009). Indeed, there are several parameters affecting the toxicity of CNTs in vivo. Metal impurities might contribute partially to the oxidative stress, and thus, careful purification of CNTs is necessary. Chemically functionalized CNTs show higher biocompatibility than pristine CNTs. Thus, Yang et al. (2012) concluded that functionalized CNTs are generally biocompatible and less toxic for the biomedical purposes. More toxicity evaluations are encouraged to give the safety threshold value of different CNTs and clarify the toxicological mechanism.

6.2.2.3 Human Toxicity of Carbon Nanotubes

As applications of functionalized CNTs linked with therapeutic molecules are still not assayed in humans for clinical studies, most publications found in the literature suggested that pristine CNTs could be the source of occupational lung diseases in workers of CNT industries such as asbestos pathology previously observed in humans (Yang et al., 2012; Lacerda et al., 2006; Donaldson et al., 2006). Based on several rodent studies in which test dusts were administered intratracheally or intrapharyngeally to assess the pulmonary toxicity of manufactured CNTs, these authors concluded that CNTs were capable of producing inflammation, epithelioid granulomas, fibrosis, and biochemical changes in the lungs (Shvedova et al., 2009; Fisher et al., 2012; Ursini et al., 2012; Lacerda et al., 2006; Takanashi et al., 2012; Donaldson et al., 2006; Poland et al., 2008). Ali-Boucetta et al. (2013) reported that the asbestos-like reactivity and pathogenicity reported for long, pristine nanotubes can be completely alleviated if their surface is modified and their effective length is reduced as a result of chemical treatment, such as with tri(ethylene glycol). However, opinions about the potential hazards of exposures to pristine CNTs and their residual metal impurities are still discussed. Moreover, the apparent similarity between MWCNTs and asbestos fibers has raised many questions about their safety profile. Despite their immense potential benefits in therapeutics, the toxicity of CNTs is a major concern that needs to be more clearly explained and apprehended.

6.2.3 Toxicity and Biosafety of Graphene and Graphene Derivatives

The cytotoxicity of graphene and its derivatives is also highly dependent on surface functionalization, in that the pristine and reduced GO with fewer surface functional groups tend to be more toxic than GO, as suggested by a number of in-depth studies (Liao et al., 2011; Zhang et al., 2010a,b). The observed cytotoxicity of pristine graphene is believed to be the result of two signaling pathways: the mitogen-activated protein kinase (MAPK) pathway and the transforming growth factor beta (TGF-β) pathway. These two signaling pathways both lead to the upregulation of proapoptotic proteins and initiate the apoptosis process that results in cell death under oxidative stress (Li et al., 2012). In striking contrast, it has been reported that functionalized graphenes, including GO and GO modified with many different hydrophilic macromolecules, can significantly reduce the cytotoxicity. It has been suggested that the GO has distinct advantages of reduced toxicity over CNTs, owing to the excellent aqueous solubility without the need to use surfactant to debundle and disperse in water, as well as the lack of metal catalyst impurities that are usually present in CNTs and could be the cause of oxidative stress related to CNTs (Loh et al., 2010). These arguments have been validated in a series of studies showing little or no cytotoxic effect of plain and modified GO at relatively high concentrations. For example, the as-oxidized plain GO without further conjugation to the surface groups does not exhibit any statistically significant reduction of A549 cell viability until an incubation concentration of 50 mg L^{-1} is reached (Chang et al., 2011). It is reported that the PEGylated nano-GO shows little cytotoxicity up to a cell incubation concentration of >100 mg L^{-1}, in which branched PEG chains are covalently linked to the −COOH groups in GO (Liu et al., 2008; Sun et al., 2008). Reduced GO with a decreased number of available functional groups can be noncovalently PEGylated with the branched C18-PMH-mPEG through hydrophobic interaction, and a low cytotoxicity has been found with an IC50 value of 80−90 mg L^{-1} to different cancer cell lines (Robinson et al., 2011). Besides our group's PEGylation methodology to reduce cytotoxicity, a number of other strategies of surface functionalization have been used by others to reduce the cytotoxicity of GO. Proteins, including the unpurified fetal bovine serum

(FBS) proteins and bovine serum albumin (BSA, which is the major component in FBS) (Mu et al., 2012), have been used to coat the surface of GO via nonspecific binding, and the cytotoxicity of GO-protein complex is found to be dependent on the protein concentration, where a higher concentration of proteins results in mitigated cytotoxicity of GO (Hu et al., 2011). Other surface-passivating molecules and polymers, including chitosan (Liao et al., 2011), dextran (Zhang et al., 2011a,b,c), and peptide (Bhunia et al., 2011), have been used to coat GO through either electrostatic interaction or covalent bonding between GO and protein. Reduced hemolytic activities and cell toxicity after surface coating suggest a biocompatible protective layer on the GO sheet to reduce the toxic effect on the exposed cells.

It is very important to investigate the physicochemical interaction of the NPs with in vitro and in vivo organelles before applying or considering them for biological application. Because graphene has been primarily considered as an electronic material, many studies have established graphene for bio-applications such as tissue engineering, drug delivery, and stem cell research. However, extensive observations are required in both in vivo and in vitro to investigate their cell and biomolecular interactions. Graphene is composed of only carbon atoms; however, GO and graphene quantum dots (GQDs) contain oxygen due to oxidation. Although few in vitro and in vivo toxicology studies have been reported previously, there has not been much focus on the biochemistry and histological impact (Josefsen et al., 2012; Chong et al., 2014). It has been conducted an extensive toxicity evaluation experiment to conduct a deep investigation based on biochemical and histological observations in GQDs-treated animals.These observations do not reveal any significant toxicity exerted by GQDs in vitro and in vivo. GO has several advantages over graphite or graphene, such as its dispersion in aqueous media, which is essential for biological application. GO contains hydrophilic functional groups, which enable chemical modification and functionalization. The in vivo studies using GO are based on the appraisal of bioaccumulation and excretion. The route of administration is also one of the important parameters to be considered in the case of the toxicity of nanomaterials. Because of the increasing importance of GO, there is a need for more detailed and accurate in vitro and in vivo studies regarding the toxicity of the GO. Wang et al. reported that GO could induce dose- and time-dependent cytotoxicity and can also enter into the cytoplasm and nucleus, decreasing cell adhesion and inducing cell coating and apoptosis (Wang et al., 2010; Zhang et al., 2011a,b,c; Chang et al., 2011). Another group reported that GO shows less toxicity in fibroblast HeLa cells over other carbon materials such as MWCNT and nanodiamond (Zhang et al., 2012). In addition to GO, GO-based polymer nanocomposites were also found to show toxicity on bacterial cells (Mejias et al., 2012). The size-dependent toxicity of graphene nanoflakes were investigated using a cell-based electrochemical impedance sensor, which depends on an interdigitated indium tin oxide electrode. Their results showed that the increased toxicity with smaller graphene nanoflakes can be used for electrochemical impedance sensing, optical imaging of cells, and bioassays (Yoon et al., 2014). Another graphene derivative, graphene nanowalls, posed greater toxicity on their contact with the bacterial cell membrane leading to the efflux of RNA from the cells. GO and some of its derivatives, such as oxygenated and carboxylated GO nanomaterials, showed toxicity in human cancer cells by an MTT assay (Wu et al., 2011; Mullick et al., 2013). Zebrafish is considered as one of the most used animal models to evaluate the in vivo toxicity of graphene-related materials. One of the research groups reported that MWCNTs, GO, and reduced GO do not show high toxicity to zebrafish embryos, but had some sublethal effects on their heart rate, hatching rate, and the length of their larvae (Liu et al., 2014). Nanosize GO and reduced GO showed lower toxicity in biomedical areas with higher photothermal effects (Table 6.2) (Robinson et al., 2011).

6.2.3.1 Biological Effect of Graphene Derivatives

Although not many, important studies have been conducted to understand the mechanism of interaction between graphene and biomolecules, especially intracellular organelles. The study reported by Li et al. (2012) showed that the commercially available pristine graphene increased the generation of ROS and decreased mitochondrial membrane potential, thus greatly affecting the immune system. As pristine graphene increases intracellular ROS, it triggers apoptosis through a mitochondrial pathway. In this study, authors selected murine RAW 264 and demonstrated macrophages triggered cell death, which was evaluated by cell signaling pathways such as the MAPK-related and TGF-β—related pathways (Li et al., 2012). Another report includes the biological effects of pristine graphene in primary murine and immortalized macrophages. The investigation reported that the secretion of cytokines (Th1/Th2, IL-1a, IL-6, IL-10, TNF-α, and GM-CSF) and chemokines (MCP-1, MIP-1a, MIP-1b, and RANTES) were increased due to pristine graphene (Zhou et al., 2012). The observations revealed that the graphene activated the TLR-mediated and NF-kB-dependent transactions. The report also demonstrated that the graphene remodeled actin assembly, thus altering the morphology of native macrophages and resulted in cells losing their adherence with the extracellular matrix. Although the in vitro studies in primary cells demonstrated that graphene induced apoptosis and attenuated phagocytosis, in vivo studies are required to obtain comprehensive information about these limitations (Fig. 6.1).

6.2.3.2 Size- and Dose-Dependent Therapeutic Effect of Graphene

NPs (crystal, semicrystal, semiconductors, or metals) with sizes less than 5 nm mostly and equally accumulate in the liver and kidney before finally being

Table 6.2 Toxicity of Graphene Derivatives In Vitro and In Vivo

Graphene Derivatives	Study	Model (Cell Line/Animal)	Observation	References
Graphene quantum dots (GQDs)	In vitro, in vivo	HeLa cells/female BALB/c mice	No apparent in vitro and in vivo toxicity of GQD, resulting from its small size and high oxygen content compared with that of the widely used Graphene oxide (GO)–PEG	Depan et al. (2014)
GQDs	In vitro, in vivo	KB, MDA-MB231, and A549 cells/ BALB/c nude mice	No acute toxicity or morphological changes of carboxylated GQDs were noted in either system at the tested exposure levels	Josefsen et al. (2012)
GO	In vitro, in vivo	Human fibroblast cells/Kunming mice	GO may induce severe cytotoxicity and lung diseases	Wang et al. (2010)
GO	In vivo	Kunming mice	Higher doses of GO showed toxicity in mice organs	Zhang et al. (2011a,b,c)
GO	In vitro	A549 cells	The effect of GO on A549 cells is dose and size related	Chang et al. (2011)
GO	In vitro	HeLa cells	GO toxic in HeLa cells	Zhang et al. (2012)
GO composite	In vitro	*subtilis*, *Rhodococcus opacus*, *Cupriavidus metallidurans*, CH_4 and NIH 3T3 fibroblast cells	Lower toxicity in bacterial and mammalian cells	Mejias et al. (2012)
Graphene nanoflakes	In vitro	HeLa cells	Evaluate size-dependent toxicity of graphene nanoflakes	Yoon et al. (2014)
GO nanowalls	Bacterial activity	*Escherichia coli* and *S. aureus*	Cell membrane of the bacteria was effectively damaged by the direct contact of the bacteria	Peng et al. (2012)
Oxidized graphene nanoribbons	In vitro	HeLa, MCF-7, SKBR3 and NIH 3T3 cells	Oxidized graphene nanoribbons showed cytotoxic effects more than GO	Mullick et al. (2013)
GO and carboxyl graphene nanoplatelets	In vitro	Hep G2 cells	GO and carboxyl graphene nanoplatelet-treated cells demonstrated toxicity in cancer cells	Wu et al. (2011)
Reduced GO	In vitro	Wild-type zebrafish	Toxicity to zebrafish embryos and sublethal effects on the heart rate, hatching rate, and the length of larvae	Liu et al. (2014)
Nano-GO and nano-reduced GO	In vitro	U87MG human glioblastoma cells	Nano-GO and nano-rGO seemed to show similar levels of toxicity on breast cancer cells	Robinson et al. (2011)

PEG, *polyethylene glycol;* rGO, *reduced graphene oxide.*

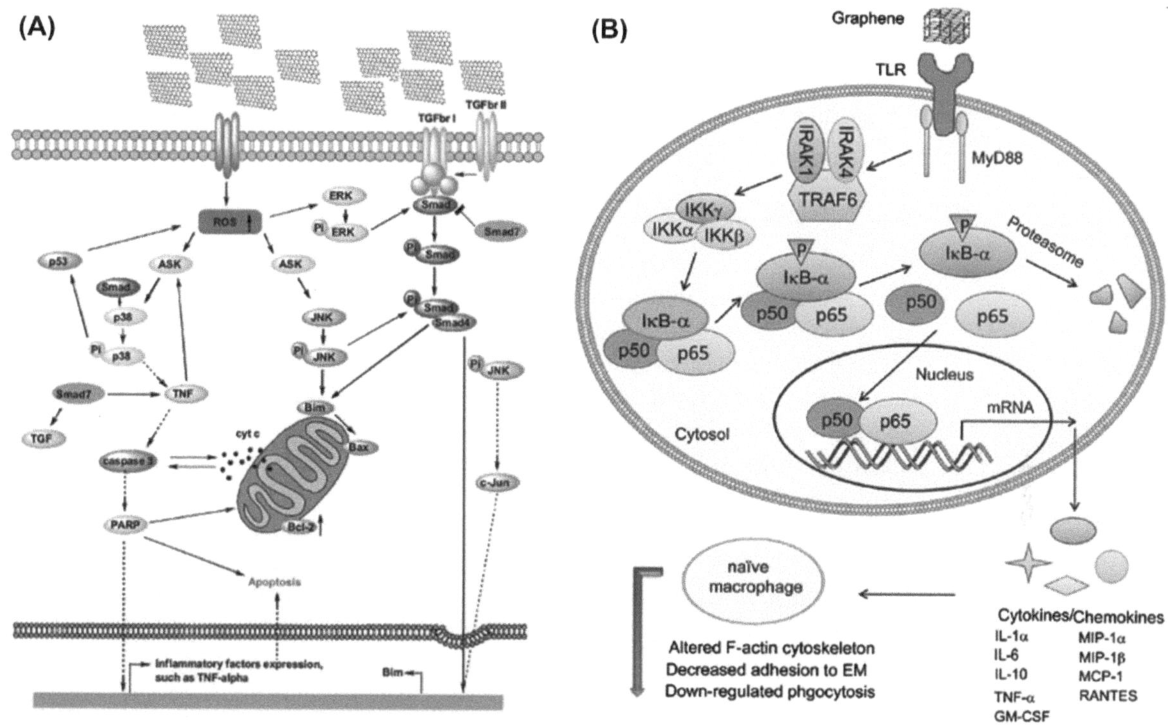

FIGURE 6.1

(A) Signaling pathway of cell apoptosis involved in pristine graphene. This scheme shows that cell apoptosis is caused through reactive oxygen species (ROS)—activated mitogen-activated protein kinase and transforming growth factor beta (TGF-β) pathways. (B) Signaling pathway of macrophage activation stimulated by graphene nanosheets. *Figure adapted with permission from Li, Y., et al., 2012. The triggering of apoptosis in macrophages by pristine graphene through the MAPK and TGF-beta signaling pathways. Biomaterials 33, 402–411 and Zhou, H. et al., 2012. The interactions between pristine graphene and macrophages and the production of cytokines/chemokines via TLR- and NF-κB-related signaling pathways. Biomaterials 33, 6933–6942 respectively.*

excreted through the kidney, whereas previous studies have proved that larger particles accumulate in the liver. Unfortunately, no in-depth investigations have been carried out to understand the effect of larger NPs and their effects in the liver and/or other organs. A very recent and advanced study reported by Volarevic et al. (2014) demonstrated that larger GQDs with a size of 40 nm are highly accumulated in the liver, which can alleviate immune-mediated liver damage. In addition, the rate of liver accumulation is higher for higher doses (50 mg kg^{-1}) compared to the lower doses. Although many previous studies have shown that graphene derivatives, especially GQDs with smaller diameters (5–10 nm), interfere with regular cellular mechanisms through interactions with intracellular pathways and induce apoptosis and reduce immunity, this study on GQDs with larger diameters shows complementary results. Nevertheless, both in vitro and in vivo studies show that GQDs with diameters of 40 nm play a critical role as a therapeutic agent to be used to treat liver inflammation/hepatitis (Volarevic et al., 2014).

6.3 CONCLUSION

This chapter summarized biological safety progress of the typical carbon NPs. Here we discussed that the results of the experiments performed up to date on the toxicity of carbon NPs are still not univocal and need to be viewed as the basis for future investigations. Some emerging concepts of nanotoxicology can be identified from these data. Toxicity of carbon NPs depends on several different factors such as their size, shape, the surface characteristics, and the amount of the substances present in the particle preparations. Based on our results, we hypothesize that the lower inflammatory potential of pure SWNTs and fullerenes as compared to graphite particles and not pure and cut-opened SWNTs toward mononuclear cells in vitro could be explained by the low amount of catalysts and graphite into the C-particle preparations tested. The higher inflammatory potential of graphite particles and not pure and open SWNTs may be due to the presence of a lot of dangling carbon bonds on their surface that are highly chemically reactive. Furthermore, we can speculate that the inflammatory effects on the murine airway cells, observed in some of the experiments reported in these review, could be consistent with the presence of graphite particles and/or metals into the carbon nanotube preparations tested. Anyway, this hypothesis need to be confirmed and a more detailed physicochemical characterization of these C-nanomaterials must be assessed to evaluate quantitatively and qualitatively the chemical reactive sites existing on their surface. Thus we believe that, at present, speculating about carbon NPs toxicity is still inconclusive. Before any investigation is performed and conclusions about their biocompatibility drawn, we need to know how the particles have been synthesized and dispersed, in which amount metal catalysts and graphite are present in the preparation, their surface modifications and functionalization, their surface chemistry (coating), and chemically reactive sites (free radicals). Additional considerations for assessing safety of carbon NPs include a careful selection of appropriate doses/concentrations related to the cell types and tissue species. The toxicity of C-nanomaterials needs to be understood in the framework of the material characterization. If scientists do not understand the material from a physical and chemical perspective, they cannot interpret exposure or toxicity measurements.

An interdisciplinary team approach is imperative for nanotoxicology research to reach an appropriate risk assessment in terms of toxicity of C-nanoderivatives to the biological environment.

Spectacular benefits of carbon nanotubes during their recent applications in different areas of pharmacy and medicine. The discovery of this bio-nanotechnology has opened new alternatives more effective than the ancient drug delivery methods because CNTs can pass through cell membranes, carrying drugs, genes, biomolecules, vaccines, and so forth deep into the target cells or organs previously unreachable. Another novel approach is the use of collagen CNT materials as scaffolds in tissue generation and artificial implants because CNTs resist biodegradation and are a powerful engineering candidate over other existing materials used to repair defective organs. Besides, CNTs combined with biosensors or other materials have proven excellent implements for the therapeutic monitoring and the diagnosis of diseases as well as for the analysis of drugs in different areas. It is also advisable to develop the free radical scavenger potential of functionalized CNTs for health maintenance. Overall, this nanotechnology could revolutionize the therapeutic concepts in the future and give a glimmer of hope for the treatment of many incurable diseases.

However, despite many surprising results of CNTs obtained during the beginning of this research field, there are still tremendous opportunities to be explored and significant challenges and risks to be solved. Therefore, more imagination and innovation are needed to elaborate different new forms of CNTs and their conjugates with high efficacy and safety for medicinal use in the future. For example, future preparation of CNTs should be attached with new sensitive markers so that they could directly reach the target cells or scientists could easily drive them from the exterior to the target organ to avoid side effects on the other healthy tissues.

At the same time, many hurdles of this nanotechnology have to be solved or elucidated. The human toxicity of different forms of CNTs for short- and long-term treatment is the first vital concern when they will be assayed in clinic. Some preliminary toxicological studies in vitro and in animal that were recently performed have shown controversial results. Carefully optimizing the physicochemical parameters to minimize the toxicity of CNTs is highly favorable. More toxicological investigations of different forms of CNTs from pristine CNTs to functionalized CNTs and their conjugates are highly recommended before they could be effectively used in clinical study and then marketed worldwide.

References

Aillon, K.L., et al., 2009. Effects of nanomaterial physicochemical properties on in vivo toxicity. Adv. Drug Deliv. Rev. 61, 457–466.

Albanese, A., et al., 2012. The effect of nanoparticle size, shape, and surface chemistry on biological systems. Annu. Rev. Biomed. Eng. 14, 1–16.

Ali-Boucetta, H., et al., 2013. Asbestos-like pathogenicity of long carbon nanotubes alleviated by chemical functionalization. Angew. Chem. 52 (8), 2274–2278.

Almeida, J.P., et al., 2011. In vivo biodistribution of nanoparticles. Nanomedicine (Lond.) 6, 815–835.

Arvizo, R.R., et al., 2012. Identifying new therapeutic targets via modulation of protein corona formation by engineered nanoparticles. PLoS One 7 (3) e33650:1–8.

Baierl, T., et al., 1996. Comparison of immunological effects of Fullerene C_{60} and raw soot from Fullerene production on alveolar macrophages and macrophage like cells in vitro. Exp. Toxicol. Pathol. 48, 508–511.

Baker, S.N., et al., 2010. Luminescent carbon nanodots: emergent nanolights. Angew. Chem. Int. Ed. 49, 6726–6744.

Barathmanikanth, S., et al., 2010. Anti-oxidant effect of gold nanoparticles restrains hyperglycemic conditions in diabetic mice. J. Nanobiotechnology 8, 16:1–15.

Bhunia, S.K., et al., 2011. Peptide-functionalized colloidal graphene via interdigited bilayer coating and fluorescence turn-on detection of enzyme. ACS Appl. Mater. Interfaces 3, 3335–3341.

Casey, A., et al., 2008. Single walled carbon nanotubes induce indirect cytotoxicity by medium depletion in A549 lung cells. Toxicol. Lett. 179 (2), 78–84.

Cassee, F.R., et al., 2011. Exposure, health and ecological effects review of engineered nanoscale cerium and cerium oxide associated with its use as a fuel additive. Crit. Rev. Toxicol. 41, 213–229.

Chang, Y., et al., 2011. In vitro toxicity evaluation of graphene oxide on A549 cells. Toxicol. Lett. 200 (3), 201–210.

Cheng, J., et al., 2011. Poly(ethylene glycol)-conjugated multi-walled carbon nanotubes as an efficient drug carrier for overcoming multidrug resistance. Toxicol. Appl. Pharmacol. 250, 184–193.

Cherukuri, P., et al., 2004. Near-infrared fluorescence microscopy of single-walled carbon nanotubes in phagocytic cells. J. Am. Chem. Soc. 126, 15638–15639.

Chong, Y., et al., 2014. The in vitro and in vivo toxicity of graphene quantum dots. Biomaterials 35, 5041–5048.

Dastjerdi, R., et al., 2010. A review on the application of inorganic nano-structured materials in the modification of textiles: focus on anti-microbial properties. Colloids Surf. B Biointerfaces 79, 5–18.

De Lima, R., et al., 2012. Silver nanoparticles: a brief review of cytotoxicity and genotoxicity of chemically and biogenically synthesized nanoparticles. J. Appl. Toxicol. 32, 867–879.

Depan, D., et al., 2014. The synergistic effect of a hybrid graphene oxide–chitosan system and biomimetic mineralization on osteoblast functions. Biomater. Sci. 2, 264–274.

Donaldson, K., et al., 2006. Carbon nanotubes: a review of their properties in relation to pulmonary toxicology and workplace safety. Toxicological Sci. 92 (1), 5–22.

Dumortier, H., et al., 2006. Functionalized carbon nanotubes are non-cytotoxic and preserve the functionality of primary immune cells. Nano Lett. 6 (7), 1522–1528.

Eifler, A.C., Thaxton, C.S., 2011. Nanoparticle therapeutics: FDA approval, clinical trials, regulatory pathways, and case study. Methods Mol. Biol. 726, 325–338.

El-Hady, M.M., et al., 2013. Flame retardancy and UV protection of cotton based fabrics using nano ZnO and polycarboxylic acids. Carbohydr. Polym. 92, 400–406.

Farouk, A., et al., 2013. Preparation of multifunctional cationized cotton fabric based on TiO_2 nanomaterials. Int. J. Biol. Macromol. 61C, 230–237.

Firme III, C.P., et al., 2010. Toxicity issues in the application of carbon nanotubes to biological systems. Nanomedicine 6, 245–256.

Fisher, C., et al., 2012. Applications and nanotoxicity of carbon nanotubes and graphene in biomedicine. Review. J. Nanomaterials 2012, 315185:1–19.

Folkmann, J.K., et al., 2009. Oxidatively damaged DNA in rats exposed by oral gavage to C_{60} fullerenes and single-walled carbon nanotubes. Environ. Health Perspect. 117 (5), 703–708.

Fraczek, A., et al., 2008. Comparative in vivo biocompatibility study of single- and multi-wall carbon nanotubes. Acta Biomaterialia 4 (6), 1593–1602.

Fröhlich, E., et al., 2012. The role of surface charge in cellular uptake and cytotoxicity of medical nanoparticles. Int. J. Nanomedicine 7, 5577–5591.

Ganguly, K., et al., 2011. Impaired resolution of inflammatory response in the lungs of JF1/Msf mice following carbon nanoparticle instillation. Respir. Res. 12, 94:1–11.

Ge, Y., et al., 2011. Proteome profiling reveals potential toxicity and detoxification pathways following exposure of BEAS-2B cells to engineered nanoparticle titanium dioxide. Proteomics 11, 2406–2422.

Han, Y., et al., 2012. In vitro toxicity of multi-walled carbon nanotubes in C6 rat glioma cells. NeuroToxicology 33 (5), 1128–1134.

Haniu, H., et al., 2011. Toxicoproteomic evaluation of carbon nanomaterials in vitro. J. Proteomics 74, 2703–2712.

Hebeish, A.A., et al., 2013. TiO_2 nanowire and TiO_2 nanowire doped Ag-PVP nanocomposite for antimicrobial and self-cleaning cotton textile. Carbohydr. Polym. 91, 549–559.

Herzog, E., et al., 2009. SWCNT suppress inflammatory mediator responses in human lung epithelium in vitro. Toxicol. Appl. Pharmacol. 234, 378–390.

Hirano, S., et al., 2008. Multi-walled carbon nanotubes injure the plasma membrane of macrophages. Toxicol. Appl. Pharmacol. 232, 244–251.

Hong, G.S., et al., 2012. Multifunctional in vivo vascular imaging using near-infrared II fluorescence. Nat. Med. 18, 1841–1846.

Horie, M., et al., 2012. In vitro evaluation of cellular response induced by manufactured nanoparticles. Chem. Res. Toxicol. 25, 605–619.

Hoshino, A., et al., 2011. Toxicity of nanocrystal quantum dots: the relevance of surface modifications. Arch. Toxicol. 85, 707–720.

Hosseini, A., et al., 2013. Antiapoptotic effects of cerium oxide and yttrium oxide nanoparticles in isolated rat pancreatic islets. Hum. Exp. Toxicol. 32, 544–553.

Hu, W.B., et al., 2011. Protein corona-mediated mitigation of cytotoxicity of graphene oxide. ACS Nano 5, 3693–3700.

Huczko, A., et al., 1999. Fullerenes: experimental evidence for a null risk of skin irritation and allergy. Full. Sci. Technol. 7, 935–939.

Jacobsen, N.R., et al., 2008. Genotoxicity, cytotoxicity, and reactive oxygen species induced by single-walled carbon nanotubes and C(60) fullerenes in the fe1-mutatrade markmouse lung epithelial cells. Environ. Mol. Mutagen 49, 476–487.

Jia, G., et al., 2005. Cytotoxicity of carbon nanomaterials: single-wall nanotube, multi-wall nanotube, and fullerene. Environ. Sci. Technol. 39, 1378–1383.

Josefsen, L.B., Boyle, R.W., 2012. Unique diagnostic and therapeutic roles of porphyrins and phthalocyanines in photodynamic therapy, imaging and theranostics. Theranostics 2, 916–966.

Kam, N.W.S., et al., 2004. Nanotube molecular transporters: internalization of carbon nanotube-protein conjugates into mammalian cells. J. Am. Chem. Soc. 126, 6850–6851.

Kam, N.W.S., et al., 2005a. Carbon nanotubes as multifunctional biological transporters and near-infrared agents for selective cancer cell destruction. Proc. Natl. Acad. Sci. U.S.A. 102, 11600–11605.

Kam, N.W.S., et al., 2005b. Carbon nanotubes as intracellular protein transporters: generality and biological functionality. J. Am. Chem. Soc. 127, 6021–6026.

Karakoti, A.S., et al., 2006. The potential toxicity of nanomaterials-the role of surfaces. Overview Surf. Modification Bioapplications (JOM) 58, 77–82.

Kim, S.T., et al., 2013. The role of surface functionality in determining nanoparticle cytotoxicity. Acc. Chem. Res. 46, 681–691.

Kosuge, H., et al., 2012. Near infrared imaging and photothermal ablation of vascular inflammation using single-walled carbon nanotubes. J. Am. Heart Assoc. 1 (6) e002568:1–8.

Kumar, V., et al., 2012. Evaluating the toxicity of selected types of nanochemicals. Rev. Environ. Contam. Toxicol. 215, 39–121.

Lacerda, L., et al., 2006. Carbon nanotubes as nanomedicines: From toxicology to pharmacology. Adv. Drug Deliv. Rev. 58, 1460–1470.

Lai, Z.W., et al., 2012. Emerging techniques in proteomics for probing nano-bio interactions. ACS Nano 6, 10438–10448.

Lanone, S., et al., 2006. Biomedical applications and potential health risks of nanomaterials: molecular mechanisms. Curr. Mol. Med. 6, 651–663.

Larguinho, M., et al., 2012. Gold and silver nanoparticles for clinical diagnostics - from genomics to proteomics. J. Proteomics 75, 2811–2823.

Li, J.J., et al., 2011. Genomic instability of gold nanoparticle treated human lung fibroblast cells. Biomaterials 32, 5515–5523.

Li, Y., et al., 2012. The triggering of apoptosis in macrophages by pristine graphene through the MAPK and TGF-beta signaling pathways. Biomaterials 33, 402–411.

Liao, K.H., et al., 2011. Cytotoxicity of graphene oxide and graphene in human erythrocytes and skin fibroblasts. ACS Appl. Mater. Interfaces 3, 2607–2615.

Liu, Z., et al., 2008. PEGylated nanographene oxide for delivery of water-insoluble cancer drugs. J. Am.Chem. Soc. 130, 10876–10877.

Liu, X.T., 2014. Toxicity of multi-walled carbon nanotubes, graphene oxide, and reduced graphene oxide to zebrafish embryos. Biomed. Environ. Sci. 27, 676–683.

Loh, K.P., et al., 2010. Graphene oxide as a chemically tunable platform for optical applications. Nat. Chem. 2, 1015–1024.

Manna, S.K., et al., 2005. Single-walled carbon nanotube induces oxidative stress and activates nuclear transcription factor-kappa B in human keratinocytes. Nano Lett. 5, 1676–1684.

Manus, L.M., et al., 2010. Gd(III)-nanodiamond conjugates for MRI contrast enhancement. Nano Lett. 10, 484–489.

Mejias Carpio, I.E., et al., 2012. Toxicity of a polymer–graphene oxide composite against bacterial planktonic cells, biofilms, and mammalian cells. Nanoscale 4, 4746–4756.

Menchón, C., et al., 2012. Gold nanoparticles supported on nanoparticulate ceria as a powerful agent against intracellular oxidative stress. Small 8, 1895–1903.

Moussa, F., et al., 1995. The influence of C_{60} powders on cultured human leukocytes. Full. Sci. Technol. 3, 333–342.

Moussa, F., et al., 1997. Preliminary study of the influence of cubic C-60 on cultured human monocytes: lack of interleukin-1 beta secretion. Full. Sci. Technol. 5, 503–510.

Mu, Q.X., et al., 2012. Size-dependent cell uptake of protein-coated graphene oxide nanosheets. ACS Appl. Mater. Interfaces 4, 2259–2266.

Mullick Chowdhury, S., et al., 2013. Cell specific cytotoxicity and uptake of graphene nanoribbons. Biomaterials 34, 283–293.

Nohynek, G.J., et al., 2008. Nanotechnology, cosmetics and the skin: is there a health risk? Skin Pharmacol. Physiol. 21, 136–149.

Oberdorster, E., 2004. Manufactured nanomaterials (fullerenes, C_{60}) induce oxidative stress in the brain of juvenile largemouth bass. Environ. Health Perspect. 112, 1058–1062.

Oberdorster, G., et al., 2004. Translocation of inhaled ultrafine particles to the brain. Inhal. Toxicol. 16, 437–445.

Oesch, F., et al., 2012. Genotoxicity investigations on nanomaterials. Arch. Toxicol. 86, 985–994.

Pacurari, M., et al., 2008. Raw single-wall carbon nanotubes induce oxidative stress and activate MAPKs, ap-1, NF-kappab, and akt in normal and malignant human mesothelial cells. Environ. Health Perspect. 116, 1211–1217.

Pan, Y., et al., 2012. Cytotoxicity of gold nanoparticles. Methods Enzymol. 509, 225–242.

Park, B., et al., 2008. Hazard and risk assessment of a nanoparticulate cerium oxide-based diesel fuel additive - a case study. Inhal. Toxicol. 20, 547–566.

Peng, J., et al., 2012. Graphene quantum dots derived from carbon fibers. Nano Lett. 12, 844–849.

Poland, C.A., et al., 2008. Carbon nanotubes introduced into the abdominal cavity of mice show asbestos-like pathogenicity in a pilot study. Nat. Nanotechnol. 3 (7), 423–428.

Prasad, R.Y., et al., 2013. Titanium dioxide nanoparticles activate the ATM-Chk2 DNA damage response in human dermal fibroblasts. Nanotoxicology 7, 1111–1119.

Rancan, F., et al., 2002. Cytotoxicity and photocytotoxicity of a dendritic C (60) mono-adduct and a malonic acid C (60) tris-adduct on Jurkat cells. J. Photochem. Photobiol. B 67 (3), 157–162.

Ray, S.C., et al., 2009. Fluorescent carbon nanoparticles: synthesis, characterization, and bioimaging application. J. Phys. Chem. C 113, 18546–18551.

Robinson, J.T., et al., 2011. Ultrasmall reduced graphene oxide with high near-infrared absorbance for photothermal therapy. J. Am. Chem. Soc. 133, 6825–6831.

Sayes, C.M., et al., 2004. The differential cytotoxicity of water-soluble fullerenes. Nano Lett. 4, 1881–1887.

Sayes, A.M., et al., 2005. Nano-C_{60} cytotoxicity is due to lipid peroxidation. Biomaterials 26, 7587–7595.

Schmid, K., et al., 2008. Use of nanoparticles in Swiss industry: a targeted survey. Environ. Sci. Technol. 42, 2253–2260.

Schrand, A.M., et al., 2007. Are diamond nanoparticles cytotoxic? J. Phys. Chem. B 111, 2–7.

Selvam, S., et al., 2012. Antibacterial effect of novel synthesized sulfated β-cyclodextrin crosslinked cotton fabric and its improved antibacterial activities with ZnO, TiO_2 and Ag nanoparticles coating. Int. J. Pharm. 434, 366–374.

Shvedova, A.A., et al., 2009. Mechanisms of pulmonary toxicity and medical applications of carbon nanotubes: two faces of Janus? Pharmacol. Ther. 121 (2), 192–204.

Shvedova, A.A., et al., 2012. Mechanisms of carbon nanotube-induced toxicity: focus on oxidative stress. Toxicol. Appl. Pharmacol. 261, 121–133.

Sun, X.M., et al., 2008. Nano-graphene oxide for cellular imaging and drug delivery. Nano Res. 1, 203–212.

Takahashi, N., et al., 2006. Biocompatibility of modified ePTFE for an artificial dura mater. Nucl. Instrum. Methods Phys. Res. B 242, 61–64.

Takanashi, S., et al., 2012. Carcinogenicity evaluation for the application of carbon nanotubes as biomaterials in rasH2 mice. Scientific Rep. 2, 498:1–7.

Tao, H.Q., et al., 2012. In vivo NIR fluorescence imaging, biodistribution, and toxicology of photoluminescent carbon dots produced from carbon nanotubes and graphite. Small 8, 281–290.

Teeguarden, J.G., et al., 2011. Comparative proteomics and pulmonary toxicity of instilled single-walled carbon nanotubes, crocidolite asbestos, and ultrafine carbon black in mice. Toxicol. Sci. 120, 123–135.

Tenzer, S., et al., 2011. Nanoparticle size is a critical physicochemical determinant of the human blood plasma corona: a comprehensive quantitative proteomic analysis. ACS Nano 5, 7155–7167.

Tsuchiya, T., et al., 1996. Novel harmful effects of [60]fullerene on mouse embryos in vitro and in vivo. FEBS Lett. 393, 139–145.

Uhrich, K.E., et al., 1999. Polymeric systems for controlled drug release. Chem. Rev. 99, 3181–3198.

Ungurenasu, C., et al., 2000. Highly stable C-60/poly-(vinylpyrrolidone) charge-transfer complexes afford new predictions for biological applications of underivatized fullerenes. J. Med. Chem. 43, 3186–3188.

Ursini, C.L., et al., 2012. Study of cytotoxic and genotoxic effects of hydroxyl-functionalized multiwalled carbon nanotubes on human pulmonary cells. J. Nanomaterials 2012, 815979:1–9.

Ventola, C.L., et al., 2012. The nanomedicine revolution: part 3: regulatory and safety challenges. P & T 37, 631–639.

Volarevic, V., et al., 2014. Large graphene quantum dots alleviate immune-mediated liver damage. ACS Nano 8, 12098–12109.

Walczyk, D., et al., 2010. What the cell "sees" in bionanoscience. J. Am. Chem. Soc. 132, 5761–5768.

Wang, H., et al., 2004. Biodistribution of carbon single-wall carbon nanotubes in mice. J. Nanosci. Nanotechnol. 4, 1019–1024.

Wang, K., et al., 2010. Biocompatibility of graphene oxide. Nanoscale Res. Lett. 6, 8:1–8.

Wick, P., et al., 2007. The degree and kind of agglomeration affect carbon nanotubes cytotoxicity. Toxicol. Lett. 168, 121–131.

Wolff, D.J., et al., 2000. Inhibition of nitric oxide synthase isoforms by tris-malonyl-C_{60}-fullerene adducts. Arch. Biochem. Biophys. 378, 216–223.

Wolff, D.J., et al., 2001. C_{60}-Fullerene monomalonate adducts selectively inactivate neuronal nitric oxide synthase by uncoupling the formation of reactive oxygen intermediates from nitric oxide production. Biochemistry 40, 37–45.

Wu, J., et al., 2011. Organic light-emitting diodes on solution-processed graphene transparent electrodes. ACS Nano 4, 43–48.

Yamago, S., et al., 1995. In vivo biological behavior of a water-miscible fullerene: 14C labeling, absorption, distribution, excretion and acute toxicity. Chem. Biol. 2 (6), 385–389.

Yamakoshi, Y., et al., 2003. Active oxygen species generated from photoexcited fullerene (C_{60}) as potential medicines: $O_2^{-\cdot}$ versus $^1O_2$. J. Am. Chem. Soc. 125, 12803–12809.

Yang, S.T., et al., 2008. Long-term accumulation and low toxicity of single-walled carbon nanotubes in intravenously exposed mice. Toxicol. Lett. 181, 182–189.

Yang, S.T., et al., 2009. Carbon dots as nontoxic and high-performance fluorescence imaging agents. J. Phys. Chem. C 113 (42), 18110–18114.

Yang, Z., et al., 2010. Pharmacological and toxicological target organelles and safe use of single-walled carbon nanotubes as drug carriers in treating Alzheimer disease. Nanomed. Nanotechnol. Biol. Med. 6, 427–441.

Yang, S., et al., 2012. Pharmacokinetics, metabolism and toxicity of carbon nanotubes for biomedical purposes. Theranostics 2 (3), 271–282.

Yoon, O.J., et al., 2014. Toxicity of graphene nanoflakes evaluated by cell-based electrochemical impedance biosensing. J. Biomed. Mater. Res. Part A 102, 2288–2294.

You, C., et al., 2012. The progress of silver nanoparticles in the antibacterial mechanism, clinical application and cytotoxicity. Mol. Biol. Rep. 39, 9193–9201.

Yu, S.J., et al., 2005. Bright fluorescent Nano diamonds: no photo bleaching and low cytotoxicity. J. Am. Chem. Soc. 127, 17604–17605.

Zhang, W.X., et al., 2005. Nanoscale environmental science and technology: challenges and opportunities. Environ. Sci. Technol. 39, 94A–95A.

Zhang, D., et al., 2010a. Long-term hepatotoxicity of polyethylene-glycol functionalized multi-walled carbon nanotubes in mice. Nanotechnology 21 (17), 175101:1–10.

Zhang, Y.B., et al., 2010b. Cytotoxicity effects of graphene and single-wall carbon nanotubes in neural phaeochromocytoma-derived PC12 cells. ACS Nano 4, 3181–3186.

Zhang, H., et al., 2011a. Quantitative proteomics analysis of adsorbed plasma proteins classifies nanoparticles with different surface properties and size. Proteomics 11, 4569–4577.

Zhang, S.A., et al., 2011b. In vitro and in vivo behaviors of dextran functionalized graphene. Carbon 49, 4040–4049.

Zhang, X., et al., 2011c. Distribution and biocompatibility studies of graphene oxide in mice after intravenous administration. Carbon 49, 986–995.

Zhang, X., et al., 2012. A comparative study of cellular uptake and cytotoxicity of multi-walled carbon nanotubes, graphene oxide, and nanodiamond. Toxicol. Res. 1, 62–68.

Zhao, Q.L., et al., 2008. Facile preparation of low cytotoxicity fluorescent carbon nanocrystals by electrooxidation of graphite. Chem. Commun. 5116–5118.

Zhou, H., et al., 2012. The interactions between pristine graphene and macrophages and the production of cytokines/chemokines via TLR- and NF-κB-related signaling pathways. Biomaterials 33, 6933–6942.

Index

'*Note:* Page numbers followed by "f" indicate figures.'

FSC
www.fsc.org
MIX
Papier | Fördert gute Waldnutzung
FSC® C083411

Kolibri 360 GmbH
Ferdinand-Jühlke-Straße 7
99095 Erfurt, Deutschland
produktsicherheit@kolibri360.de